T0185926

Critical Risks of Different Economic Sectors

Dmitry Chernov • Didier Sornette

Critical Risks of Different Economic Sectors

Based on the Analysis of More Than 500 Incidents, Accidents and Disasters

 Springer

Dmitry Chernov
Chair of Entrepreneurial Risks
ETH Zürich
Zürich, Switzerland

Didier Sornette
Chair of Entrepreneurial Risks
ETH Zürich
Zürich, Switzerland

ISBN 978-3-030-25033-1   ISBN 978-3-030-25034-8   (eBook)
https://doi.org/10.1007/978-3-030-25034-8

Cover illustrations: Igor Kostin / The Atomic Photographers – Breck P. Kent / shutterstock.com –
dvoevnore / shutterstock.com – TK Kurikawa / shutterstock.com

This Springer imprint is published by the registered company Springer Nature Switzerland AG.
The registered company address is: Gewerbestrasse 11, 6330 Cham, Switzerland

Contents

Chapter 1
Introduction

There is an increasing trend towards the generalization of theories and solutions. Politicians try to impose political models and solutions—which may have worked well in some countries—in very different societies elsewhere. Management theorists take what they see as good practice from leading companies in one country and propose universal solutions to millions of other organizations worldwide. MBA programs train universal managers capable—on paper at least—of running any company.

However, in practice, we see the disastrous consequences of such a generalized approach in many spheres. Hundreds of millions of people have suffered or died, victims of social engineering experiments that imposed alien political ideas on whole countries and even regions, and brought wholesale economic or social devastation. Companies have faced collapse and bankruptcy because they tried to implement "*modern*" and "*universal*" managerial ideas without regard for the readiness of the market, the "*business culture*" in that region or local laws on business competition.

In risk management, there has been a similar trend, where solutions effective in one industry (mainly derived from insurance, finance and heavy industry) have been transferred to the practice of organizations working in entirely different fields [1,2,3,4]. This generalization of risk management solutions originates from a desire for systematization: risk management experts seek to establish universal strategies

---

[1] John J. Hampton, Fundamentals of Enterprise Risk Management. How Top Companies Assess Risk, Manage Exposure, and Seize Opportunity, Second Edition, AMACOM, 2009

[2] Risk culture. Under the Microscope, The Institute of Risk Management, 2012, https://www.theirm.org/media/885907/Risk_Culture_A5_WEB15_Oct_2012.pdf and Risk culture. Resources for Practitioners, The Institute of Risk Management, 2012, https://www.iia.org.uk/media/329076/firm_risk_culture_-_resources_for_practitioners.pdf

[3] Extended Enterprise: Managing risk in complex 21st century organizations, The Institute of Risk Management, 2012, https://www.theirm.org/media/1155369/IRM-Extended-Enterprise_A5_AW.pdf

[4] Marcia Clemmitt, Issue: Crisis Management, Sage Business Researcher, February 9, 2017

© Springer Nature Switzerland AG 2020
D. Chernov, D. Sornette, *Critical Risks of Different Economic Sectors*,
https://doi.org/10.1007/978-3-030-25034-8_1

and a standardized approach for all enterprises based on effective, successful and logical solutions from a limited number of industries [5,6,7,8,9].

To illustrate the nature of the problem, we now briefly outline four examples of disastrous failures in different industry areas. In so doing, we would like to ask our readers a simple question: "*Is it possible that one risk executive/risk manager/risk specialist during her or his career could adequately manage the risks of all the following companies and effectively bring to an end the consequences of the following calamities?*"

Our first example comes from retail manufacturing. The company is the leading smartphone producer in the world, with a market share of 20–25% of all global smartphone shipment. Billions of customers have used its phones for many years and trust the brand: they will stick to it when upgrading their phone, and recommend it to relatives and friends. Then, the company releases a new smartphone, but before long a hundred customers have found that their phones caught fire during recharge. The problem affects less than 0.1% of the entire volume of new phones sold. But even such an apparently insignificant percentage of incidents should not mislead anyone: since the company sells tens of millions of devices globally, one would still expect tens of thousands of fires, not only at home but during car journeys or even air flights. The near-universal penetration of social media—and the actions of competitors in some markets—make each fire case a widely published event, damaging the brand in the eyes of potential customers and threatening the loyalty of existing users. Finally, under pressure from both regulators and the wave of negative publicity, the company makes a recall of all the potentially defective devices. The recall costs the company several billion dollars; but it protects the brand in the long term by putting an end to public discussion of the fires in the defective model, which were casting a shadow on every model produced by the brand.

Our second example is from heavy industry—from oil and gas extraction. The oil company in question is one of the major energy producers in the world. One of its production success stories is offshore drilling in a particular oil-rich region. For over two decades, the company has seen no significant accidents during deepwater drilling. The company then embarks on an exploratory well to a depth of 6500–1500 m below sea level and a further 4000 m beneath the seafloor. The stakes are high: the proven reserves of this field are 110 million barrels, with a potential revenue of around US$10 billion. But because of unexpectedly challenging geological conditions, the drilling team is soon running 43 days behind schedule and $58

---

[5]Risk Management Standard, IRM, 2002

[6]ISO 31000:2009, Risk management – Principles and guidelines, International Organization for Standardization

[7]A structured approach to Enterprise Risk Management (ERM) and the requirements of ISO 31000, AIRMIC, Alarm, IRM, 2010

[8]Katelyn Smith, Best Practices for Effective Corporate Crisis Management: A Breakdown of Crisis Stages Through the Utilization of Case Studies, California Polytechnic State University, 2012

[9]Enterprise Risk Management, Aligning Risk with Strategy and Performance, Executive Summary, Committee of Sponsoring Organizations of the Treadway Commission, June 2016 edition

million over budget. Everybody is in a rush, because leasing the platform is costing over \$1 million a day. One of the company's contractors conducts several tests on the cement mixture for the walls of the well, and all of them fail to confirm that the mixture will withstand the required oil pressure. But instead of informing the oil company, the cement contractor simply delivers the defective mixture to the drilling platform. The quality of the cement job is critical for the safe operation of deepwater wells: according to a study by one regulator in the field, cementing has been the most significant factor in 18 out of 39 well blowouts over 14 years in the area where the company is drilling. Unaware that cement test results have been concealed from them, and anxious to save money and time, the project managers decide to reduce the number of centralizers, which distribute cement evenly in a well, from 21 to just six. Moreover, once the well has been drilled and cemented, they continue to cut back on expenditure and time: they cancel the final acoustic test of the cement job, thinking that they are saving \$128,000 by doing so. But a few hours later, there is a blowout of oil, gas and concrete from the well, causing an explosion and a fire that leads to the decision to sink the platform. 126 crewmembers are on the rig during the accident; 11 people are killed and 17 injured. It takes 87 days to deal with the accident, and at least 4.9 million barrels of oil are discharged into the sea. This is the largest marine oil spill in the history of the industry. The company's total losses from the accident are \$46 billion—including \$18 billion on additional government fines and penalties—and within 3 months its stock market value has fallen by \$70 billion.

Our third example is from transportation. The sinking of the Titanic in 1912 had a seismic impact on transatlantic passenger transportation. But around 100 years later, the leisure cruise market is booming. The capacity of a typical modern cruise liner is several times that of the Titanic: more than 4000 tourists and 1000 crew members can be onboard at any time. The company in our example has a history of over 60 years of successful passenger shipping and cruising. But one of its captains has a reputation as a roughneck, a maverick and a reckless navigator: before the disaster we are relating here, he has taken risky and controversial decisions several times in the past. This time, during a winter cruise, the captain decides to deviate from the scheduled route to go closer to a nearby island. Later he explains that this was in order *"'to kill three birds with one stone': to please the passengers, salute a retired captain on [the island] and do a favour to the vessel's head waiter, who was from the island"*. However, the unauthorized deviation takes the ship too close to underwater rocks—it runs aground and sinks partially several hours later. The captain prefers not to sound a general emergency alarm straight after the collision, instead sending falsely reassuring reports to the emergency services. This delays the evacuation of 3206 passengers and 1023 crew members. The alarm is eventually given 45 min after the collision—but some passengers do not understand the message because of the language barrier, and stay in their cabins. Even though there are still passengers onboard, the captain, some senior officers and hundreds of ordinary crew members get off the sinking ship. Only around 70% of those on board find their way to the ship's life-rafts during the evacuation. More than 1200 people use municipal and emergency services: coastguard helicopters and vessels, and local fishing boats. During the accident, 32 people are killed and 157 injured. The cruise company lost nearly €1.5 billion in legal fees, compensation payments and salvage. Damage to the cruise company brand is so severe that, for a decade after the accident, it uses discounts to attract customers.

Our fourth example is from professional services. Audit firms trade on their names and reputations: they rely on potential clients trusting that the accounting books of companies they have audited are fair, and correspond accurately to real financial results. On the global financial services market, there is an elite group of accounting firms (referred to as the "Big Five") whose assessment is recognized as reliable by international investors. One of these audit firms has a branch in an oil-rich region where many energy companies have headquarters. Employees of this branch simultaneously provide accounting, consulting and auditing services to one of their clients. The client's management team has a very aggressive strategy, aiming to be one of the largest energy firms in a given country (and worldwide) through constant growth of their market capitalization. Representatives of the branch and executives of the client secretly come to an agreement to *"cook the client's books"*. The client hides debts and losses from its balance sheets, in order to keep their credit rating up to investment grade and keep down the cost of capital borrowing. The branch put the audit firm's very reputable name to the falsified financial reports of the client. In return, the branch receives around US$50 million annually from the client for all its services. With its falsified accounts, the client's market value grows by more than $4^1/_2$ times—$60 billion, or 70 times total earnings and six times book value—over 4 years. For several years, the scam stays uncovered, until it leads to losses of US$63.4 billions. The client's failure becomes the worst bankruptcy in corporate history at the time. Although the falsifications have only taken place in one branch of the global accounting firm, this company has been irreparably damaged: as we have noted, auditing is based mainly on trust in the name and reputation of an auditor. The firm loses its business, and its staff have to look for new positions in the remaining "Big Four" audit firms or elsewhere.

By now, to our question as to whether any one person could be a *"universal risk manager"*, the answer must be evident. Risk analysis methods in the outlined cases of a smartphone manufacturer, an oil company, a cruise company and an accounting firm are quite similar; but risk profiles, mitigation measures and the practicalities of putting an end to the consequences of specific risks coming to pass are so different that effective solutions cannot really be transferred from one case into another. We thus argue that it is quite difficult to train a universal risk executive to tackle all cases effectively because of the complete divergence of risk profiles, and the complexity and dissimilarity of risk mitigation solutions in each case.

Nevertheless, the prevalent approach to risk mitigation is still to offer a general set of processes and solutions, which are presented as being applicable to the *"average organization"*. It is widely assumed that *"executives in many firms may have treated risk management as synonymous with insurance and financial portfolio*

Table 1.1 Percentage of the contribution to GDP of the service, industry and agriculture sectors in different countries and for the world (2014 data)[a]

	Services	Industry	Agriculture
USA	78	20.7	1.3
Switzerland	73.6	25.7	0.8
Russia	63.7	32.1	4.2
China	47.8	43.1	9.1
Central African Republic	41.2	16.6	42.2
World	68.2	27.7	3.9

[a]World Bank national accounts data, and OECD National Accounts data files, 2014

management" [10]. And if a solution has proved effective in a single industry, some risk managers will try to implement it in all other industries. On the other hand, some researchers have seen the limitations of this universalist approach, and have attempted to provide industry-specific answers to certain risk management challenges [11,12,13,14].

In this context, we would thus like to call the attention of senior management of different companies and the risk management community to the existence (and consequences) of huge managerial differences between the two broad economic sectors, namely production (including agriculture) on the one hand and services on the other. Most of the organizational and managerial theory that informs the management of modern companies originates from observing the industry-dominated economy of the 1920s–1950s. Many modern companies still seem stuck with such *"universal"* managerial approaches derived from industrial practice. Without consideration of which sector their companies fall into, managers implement industrial practices in service companies and vice versa. It has become important to understand these sector differences because of the service revolution that has developed over the last 40–50 years. Services have only recently become the leading sector in terms of GDP in many countries (see Table 1.1).

---

[10]Douglas W. Hubbard, The Failure of Risk Management: Why It's Broken and How to Fix It, John Wiley & Sons, 2009, p. 22

[11]Michel Crouhy, Dan Galai, Robert Mark, Risk Management, McGraw-Hill Education, 2000, pp. 615–661

[12]Michael Regester, Judy Larkin, Risk issues and crisis management: a casebook of best practice, 2005, 3rd ed. pp. 133–162

[13]James Lam, Enterprise Risk Management: From Incentives to Controls, John Wiley & Sons, 2014, 2nd ed. pp. 277–336

[14]Nicola Martina Zech, Doctoral thesis "Crisis management within the hotel industry – a stakeholder relationship management approach", Faculty of Economics and Management, University of Latvia, Riga, 2016

The principles of operation of each of these sectors and subsectors (industries within a sector) are very different and the issue of managerial differences between economic sectors has been widely explored [[15,16,17,18,19,20,21,22,23,24,25,26,27]].

A cursory analysis of the types of accidents seen in different industries allows us to hypothesize that each sector requires its own risk profiles, specific templates for the accidents that are likely within different industries, and its own tools and measures for the mitigation of the distinctive risks of that field. We have found no publications that explain comprehensively such differences, especially in relation to general risk management and industry-specific risk profiles. Based on the analysis of more than 500 disasters, accidents and incidents—around 230 cases from the production sector (including agriculture) and around 280 cases from the service sector—we will elaborate on the differences in risk profiles and risk response actions within different sectors, in order to demonstrate that it is impossible to apply universal risk mitigation solutions in every sector and subsector.

Before going further, it is useful to make explicit the major risks considered in this book. Risk is, by definition, a potential loss. Hence, risk is characterised by three factors: (I) the area of vulnerability, (II) the likelihood, or if available the probability, of the event occurring and (III) the consequences, quantified for example by the resulting losses. As the considerable body of cases presented here attests, the main

---

[15]W. Earl Sasser, Richard Paul Olsen, D. Daryl Wyckoff, Management of service operations: text, cases, and readings, Allyn and Bacon, 1978

[16]Philip Kotler, Principles of marketing, Prentice-Hall, 1980

[17]Christopher H. Lovelock, Classifying Services to Gain Strategic Marketing Insights, Journal of Marketing, 1983, Vol. 47, No. 3, pp. 9–20

[18]James L. Heskett, Managing in the Service Economy, Harvard Business School Press, 1986

[19]James L. Heskett, Christopher Hart, Earl Sasser, Service Breakthroughs: Changing the Rules of the Game, Free Press, 1990

[20]Richard Normann, Service Management: Strategy and Leadership in Service Business, Wiley, 1991

[21]James A. Fitzsimmons, Mona J. Fitzsimmons, Service Management: Operations, Strategy, Information Technology, McGraw-Hill/Irwin, 1998

[22]Cengiz Haksever, Barry Render, Roberta S. Russell, Robert G. Murdick Service, Management and Operations, Pearson, 1999

[23]Bart van Looy, Paul Gemmel, Roland Dierdonck, Services Management: An Integrated Approach, Pearson Education, 2003

[24]Richard L. Daft, Organization Theory and Design, Cengage Learning, 2010

[25]William J Stevenson, Operations Management, McGraw-Hill, 2012

[26]Richard L. Daft, Dorothy Marcic, Understanding Management, Cengage Learning, 2013

[27]Richard L. Daft, Management, Cengage Learning, 2015

losses include: (1) health and environment related losses [28]; (2) losses due to natural and man-made accidents [29]; (3) losses related to competition, technology and economics [30]; (4) losses linked to financial markets [31]; (5) losses related to reputation and trust [32]; (6) losses due to social instabilities, conflicts, wars [33]. In this book, we would like to analyze systematically the major risks that are distinctive to each specific subsector.

In recent decades, it has become common in many industries for executives to focus mainly on delivering financial results, to stay far away from operational control and avoid deep immersion in the details of production or the provision of services. Thus, companies are increasingly led by businessmen, who understand neither the complexity of their organizations, nor the full picture of risks in the industry in question. In some instances, the entire management team at a hazardous industrial site has been replaced: a team of skilled, technically proficient and experienced managers has given way to financiers and economists, who see the

---

[28]Loss of life; losses due to products that are poorly designed, or revealed later to cause illnesses and deaths; loss and disruption of business when hidden long-term health and environmental hazards are progressively revealed; losses due to unhealthy production practice endangering the health of employees; high costs for the cleaning/rehabilitation of contaminated or destroyed habitats.

[29]Losses due to supply chain disruption, especially when a business is dependent on a small number of suppliers for key parts or components; loss due to environmental disasters, earthquakes, hurricanes, floods, landslides, pandemics...; losses due to changes of legislation or higher costs for risk mitigation following an accident (e.g. in the case of the Three Mile Island reactor meltdown); losses due to industrial failure of machines and infrastructure through aging, poor maintenance; dangerous siting, inadequate risk management, or bad luck; losses due to accidents in a given industry that affect the global public perception of that whole industry (e.g. nuclear accidents); loss due to electricity blackout in a region.

[30]Loss of profits; loss of business due to competition; loss of market share due to disruptive technology; Loss of access or of market power due to changing geopolitics; loss of competitiveness due to changing regulations; losses due to liability changing with new regulation; failure to access a market due to political protectionism on grounds of national security, more stringent regulation, or opposition from local populations; loss of ownership due to nationalization; losses due to black market business or industrial malpractice revealed many years later, either within a firm itself or on the part of its suppliers/subcontractors.

[31]Losses due to volatility in currency exchange and financial markets, impacting costs as well as cash management; losses due to interest rate changes impacting debt servicing costs.

[32]Loss of reputation or public trust due to unethical practice: exploitation of ancient forests, child labor, extreme exploitation of workers in sweatshops, exploitation of animals, discrimination against specific groups such as LGBT, particular races or religions, etc.; losses due to sub-standard performance by employees leading to low quality products, or unhygienic conditions that later threaten the health or life of customers; losses due to employee error, incompetence, or psychologically unbalanced behavior; losses due to customer bad habits or mistakes that the company nevertheless has to endorse in order not to lose trust and market share; losses due to corruption charges; losses due to theft or counterfeiting; losses due to lawsuits and class actions.

[33]Losses due to destruction by military or terrorist sabotage action; losses due to international sanctions; losses due to social disruption in a country, say due to a general strike; losses due to employee strikes and employer-employee conflicts; losses due to cyber-attacks and loss of control of a company's IT system, losses due to sabotage by employees or external agents.

plant in their charge not as a complex socio-technical system but simply as a profits generator, and prioritize financial security over safety and the public interest. If senior management see their role primarily as to maximize income, they may see little need to learn the details of the business—which leads to a situation where people with only a fragmented perception of the real situation "*on the shop floor*" are in charge of potentially dangerous projects or facilities. Such executives prefer to disclaim responsibility for the most problematic and complicated issues in an organization by delegating control over specific risks down to narrow risk specialists: health, safety and environment managers, quality control engineers, process safety managers, finance and insurance managers, insurance professionals, internal control specialists, cyber security engineers, security professionals, and so on. This approach does not allow for adequate control over critical risks in an organization, because many risk mitigation solutions are multi-functional and would require oversight and action at higher management level.

Against this backdrop, we insist that executives should keep the critical risks threatening their organization under personal control. Top management—not only highly specialized risk professionals—are responsible for proper risk management. However, executives cannot manage all risks equally. In a large organization, if information about the potential outcomes of every single risk and delivery reports about all mistakes and incidents were passed up to executives, it would lead to information overload for senior managers. But executives can and should prioritize organizational risks, in order to mitigate the occurrence of those risks that are critical for the sustainable development of an organization over the long term. Such risks should be under the direct control of executives, but monitoring and control over other less significant risks can be delegated within the organizational hierarchy (nevertheless, executives should be involved in decision-making concerning important issues for the mitigation even of less significant risks). This prioritization of risks is inevitable: no organization can mitigate all its risks to reach a zero probability of occurrence due to limitation of available resources. Organizations must prioritize to allocate their resources wisely.

As an illustration to give some substance to our proposition, let us analyze briefly the critical risks of the organizations involved in the four examples we have just outlined, in order to demonstrate risk prioritization in action.

For the global producer of smartphones, the critical risks are those that threaten the long-term loyalty of customers to its products. New phones are purchased quite frequently: over the last 20 years when mobile communication has become affordable to a mass market, the average lifetime of a smartphone has dropped to just a few years—because new phones are constantly being released and, unfortunately, because they are designed to slow down once their memories contain significant amounts of data [34]. Phone producers need to be known for making reliable phones,

---

[34]Some phone producers manufacture very reliable phones, which can work properly during many years but, unknown to the customers, they install special software in their phones, which makes them slow down so as to motivate customers to buy new more powerful phones with new features,

which will work properly without causing headaches for several years. Thus, the most critical risks for a phone producer are (I) design errors and (II) production defects. Another important challenge is negative publicity over the quality of phones, when defects threaten the safety or even the life of customers or cause dissatisfaction. All these discussions can (III) damage the brand and (IV) cast doubt on the long-term stability of the producer's market share and profits. Response measures to mitigate these risks include ongoing R&D, quality management, crisis communications, branding and customer care. These areas should be under the direct supervision of company executives. Producers also face a long list of other risks from regulation concerning the national security of telecommunications infrastructure and software in some countries, localization of production processes in different regions, reliability of the supply chain, technical certification of phones, distribution of phones to retail networks, etc. Nevertheless, such risks are less important than those posed by low production quality and brand failure—so control over non-critical risks could be delegated to specific risk professionals under the general oversight of executives.

A large oil company faces thousands of different risks, amongst which the following are critical: (I) loss of their right to extract resources in oil-rich regions because regulators, the general public and local residents do not trust the company (and the entire industry) to ensure ecologically safe production of large amounts of hydrocarbons in the long term; (II) an industrial disaster at a remote, complex and expensive production facility, whose dealing with may incur tremendous expenses for cleaning up oil spills, restoration of damaged territories over decades, payment of government penalties, compensation to victims and their families, lost equipment (even if insured) and lost profits and oil reserves; (III) reduction of market value, lawsuits from investors, deterioration of the conditions for financing the massive capital investments required for oil projects; (IV) damage to the retail brands through which the oil company sells fuel to motorists in a very competitive market, if clear evidence comes to light of irresponsibility towards environmental and production safety on the company's part.

For a cruise company, the most critical risks are: (I) failing to provide passengers safe and comfortable transportation because of crew errors, technical reasons or bad weather; (II) damage to the cruise brand, because customers usually assume that any kind of emergency on a ship reflects the attitude towards safety, level of crew preparation and quality of service across the company's whole fleet.

For an audit firm, the most dangerous risks are: (I) deliberate action by highly educated and competent staff to make false or misleading assessments of the financial position of the clients they are auditing; (II) damage to the reputation of the firm, because a single large criminal case concerning the conduct of employees from just one regional office is seen as casting doubt on the work of an entire firm worldwide.

---

just after several years of usage (Angelo Amante, Paresh Dave, Italian watchdog fines Apple, Samsung over software updates, Reuters, October 24, 2018)

In this connection, the book presents the following system of subsector analysis based on a risk prioritization approach:

- Determination of the key business and managerial features of each subsector and specific industry.
- Application of the concept of *"critical success factors"* to help determine and rank in importance the unique factors that allow organizations in a given subsector to succeed. This concept has seen widespread acceptance in strategic business planning since 1961, when it was first proposed in the Harvard Business Review by Ron Daniel, a consultant at McKinsey & Company [35]. It was developed further in the 1970s and 1980s by John Rockart, Director of the Center for Information Systems Research at the MIT's Sloan School of Management [36,37].
- Application of the concept of *"stakeholders"* to identify and assess the relative importance of the key audiences that influence the overall business of a typical organization within a particular industry. Based on our observations, research and expertise, we subjectively weight the influence that each key audience has on achieving the critical success factors for an organization (using the expert opinion of each audience). This weighting of the influence of an organization's stakeholders allows us to assess what executives should prioritize in their strategic actions.
- Presentation of the typology of commonly occurring dangerous events and accidents within each subsector in order of severity (ranking risks by severity will allow us to determine the key threats to the long-term sustainable development of organizations from each subsector).
- Accounts of the most notorious accidents occurring within each subsector, to provide clear examples of failure to meet critical success factors for that subsector or specific industry.
- Identification of the main features of each of these major accidents.
- Presentation of key measures to prevent the kind of accidents that are common within a given subsector and industry.
- Summary of specific risk mitigation features characteristic of each industry.

The book is mainly intended for executives, strategists, senior risk managers of enterprise-wide organizations and risk management experts engaged in academic or consulting work. For executives and practical risk specialists, the book will give a general picture of the risks peculiar to each sector and subsector; this will enable them to reassess the most critical risk areas in the work of their organization through our analysis of risks and disasters that have affected other companies in their

---

[35]D. Ronald Daniel, Management Information Crisis, Harvard Business Review, September–October 1961

[36]John F. Rockart, Chief Executives Define Their Own Data Needs, Harvard Business Review, March–April 1979

[37]John F. Rockart, The changing role of the information systems executive: a critical success factors perspective, Center for Information Systems Research, Sloan School of Management, MIT, April 1982

industry. For academics and consultants, the book will help to clarify the conceptual difference between the various risk management approaches in the production and service sectors.

By setting out clearly the sector differences in risk management, we aim to improve the practice of general risk assessment with regard to identifying and prioritizing industry-specific risks, and of risk control with regard to planning appropriate mitigation measures. These aspects are covered by the following provisions of ISO 31000:2009:

- "Design of a framework for managing risk" (4.3), including "Understanding the organization and its context" (4.3.1),
- "Establishing risk management policy" (4.3.2),
- "Accountability" (4.3.3),
- "Integration into organizational processes" (4.3.4),
- "Resources" (4.3.5),
- and a parts of COSO ERM:2004: "Enterprise Risk Management", specifically with regard to "Internal Environment", "Objective Setting", "Event Identification", "Risk Assessment", "Risk Response", and "Control Activities".

For the classification of sectors and subsectors (industries) within the economy, we have used, as a point of reference, the UN's International Standard Industrial Classification, which is the system adopted by the United Nations Statistics Division (Revision 4) [38], as shown in the Table 1.2.

We have selected only subsectors and industries where enterprise-wide activity takes place, and we have omitted detailed discussions of the public services—including public administration, national defense, compulsory social security, security and investigation activities, and the activities of membership organizations, extraterritorial organizations and bodies, and so on.

In our analysis of each subsector, we will give a brief account of the most prominent risks within a given industry without in-depth, specialist discussion and details—our main goal is to demonstrate the managerial differences between, and specific risks within, the different sectors and subsectors without digging too deeply into any of them. Nevertheless, some notable cases, which became pivotal for a given industry, are described at greater length to illustrate the specific risks associated with a whole subsector. These include the influence of Saudi Arabia on the oil market, the impact of shale energy production on the American environment, the consequences of the Chernobyl and Fukushima disasters on nuclear energy production and energy portfolio allocations around the world, the global recall of Perrier sparkling water, the sinking of the Costa Concordia, the failure of HealthCare.gov, the problems of zero-day vulnerabilities and "backdoors" in popular software, and others.

---

[38]International Standard Industrial Classification of All Economic Activities, Rev.4, https://unstats.
un.org/unsd/cr/registry/regcst.asp?Cl=27   and   https://unstats.un.org/unsd/publication/seriesM/
seriesm_4rev4e.pdf

Table 1.2 Categories of economic activity according to the international standard of industrial classification (Rev. 4)

	Sectoral share of total economic activity within the world economy (value added as % of GDP)
Mining, oil and gas extraction	Industry (27.7%)
Manufacturing	
Electricity, gas, steam and air conditioning supply	
Water supply; sewerage, waste management and remediation activities	
Construction	
Agriculture, forestry and fishing	Agriculture (3.9%)
Wholesale and retail trade; repair of motor vehicles and motorcycles	Services (68.2%)
Transportation and storage	
Accommodation and food service activities	
Information and communication	
Financial and insurance activities	
Real estate activities	
Professional, scientific and technical activities	
Administrative and support service activities	
Public administration and defense; compulsory social security	
Education	
Human health and social work activities	
Arts, entertainment and recreation	
Other service activities	
Activities of households as employers	
Activities of extraterritorial organizations and bodies	

In the final chapter of the book, we summarise the risk differences between sectors and survey as well similarities in risks that exist between industries in some of the subsectors, based on our analysis of the specific risk profiles of different industries.

Chapter 2
Specific Features of Risk Management in the Industrial and Agricultural Sectors

2.1 Mining, Oil and Gas Extraction

This subsector includes the following industries according to ISIC Rev. 4:

- 05—Mining of coal and lignite
- 06—Extraction of crude petroleum and natural gas
- 07—Mining of metal ores
- 08—Other mining and quarrying
- 09—Mining support service activities
- 49—Transport via pipelines (we will include this service activity, which is part of the production process of many energy companies, in the subchapter)
- 50—Tanker transportation of energy-related substances on water (as with pipeline transportation, we will include tanker transportation in this subchapter)

General Description and Key Features of Subsector and Incorporated Industries

Extraction of raw material resources includes the production of oil and gas by conventional and unconventional methods, and the underground and surface mining of coal, peat, ore, uranium, stone, sand, and other materials for use in manufacturing, utilities and construction. We have also included in this subchapter the transportation of liquid energy-related substances via pipelines and marine transport (mainly by oil and LNG tankers): these activities are usually associated with energy production, and are part of the production process of many oil and gas companies.

To keep production costs down, companies from this subsector tend to develop and manage large-scale industrial complexes in conventional resource-rich regions. Prospecting for and extracting natural resources is of crucial importance for national security, budget revenues and domestic output, and for the output of co-operating

© Springer Nature Switzerland AG 2020
D. Chernov, D. Sornette, *Critical Risks of Different Economic Sectors*,
https://doi.org/10.1007/978-3-030-25034-8_2

industries. Thus, national and regional authorities expect to have a strong influence when it comes to deciding which extraction companies (especially foreign and transnational corporations) should have access to deposits. In this subsector, there is high competition for prospecting and operating rights between rival companies. Due to the common practice of trading raw materials through commodity exchanges, selling prices for such products are highly dependent on global economic growth, demand on feedstock, global changes in energy matters, environmental policies, technological progress, and so on. However, consumers of natural resources usually have a limited choice of suppliers. Production processes are tailored to the specific characteristics of a given raw material, so there are relatively few suppliers of each material worldwide, only a few of whom will obtain permission to operate in any particular country. Production activity is usually highly localized in limited areas and restricted by national environment legislation. It is a capital-intensive process that requires access to long-term and low-cost investment capital.

Critical Success Factors for an Organization Within the Subsector

Due to the global scarcity of many raw resources, there are two main critical factors in the subsector: (I) unrestricted ability to participate in the distribution of cost-effective state natural resources in different regions, and (II) the fostering of good, stable and long-term relationships with the authorities in key resource-rich regions.

If an extraction company is developing deposits abroad, they will need to maintain a close relationship with their own native government, which will see the promotion and protection of its companies in foreign countries as an opportunity to extend its "*soft power*": many countries seek to extend their influence, less nowadays in most countries by means of military force, but by intensified economic expansion of its corporations, cultural and media penetration, and so on. Another critical factor is ensuring a low accident rate at production sites, and operating all sites in an environmentally responsible way as the basis for their long-term sustainable development: this requires the respect and cooperation of the authorities, general public, local communities and environmental protection organizations. Operational sustainability, the stability of production quality and the ability to attract and retain qualified technical staff with a conservative safety-oriented attitude and working practices will all help to ensure the safety of the production process. A constructive relationship with local communities near mineral deposits reduces the potential for conflicts with national authorities, and the likelihood of excessive public scrutiny of production activity or difficult questions about the fair distribution of profits from the exploitation of national raw resources.

Stakeholders in this Subsector [1]

- Home government and national governments of production regions—50%
- Customers −15%
- Employees—10%
- Investors—10%
- Hardware vendors and service providers—5%
- Local communities—5%
- Other—5%

Typology of Common Risks, Main Features of Major Accidents and Risk Mitigation Measures Within the Subsector

Crises Induced by Conflicts with Governments and Regulators

Extraction companies can sometimes get involved in conflicts between national governments due to geopolitical tensions between their own government and the country of operation. Other flashpoints can include controversy over taxation and fair distribution of earnings from the extraction, plans by the company to reduce or limit investment in production activity in a given region which may harm the host country's revenue, the company's wish to sell what the country sees as national assets to other players without agreement from the national government, the financing of parties which are in opposition to the officially elected government, and so on. Generally, extraction companies are in a weak and subordinate position in any conflict with a national government—which has both the legal right as a sovereign power, and the support of local people, to regulate any activity concerning the development of national mineral resources. Several important cases illustrate the kind of conflict that can develop.

The Anglo-Iranian Oil Company (the Future BP) and the Prime Minister of Iran (Iran, 1953) In August 1941, Great Britain and the USSR occupied Iran in order to prevent Nazi Germany's capture of Iranian oil fields. After this military invasion, Reza Shah, the ruler of Iran, proposed his son, Mohammad Reza Pahlavi, as a new Shah of Iran who would be loyal to the Allies. Consequently, Pahlavi ruled the country until the Iranian Revolution in 1979. In the first stage of his reign, the Shah kept a balance between the West and the Soviet Union. But he was forced to take the Western side in 1953 after a successful cover operation by the CIA and MI-6, during which Mohammad Mossadegh, the democratically elected Prime

---

[1]Our informed appraisal of the influence of each audience on a typical organization within the subsector (100% = combined influence of all audiences)

Minister of Iran, was overthrown [2]. Previously Mossadegh, who was backed by Iran's communist Tudeh Party, had nationalized the Iranian oil industry and tried to kick the Anglo-Iranian Oil Company out of the country—where the company had been working since 1908. The coup allowed the company to work in Iran for a further 25 years. Eventually, after the Iranian Revolution in 1979, all oil assets owned by foreigners were confiscated and nationalized within the National Iranian Oil Company.

The Creeping Nationalization of Saudi Aramco (Saudi Arabia, 1970s–1980s)
In 1933, Saudi Arabia's only major source of income was the hajj, the annual pilgrimage of Muslims to the holy site of Mecca. Keen to diversify budget revenues and bring in more money, the House of Saud—the kingdom's ruling family—happily agreed to grant a concession for 60 years of oil exploration to the California-Arabian Standard Oil Co, which would later become the Arabian American Oil Co (ARAMCO). The company was a 100% subsidiary of Standard Oil of California (the future Chevron) and the Texas Oil Co (the future Texaco). Finally, in 1938, the first commercial Saudi oil was produced and exported. The Second World War confirmed the importance of oil to industrial society, and Saudi Arabia had estimated reserves amounting to 20 billion barrels of it—at that time, equivalent to all of America's proven reserves. Maintaining good relations with the Saudis thus became a strategic necessity for the US, and for the energy security of the Allies in Europe and the Pacific region. In 1943, the US government decided to provide military and financial assistance to Saudi Arabia in order to strengthen Saudi national security. The Saudis confirmed their consent for the use of their ports by Allied navies, and for the construction of an American air base in Dhahran, a major administrative center of the Saudi oil industry.

The initial conditions of the ARAMCO oil exploration concession were disadvantageous to Saudi Arabia because the country had no experience of the oil business when they agreed to it [3]. After the war, there was tremendous economic growth worldwide during the recovery of Europe and Japan, and oil consumption rose to eight times pre-war levels; by the mid-1960s, the Middle East had overtaken the United States as the largest oil-producing region in the world, with an output of 8.5 million barrels a day [4]. The production costs of oil extraction in Saudi Arabia were and remain the lowest in the world—around US$2/barrel [5]. Other Persian

---

[2]Ervand Abrahamian, The Coup: 1953, the CIA, and the Roots of Modern US-Iranian Relations, New Press, New York, 2013

[3]Alexey Vasilyev, History of Saudi Arabia, 1745–1973, Trud Publishing House, Moscow, 1999, p.363

[4]Alexey Vasilyev, History of Saudi Arabia, 1745–1973, Trud Publishing House, Moscow, 1999, p.378

[5]Interview with Ali Al-Naimi, Saudi Arabian Minister of Petroleum and Mineral Resources and former CEO of Saudi Aramco, from documentary "60 Minutes: Season 41, Episode 11: The Oil Kingdom", CBS television network, December 7, 2008

Gulf countries had very low breakeven costs too, from US$1/barrel to $5/barrel [6]. In the 1950s and 1960s, production costs for Saudi oil were five times lower than those in Venezuela and ten times lower than in the United States. In 1961, the rate of return on ARAMCO's capital investment was phenomenal, standing at 81.5% when the rate across the US oil industry was no more than 10–12% [7]. Naturally, over the following decades the House of Saud was keen to raise its share in the uniquely profitable ARAMCO. They could also see their interest in reducing the influence of Americans on the Saudi oil business, by bringing in other international oil companies who would then offer more profitable oil concessions.

In 1960, the Organization of the Petroleum Exporting Countries (OPEC) was established by the governments of Iran, Iraq, Kuwait, Saudi Arabia and Venezuela in order to ensure the steady rise of oil prices and reduce the power of the "Seven Sisters" (the seven private Western oil companies, which between them controlled a majority of the world's petroleum reserves). Later Abu Dhabi, Qatar, Libya, Algeria, Indonesia, Nigeria, Ecuador, and Gabon joined the OPEC cartel too. OPEC aimed to coordinate oil production in every participating country, in order to manage oil prices and put constant pressure on the "Seven Sisters" to increase the native governments' share in export revenues. In 1965, the Saudi government began to receive 50% of ARAMCO profits. The tipping point in the struggle for getting a better share of oil profits came in 1967, when Israel attacked and defeated Egypt, Syria and Lebanon in the Six-Day War. Israel occupied the Gaza Strip and the Sinai Peninsula (part of Egypt), the West Bank, including East Jerusalem (Jordan) and the Golan Heights (Syria). Israel had vigorous support from the United States and other Western countries. It was incumbent on Saudi Arabia, having positioned itself as the defender of all Arabs and Muslims, to provide political, moral and financial help to the affected nations. And in spite of its close relations with the West, Saudi was also obliged to impose a 3-month oil embargo against the US, Britain, and West Germany for their support of Israel: they had to be seen to demonstrate the unity of Muslims against the occupation of East Jerusalem, site of the third holiest place in Islam, the Al-Aqsa Mosque. However, the embargo did not have a significant impact on oil price or on Western economies.

After the 1969 revolution in Libya led by Colonel Qaddafi, the new government there demanded that Western oil companies remit 54–55% of their profits to the Libyan people. When the companies refused to oblige, the Libyan government reduced the authorized oil output. In the end the companies had no choice but to accept new rules, because Libya provided 20% of the total oil imports to Western Europe. This event triggered a domino effect among other OPEC countries [8].

---

[6]Energy Security: Evaluating US Vulnerability to Oil Supply Disruptions and Options for Mitigating Their Effects, US Government Accountability Office, December 12, 1996, p.20, http://www.gao.gov/archive/1997/rc97006.pdf

[7]Alexey Vasilyev, History of Saudi Arabia, 1745–1973, Trud Publishing House, Moscow, 1999, p.383

[8]Alexey Vasilyev, History of Saudi Arabia, 1745–1973, Trud Publishing House, Moscow, 1999, p.455

The excuse for the next oil price increase was an economic decision by Richard Nixon, which was caused by the Vietnam War and ended the post-war "Golden Age" economic cycle of the 1940s–1970s. In August 1971, the US President suspended dollar to gold exchange, a fundamental principle of the post-war monetary system established by the Bretton Woods Conference in 1944, which determined the dollar as a global reserve currency backed by gold with a fixed convertible rate. This decision provoked reciprocal action from OPEC: the proposal either to calculate oil in gold and not in dollars, or to recalculate oil prices in devalued dollars. For the period between 1970 and September 1973, the price of oil rose by 72%, while US oil imports rose from 2.2 million barrels a day in 1967 to 6.2 million barrels a day in 1973 [9].

For the United States, which became the largest oil importer in the world after its domestic oil production peaked in 1970 at 9.64 million barrels a day, additional expenses on oil meant a deterioration of trade balance and economic downturn. Nevertheless, in negotiations in the summer of 1973, Saudi Arabia offered to provide 20 million barrels/day by 1980 to the United States if they would withhold any further military and financial support to Israel. If they did not, it would be very difficult for the Saudis to supply the US even at the existing level. The US did not respond to these demands. The Israeli occupation of Arab land, and their humiliation of the Arab nations by the stunning defeat inflicted during the 1967 Six-Day War, required some response from Arab politicians. Consequently, on 6–25 October 1973, Egyptian and Syrian forces attacked Israeli positions on the Sinai Peninsula and the Golan Heights, but without significant success. During the war, the United States re-supplied Israel. This fact became the pretext for another 6-month oil embargo imposed by the Arab producers against the United States, the Netherlands, Canada, Japan and the United Kingdom. The embargo was intended to motivate the target countries to provide the required pressure on Israel to make it return the occupied territory. Colonel Qaddafi was first, declaring an oil embargo against the West and immediately doubling the price of Libyan oil; several days later, all the Arab members of OPEC as well as Egypt, Syria and Tunisia supported this decision. For Saudi Arabia, the leader of Sunni Muslims, the issue of returning the occupied territory symbolized the larger fight for Islam in the Middle East. The Saudi King Faisal, in discussion with US Secretary of State Henry Kissinger, urged him: *"Can't you help me? Can't you give me Jerusalem?"* [10]. But after a series of discussions with the Arab oil states, Kissinger announced that the use of petroleum as a weapon to influence the outcome of the Arab-Israeli conflict had little merit in reality [11].

---

[9] Andrew Scott Cooper, The Oil Kings: How the US, Iran, and Saudi Arabia Changed the Balance of Power in the Middle East, Simon and Schuster, New York, 2011, pp.90, 110

[10] Andrew Scott Cooper, The Oil Kings: How the US, Iran, and Saudi Arabia Changed the Balance of Power in the Middle East, Simon and Schuster, New York, 2011, p.131

[11] Mahmoud A. El-Gamal, Amy Myers Jaffe, Oil, Dollars, Debt, and Crises: The Global Curse of Black Gold, Cambridge University Press, 2010, p.7

The oil embargo not only met the expectations of the Arab public for a serious demonstrative response to the Israeli occupation, but also satisfied the desire of the Arab elite to redistribute wealth from the industrialized West to the Middle East and North Africa. Sheikh Yamani, the Saudi Minister of Petroleum and Mineral Resources, declared: *"The moment has come. We are masters of our own commodity"* [12]. As a result, by March 1974 oil prices had been jacked up more than 2.5 times—from US$4.90/barrel [$25 in 2018 prices] to US$12/barrel [$61 in 2018 prices] [13]. All commercial shipment of oil from the Middle East to the United States halted—and at this point more than 25% of total US oil imports came from the Middle East. This oil embargo had grave and far-reaching consequences for Western military and economic security. It tipped the United States into economic recession between November 1973 and March 1975. In 1974, inflation there reached an annual rate of 12%; in France and Belgium it reached 16%, in Great Britain 18%, in Japan 24% and in Italy 25%. US economic commentator Alan Greenspan admitted that banks on Wall Street were at risk of collapse if OPEC raised the price of oil [14]. The United States were even forced to ask their sworn enemy the Soviet Union to start exchanging American wheat for Soviet oil to cover the oil deficit.

The nationalization of the assets of the "Seven Sisters" in several Middle-Eastern countries resulted in a 74% cut in their production, from 25.5 million barrels/day in 1973 to 6.7 million barrels/day in 1982 [15]. Before the 1973 crisis, western oil companies owned 85% of world oil reserves. After the crisis, the OPEC countries became the leading power in the world energy business: by the 2000s, OPEC members already held 78% of the world's proven oil reserves [16]. According to the World Bank, by 2010, state-owned companies possessed 90% of the world's conventional oil reserves and provided 75% of total world oil output [17]. The oil embargo enabled Saudi Arabia not only to raise prices (in all fairness, Saudi Arabia was one of the most moderate countries within OPEC), but also to become a part owner of ARAMCO. The threat of nationalization forced a pool of American companies operating there to accept changes in their shareholding structure. The Saudis got 25% with the agreement to reach 51% by the beginning of 1983; nowadays, ARAMCO is a 100% Saudi state-owned company.

---

[12]Peter Maass, Crude World: The Violent Twilight of Oil, Random House LLC, New York, 2009, p. 169

[13]Andrew Scott Cooper, The Oil Kings: How the US, Iran, and Saudi Arabia Changed the Balance of Power in the Middle East, Simon and Schuster, New York, 2011, pp.123, 149

[14]Andrew Scott Cooper, The Oil Kings: How the US, Iran, and Saudi Arabia Changed the Balance of Power in the Middle East, Simon and Schuster, New York, 2011, p.10

[15]Antoine Ayoub, Oil: Economics and Politics, Energy Studies Review, vol. 6, no 1, Ontario, 1994, p. 52

[16]Gal Luft, Anne Korin, Energy Security Challenges for the twenty-first Century: A Reference Handbook, ABC-CLIO, Santa Barbara, 2009, p.83

[17]Peter Voser, End of the oil boom? Notenshtein Dialogue, St Gallen, December 2013, p.2

After this crisis, the *"petrodollar recycling machine"* was created. Under this long-standing arrangement, foreign oil producers—mainly Saudi Arabia—buy US treasuries in exchange for oil at high prices nominated in US dollars.

Chinese Attempted Acquisition of UNOCAL (United States, 2005) American politicians severely criticized the Saudis for their enforced restriction of the market share of American oil companies in the Arabian Peninsula; but at home these same politicians had similar motives in preventing the acquisition of UNOCAL (the Union Oil Company of California) by the Chinese state-controlled CNOOC (China National Offshore Oil Company Ltd.). In 2005, CNOOC was ready to offer the highest bid for UNOCAL: US$18.5 billion. However, under strong political pressure from Washington on grounds of the perceived threat to American national security, the Chinese energy company was forced to withdraw its offer. A majority of the US House of Representatives voted to veto the sale of such a large slice of American energy infrastructure to a Chinese firm, and called on the Bush administration to block the deal. Unsurprisingly, CNOOC took a dim view of the situation: *"The unprecedented political opposition [from Washington] that followed the announcement of our proposed transaction ... was regrettable and unjustified"* [18]. The Chinese foreign ministry commented: *"we encourage the US to allow normal trade relations to take place without political interference"* [19]. Ultimately, UNOCAL was sold to Chevron for $17.9 billion, allowing the American energy asset to remain under the control of an American energy major.

Let us summarize what we can deduce from these examples of crises induced by conflict with governments. The governments of countries with rich natural resources usually have huge support from their citizens for extensive control over national mineral deposits, up to and including nationalization measures; thus, the owners of extraction companies loyal to their governments may make billions in profits, while those who are arrogant and disloyal can expect prison, financial ruin and exile (as in the case of the Yukos oil company in Russia). It is in the interest of any extraction company to steer clear of conflict in their relations with the national government. Therefore, the company should follow national security priorities, pay taxes transparently and in full, participate in national economic development by working hard to exploit domestic deposits to contribute to the national budget, order any equipment or infrastructure from domestic suppliers when possible, recruit and train local staff and support local communities, invest in safety, apply strict environmental rules, and so on. They should avoid taking any official position on the domestic or foreign policy of the country where they are developing deposits. They should refrain from supporting opposition parties against the official government, and from any questionable schemes to try and reduce their tax base.

---

[18]CNOOC Withdraws Unocal Bid, Xinhua News Agency August 3, 2005

[19]David Barboza, Andrew Ross Sorkin, Chinese Oil Company Offers $18.5 Billion for Unocal, The New York Times, June 22, 2005

Geopolitical Struggle

The energy business is influenced by the geopolitical struggle between global and regional powers. An extraction company may have an excellent relationship with the national government in a resource-rich host country, but their business could still be affected by any international geopolitical crisis that affects the country, whether through local guerilla war against the national government or by acts of terrorism or sabotage. There are several prominent examples of such occurrences.

Saudi Global Oil Interventions Caused by the Soviet War in Afghanistan, the Low Oil Price Regime, the War on Terror and the Transition to High Oil Prices (Worldwide, 1980s–2000s) The 1973 oil embargo and 1979 Islamic Revolution provoked a phenomenal jump in oil prices—from US $3.3/barrel in 1973 [$18.6 in 2018 prices] to $37/barrel in 1980 [$112 in 2018]. With most of their former production taken over by the OPEC countries' state-ruled national oil companies, the "Seven Sisters" shifted their focus onto mineral prospecting in the North Sea (Norway and Great Britain), Prudhoe Bay and the Trans-Alaska pipeline (USA), Alberta and Newfoundland (Canada), and the US Gulf of Mexico.

In the 1970s, the Soviet Union discovered giant oil and gas fields in Western Siberia with a low-cost breakeven point of around US$0.80/barrel [$4.5 in 2018 prices] [20]. The rise in oil prices after the 1973 oil embargo was favorable for the income of the Soviet Union. Then, in December 1979, the USSR intervened in Afghanistan. The United States and the House of Saud used this as grounds for the declaration of a "Sacred Jihad"—a "Holy War" against infidels (in this case the Red Army) in a Muslim land (Afghanistan). The overall expenses of the US on the Afghanistan operation amounted to around $3 billion of American taxpayers' money [21]. Saudi and the US had two clear goals in supporting up to 25,000 mujahideen in Afghanistan. Sending thousands of religious radicals into the fire of the Red Army would reduce the pressure on Muslim regimes loyal to America; it could also mire the Soviet Union in Afghanistan for years (nearly a decade as it turned out), in their own equivalent of America's expensive, drawn-out and ultimately disastrous war in Vietnam. American officials tried to convince the House of Saud to reduce oil prices in order to hurt the economy of the USSR [22], which was heavily oil-dependent, with the share of oil and gas in Soviet export revenues having jumped from 16.2% in 1960 to 54.4% in 1985 [23]. Saudi Arabia was the largest oil producer in OPEC, contributing 40% of the cartel's total production with the capacity to supply an

---

[20]Gennady Shmal, Energy heart of Russia. 60 years of West-Siberian oil and gas province, Drilling and Oil Magazine, Moscow, November 2013

[21]Craig Unger, House of Bush, House of Saud: The Secret Relationship Between the World's Two Most Powerful Dynasties, Simon and Schuster, New York, 2004, p.97

[22]Peter Schweizer, Victory: The Reagan Administration's Secret Strategy that Hastened the Collapse of the Soviet Union, Atlantic Monthly Press, New York, 1994, p. 27–31

[23]Stepan Sulakshin et al., Strategies of Russia in historical and global perspectives. Directmedia, Moscow, 2009, p. 88

additional 2–3 million barrels/day, and thus had a strong influence on the determination of oil prices.

By the 1980s, over-supply from new fields was saturating the market. This situation was exacerbated by Saudi Arabia's decision in late 1985 to resume oil production at full capacity, after a period in which they had restricted production to limit supply. By 1986, oil prices had dropped back down to just US$14/barrel [$33 in 2018 prices] [24] and they wobbled around this level for almost 20 years. (There was a local price jump in 1990–1991, when Iraq invaded Kuwait, but prices dropped again after the defeat of Saddam in the Desert Storm operation). As intended, the Saudi action had an impact on Soviet state finances. But it also eroded the profits of Western oil producers with a high break-even in Alaska, the North Sea, the US Gulf of Mexico and others. In 1986, George H. W. Bush, US Vice President during Reagan's tenure, had to persuade the Saudis that the oil glut was making West Texas oil companies unprofitable, working as they were with far higher production costs than those involved in producing Saudi light crude. It is noteworthy that Arbusto, Bush Exploration, Spectrum 7 and Harken Energy—small oil companies owned by George W. Bush—were all drilling wells in Texas in the 1980s–1990s, but the wells could only be run at a loss [25]. And after the collapse of the Soviet Union, the American petroleum industry and military-industrial complex suffered not only from low oil prices, but from a huge reduction in US government expenditure on weapons. Oil prices in particular were continuing to fall: after a slight rally at the end of the 1980s, the average annual oil price dropped from US$23/barrel [$45 in 2018 prices] in 1990 to $12/barrel [$20 in 2018 prices] in 1998. The American shale revolution—which eventually occurred in 2005–2014—could have started in the 1980s, but low energy prices did not allow it.

Natural gas prices had fallen too: the average nominal natural gas price at the wellhead during the 1980s was US$72/1000 m^3 [$168 in 2018 prices], but by the 1990s, it had dropped to $67/1000 m^3 [$110 in 2018 prices] [26]. To put this in context, the average production costs of natural gas from shale formations within the United States were US$140–$230/1000 m^3 at the wellhead in 2011–2013 [27,28,29,30,31]; by

---

[24]Oil: Crude oil prices 1861–2009, BP Statistical Review of World Energy 2010, June 2010

[25]Craig Unger, House of Bush, House of Saud: The Secret Relationship Between the World's Two Most Powerful Dynasties, Simon and Schuster, New York, 2004, p. 117

[26]US Natural Gas Wellhead Price (1922–2013), US Energy Information Administration, April 2014

[27]During our calculations we use following conversion rates of natural gas from American metrics into European one: one million British Thermal Units (MMbtu) = 1000 cubic foots (Mcf), 35,000 cubic foots (Mcf) = 1000 cubic meters (m^3) of natural gas. We use factor 35 for conversation price of 1MMbtu into 1000 m^3

[28]The Future of Natural Gas, Appendix 2D: Shale Gas Economic Sensitivities, MIT Energy Initiative, 2011, p.2

[29]First 5 years of "shale gas revolution". What we now know for sure? Centre for Global Energy Markets of Energy Research Institute of the Russian Academy of Sciences, November 2012, pp.25–26, 32

[30]Euan Mearns, What is the real cost of shale gas? Energy Matters, November 28, 2013

[31]Ivan Sandrea, US shale gas and tight oil industry performance: challenges and opportunities, The Oxford Institute for Energy Studies, March 21, 2014, p. 4

contrast, the Russian conventional natural gas giant Gazprom's average production costs were US\$37–38/1000 m^3 in 2013–2014 [32]. Thus, in the 1990s, 90% of gas produced in the United States was still from low-cost conventional gas fields, but the output of these traditional deposits gradually fell as the fields became depleted.

The massive surge of oil and natural gas production in the US occurred only in 2005–2014 when higher energy prices made gas production from shale formations profitable, and environmental restrictions on fracking were lifted by the George W. Bush administration. The main causes of the oil price hike were the War on Terror and the invasion of Iraq. The US, allied and Saudi support for the mujahideen movement in its campaign against the Soviets in Afghanistan during the 1980s led to the emergence of Al Qaeda and ultimately to the 9/11 attacks in 2001 [33]. The beginning of the War on Terror in response to 9/11, and the subsequent invasion of Iraq, became the tipping points for a change in the trend of oil prices—from an annual average of US\$24/barrel in 2001 [\$34 in 2018 prices] to \$97/barrel in 2008 [\$113 in 2018 prices] .

Nowruz Oil Field Spills and Oil Tanker War (Iran, 1983) Following the Iranian Revolution in 1979, Shiite radicals took possession of the vast arsenal of American weapons that the US had sold to the Shah during his vigorous militarization program in the 1970s. Ayatollah Khomeini, who had already declared a strongly anti-Saudi stance, was now in a position where he could order an invasion by Iranian revolutionaries of the Saudis' Eastern Province (where the main Saudi oil fields are located) and Bahrain, both areas with a Shiite majority. The Americans and Saudis urgently courted Saddam Hussein, the leader of Iraq, to bring him from a pro-Soviet orientation onto their side [34]. According to some sources [35], they paid him around US \$30 billion to persuade him to intervene in Iran, and there was continuous financial, military and technical support by the US to Iraq over the whole duration of the war [36]. Saddam duly attacked Iran; the resulting war continued from September 1980 to August 1988, and caused up to a million casualties with no obvious benefit to either side. Nevertheless, the two largest and most militarily powerful nations of the Persian Gulf, either of which could easily have occupied the far weaker Saudi Arabia, were themselves severely weakened by this pointless war.

---

[32]Mikhail Korchemkin, East European Gas Analysis, September 11, 2014, http://www.eegas.com/rep2014q1-cost_e.htm

[33]Mapping militant organisations, Stanford University, August 18, 2015 (http://web.stanford.edu/group/mappingmilitants/cgi-bin/groups/view/21, retrieved 25 July 2018)

[34]Dobbs, M. (2015). U.S. Had Key Role in Iraq Buildup. Washington Post. Retrieved 25 July 2018, http://www.washingtonpost.com/archive/politics/2002/12/30/us-had-key-role-in-iraq-buildup/133cec74-3816-4652-9bd8-7d118699d6f8

[35]Craig Unger, House of Bush, House of Saud: The Secret Relationship Between the World's Two Most Powerful Dynasties, Simon and Schuster, New York, 2004, p.66

[36]Fatemeh Mohammadi, 20 things the U.S. did to help Saddam against Iran, Nov 4, 2015, http://english.khamenei.ir/news/2168/20-things-the-U-S-did-to-help-Saddam-against-Iran,   retrieved 25 July 2018

The war had an impact on the energy sector of Iran. For example, in early 1983 there were several oil spills on the Nowruz field. In February that year, an oil tanker collided with one of the offshore platforms of the field, which later caused the riser to collapse into the wellhead, releasing 1500 barrels/day. The ongoing war prevented the leakage from being tackled quickly; and in March the platform was attacked by Iraqi air forces, setting fire to the huge oil slick that already surrounded the stricken wellhead. It took around 6 months for the well to be finally capped by Iranian engineers—11 people died during the repair. Later that same month, another platform was attacked and destroyed by Iraqi helicopter gunships; the attack created a further burning oil spill of about 5000 barrels/day [37]. With the war raging on, it took more than 2 years until the spill was cleared up; estimates of the amount of oil released into the sea during this period range from 0.7 million to 1.9 million barrels. At the beginning of the war, Iraq declared that it would destroy any oil tanker entering an Iranian port to pump Iranian oil for export. True to their threat, the Iraqis carried out air attacks on the Turkish oil tanker Atlas I, the Greek tanker Scapmount, and the Liberian Neptunia.

The Gulf War and the Destruction of Kuwaiti Oil Fields (Kuwait, 1991) In the aftermath of the Iraq-Iran war, the Iraqi economy was shattered by military expenses and needed high oil prices; but the strategy of Kuwait and Saudi Arabia to sell large amounts of oil at low prices in order to please the United States—the only remaining superpower in the world—contradicted the interests of post-war Iraq. The production of the Rumaila oil field on the Iraqi–Kuwait border was shared between the two countries. Kuwait allegedly made side-drilling under Iraqi soil and were pumping oil from Iraqi side fuelling low oil price. There was little prospect of oil prices rising any time soon, so in 1990 Saddam decided to invade Kuwait in order to halt Kuwaiti oil production and provoke the oil price jump he so badly needed. Between the end of July and August 24, 1990, the world price of crude oil climbed from about US$16/barrel [$30 in 2018 prices] to more than $28/barrel [$53 in 2018 prices]. The price escalated further in September, reaching about US$36/barrel [$69 in 2018 prices] [38]. The intervention of the Iraqi army—the most powerful army in the Middle East—in the former Iraqi province of Kuwait posed a threat to Saudi Arabia, which could be occupied by Iraq in just 3 days. The situation for the Saudis was aggravated by the fact that many Arab countries supported the actions of Saddam Hussein against pro-Saudi Kuwait and Saudi Arabia itself. And if Iraq, already an oil-rich country in its own right, had conquered both Kuwait and the Eastern Province of Saudi Arabia, Saddam would have controlled 40% of world oil reserves [39].

---

[37]Nowruz Oil Field, Emergency Response Division, Office of Response and Restoration, US National Ocean Service, National Oceanic and Atmospheric Administration, https://incidentnews. noaa.gov/incident/6262

[38]Petroleum Chronology Of Events 1970–2000, US Energy Information Administration, May 2002

[39]Craig Unger, House of Bush, House of Saud: The Secret Relationship Between the World's Two Most Powerful Dynasties, Simon and Schuster, New York, 2004, p.144

During Operation Desert Storm in January–February 1991, US-led coalition forces from 34 nations won Kuwait back. During the retreat, Iraqi forces mined and exploded 600 Kuwaiti oil wells, but this diversion did not have a significant impact on falling oil prices. Moreover, the ensuing UN sanctions against Iraq led to the elimination of Iraqi oil from the world market—to the pleasure of Saudi Arabia and Iran, who compensated for Iraqi oil exports by stepping up their own production. The Kuwaiti oil business became a hostage of the great geopolitical game in the Gulf. The oil spillage from the exploded oil wells was the largest in human history: the approximate discharge was ten million barrels, as compared to around 4.9 million barrels in 2010 from the destroyed Macondo well drilled by the Deepwater Horizon rig in the Gulf of Mexico. Kuwait was forced to bring in private oil rig firefighters from abroad and pay them more than US$1 billion.

Attack on the Limburg Oil Tanker (Yemen, 2002) Al Qaeda attacked the Limburg oil tanker in the Gulf of Aden, resulting in a marine oil leakage of approximately 90,000 barrels.

Terrorist Attacks Against Foreign Workers in the Saudi Energy Industry (Saudi Arabia, 2004) 22 people were killed and 25 injured in an attack by Islamic militants on the Arab Petroleum Investments Corporation building, the Petroleum Centre and a housing complex for foreign workers. Just a month before, a similar attack had occurred at Yanbu petrochemical plant, leaving seven expat workers dead.

Continuous Sabotage of Oil Pipelines (Turkey-Northern Iraq, Columbia) The Kirkuk-Ceyhan oil pipeline was developed to transfer oil from Northern Iraq to the sea at the Turkish port of Ceyhan. However, the Kurdistan Workers Party disrupted operations on the pipeline in Turkey, and Sunni Muslim militants frequently did the same in Northern Iraq. Designed to run at a capacity of 1.6 million barrels/day, the Kirkuk-Ceyhan pipeline was only delivering 0.54 million barrels/day in June 2011, and the flow went down to just 0.16 million in July 2013 due to constant attacks on the pipeline [40]. In 2007, Massoud Barzani, leader of the Iraqi Kurds, estimated the total losses incurred by sabotage and theft at around $12 billion a year of oil export revenue; from 2003 to 2007 alone, the Iraqi oil industry was subjected to over 500 attacks [41]. A similar situation occurred in Columbia, where the Cano Limon-Covenas Trans-Andean oil pipelines were attacked by the National Liberation Army of Colombia (ELN) and the Revolutionary Armed Forces of Colombia (FARC).

Libyan Civil War and Terrorist Attack on in Amenas Gas Plant in Algeria (2011–2013) During a reign of more than 40 years, Muammar Gaddafi steered a clear pro-Libyan course regarding relations with the Western petroleum industry. He was the first of the Middle Eastern leaders to change the tax regime for major companies in

---

[40]Iraq oil exports hit 16-month low despite higher production, AFP, August 8, 2013

[41]Gal Luft, Anne Korin, Energy Security Challenges for the twenty-first Century: A Reference Handbook, ABC-CLIO, 2009, p. 21

favor of his country, demanding the partial nationalization of their assets and imposing sanctions against some Western countries for their support of Israel in 1973.

The rise of Islamic radical groups in Tunisia and Egypt, and the confusion of security forces in these countries after the dethronement of Ben Ali and Hosni Mubarak, led to a great influx of Muslim radicals into Libya to join the civil war against Colonel Gaddafi. Moreover, Al-Qaeda's North African wing pledged to help the anti-Gaddafi uprising [42]. When the Arab Spring seemed to be threatening the regime in Libya, it was NATO, starting with France and the US, Qatar and the United Arab Emirates who decided they should also intervene in the Libyan civil war. Ultimately—after October 2011, when Gaddafi was tortured to death—Libya became a failed state due to the uncontrolled rise of jihadist groups, constant internecine fights between local tribes, and the absence of a full-fledged government because security was too weak for fair elections. In total, the anti-Gaddafi uprising and the subsequent war cost more than 30,000 Libyan lives. Strikes by oil workers, and blockages of seaport facilities for oil export, caused a massive reduction of oil exports, cutting Libya's oil and natural gas output by 60–90%: in 2010, Libya produced 1.8 million barrels/day, but by August 2011 the level had fallen to around 0.45 million barrels/day and by August 2013 to only 0.3 million barrels [43,44,45].

In 2013, there were also disruptions of the Green Stream pipeline from Libya to Italy due to combat between groups of former rebel fighters. This natural gas submarine pipeline carried 12% of Italian gas imports and had an annual capacity of up to 10 billion m^3 [46,47]. In January 2013, radicals under the leadership of Mokhtar Belmoktar, military commander of Al-Qaeda in the Maghreb, invaded Algeria from "*liberated*" Libya and attacked the In Amenas gas plant (which was jointly operated by Sonatrach, the Algerian state oil company, BP and the Norwegian Statoil). Algeria is Europe's third-largest gas supplier after Russia and Norway, providing around a third of the gas used by Italy and Spain. The outage of the In Amenas gas plant contributed to a 10% drop in Algeria's hydrocarbons production in the first half of 2013. Algerian gas supplies were considered so unstable that Rome and Madrid were seeking ways to reduce their dependence on Algerian gas [48]. The interruption of Libyan and Algerian natural gas supplies was one reason why Europe

---

[42] Al Qaeda's North African wing says it backs Libya uprising, CNN Wire Staff, February 24, 2011

[43] Libya is a major energy exporter, especially to Europe, US Energy Information Administration, March 21, 2011

[44] Libyan crude oil production levels influence international crude oil markets, US Energy Information Administration, November 26, 2013

[45] Matthew Philips, Libya's Oil Industry Is in Trouble, Bloomberg Businessweek, August 14, 2013

[46] Ali Shuaib, Marie-Louise Gumuchian, Libya stops gas exports to Italy after militia fight, Reuters, March 3, 2013

[47] Libyan Berbers shut gas pipeline to Italy, cut major income source, Reuters, November 11, 2013

[48] One year on from In Amenas attack, The Economist Intelligence Unit, January 14th 2014

bought an additional 23 billion m^3 of gas from Russia in 2013 compared with the previous year [49].

Western Sanctions Against the Russian Energy Sector Caused by the Ukrainian Crisis (Since 2014) The sanctions imposed on Russia following the Ukrainian crisis are intended to reduce the country's long-term oil production, because these sanctions restrict the access of Russian energy companies to the financial resources of the West, and to sophisticated Western equipment for deepwater oil drilling in the Arctic seas and shale oil production. The majority of Russian conventional oil fields will become depleted in the foreseeable future, and the absence of new oil fields will gradually reduce the production output of one of the largest oil producer in the world.

A clear example of the implementation of these sanctions is the case of Rosneft. At the beginning of August 2014, Rosneft, together with ExxonMobil, announced the launch of drilling in the Kara Sea, the Russian part of the Arctic ocean, with the aim of discovering the northernmost oil in the world. Rosneft's CEO declared: "*We hope to open a new Kara Sea oil province, which will, according to expert estimates, in terms of resources exceed... oil and gas provinces such as the Gulf of Mexico, the Brazilian shelf, the shelf of Alaska and Canada, and will be comparable to the resource base of Saudi Arabia*" [50]. Several weeks later, Rosneft and ExxonMobil found positive oil discharge during drilling, which led to estimates of 100 million tons, or about 750 million barrels, of high-quality light oil and 338 billion m^3 of gas [51]. But due to Western sanctions, the drilling in the Kara Sea was soon suspended, because Western offshore oil drilling equipment had been used to discover the new province [52]. Previously, Rosneft had been expecting to invest US$400 billion in the Arctic over the next 15 years on developing the new oil province, with vast resources they had boldly compared to the developed reserves of Saudi Arabia [53]. Western sanctions could also have postponed the launch of several Russian LNG developments in the Far East of Russia, another joint project between Rosneft and ExxonMobil, and of the Yamal LNG project in the Arctic, a joint effort between the Russian Novatek, the French Total and the Chinese National Petroleum Company, because access to loans from the West has become more complex; ultimately,

---

[49]Nikolay Pakhomov, Oil and gas: energy becomes a political football, The Telegraph, April 29, 2014

[50]Press release of OJSC OC Rosneft: Drilling in the Kara Sea Started, FT Company Announcement, August 11, 2014; Putin gave a start drilling in the project of Rosneft and ExxonMobil, RIA Novosti, August 9, 2014

[51]Jack Farchy, Ed Crook, Rosneft and ExxonMobil strike oil in Arctic well, The Financial Times, September 27, 2014

[52]Alan Katz, Joe Carroll, Mikael Holter and Stephen Bierman Exxon Said to Halt Arctic Oil Well Drilling on Sanctions, Bloomberg, September 19, 2014

[53]Jack Farchy, Ed Crook, Rosneft and ExxonMobil strike oil in Arctic well, The Financial Times, September 27, 2014

the Russian government and Chinese banks had to step in and became the lenders for these projects [54].

Drone Attack on Key Saudi Aramco Processing Facility (Saudi Arabia, 2019)
In September 2019, facilities which process the majority of Saudi Aramco's oil were attacked, probably by drones and missiles. Iranian-backed Houthi rebels in Yemen took responsibility for this act, demanding that Saudi Arabia halt its military intervention carried out in Yemen since 2014. The short-term impact of the attack was overwhelming: it halved the country's oil production, cutting it by up to 5.7 million barrels per day, or around 5% of the world's daily production. Production of natural gas was also reduced. The attack provoked the largest technical breakdown in the history of Aramco, forcing the Saudi state oil company to buy oil abroad in order to fulfil its oil export contracts, and make urgent repairs to the affected infrastructure. The longer-term impact was also considerable: it showed the vulnerability of global energy infrastructure to attacks by drones, a cheap and easily available new air weapon which can be launched and guided from anywhere, making them very hard to detect and neutralize.

Companies which intend to start working in a resource-rich country should research and understand the history of the region, and the occurrence of inter- and intra-state wars. They should keep track of the development of local political systems, the potential stability or otherwise of the current regime, the intentions of opposition groups and their attitude towards the company and its native country, the complexity of tribal relations, influential authors and commentators, and potential motivations for terrorism or sabotage. Without such understanding, they cannot realistically assess the risks of developing energy infrastructure in the region. It is not excessive for an extraction company to discuss the possibility of investment in a new country with representatives of its native government—including foreign affairs, military and intelligence—in order to enlist their support in promoting and protecting the company in the new region.

External Economic Shocks and Output Competition

In this subsector of mining, oil and gas extraction, prices are agreed upon on the majority of raw materials through commodity exchanges, so companies have to assess carefully the global economic situation, and the motivations and actions of different resource-rich states, to protect their market shares and minimize the threat to their business from possible dumping or other hostile competitive action.

Global Financial and Economic Crisis (Worldwide, 2008–2009) During the 2000s, a real estate bubble was forming in the US [55], which burst during

---

[54]Russia's Yamal LNG to get 150 bln rbls from state in Q1 – report, Reuters, December 18, 2014

[55]W.-X. Zhou, D. Sornette, Is There a Real-Estate Bubble in the US?, Physica A: Statistical Mechanics and its Applications, 2006, 361, pp.297–308

2007–2008. More than US$17 trillion of household wealth was wiped out within 21 months. The American subprime mortgage crisis triggered a global financial and economic crisis in 2008–2009 [56], which caused the worst recession for over 50 years. Total stock market losses exceeded US$30 trillion worldwide [57]. To prevent a total collapse of the world financial system, governments imperiled trillions of taxpayers' money on bailouts of private financial institutions, which were "*too big to fail*". This global salvage operation disrupted the stability of government finance not only in the USA, but also in many European countries. The annual US federal deficit grew from US$161 billion in 2007 to $1.4 trillion in 2009 [58]; and total public debt—the total amount owed by the federal government, including debts from intra-governmental holdings—increased by US$3.5 trillion, from $8.8 trillion in the middle of 2007 to $12.3 trillion at the end of 2009 [59]. Due to the crisis and associated slowdown of economic growth, the price of Brent crude oil fell from US$147/barrel to around $40/barrel in just 6 months. The oil price drop triggered price collapses on other commodities including natural gas, coal, ferrous ore, and nonferrous metals, hitting countries like Russia, Canada and Australia which have high resource export revenues. Nevertheless, during the quantitative easing that was part of the US response to the 2008/2009 financial and economic crisis, commercial banks received large amounts of dollars from the US Federal Reserve and invested them in stocks and commodities, while they shied away from providing credit for real American businesses. There was also a global reduction in interest rates, easier access to credit and a concomitant devaluation of the dollar resulting from the quantitative easing measures, which led to a fast recovery in commodity prices between 2010 and 2014.

Saudi Oil War Against us Shale Producers (Worldwide, 2013–2016) Rising oil prices between the mid-2000s and mid-2010s initiated the development of unconventional oil and gas resources in the United States and Canada: oil sands, extra heavy oil, offshore ultra-deepwater drilling, coalbed methane, shale gas and shale oil. Natural gas and oil produced from shale strata in the United States dramatically changed the structure of the US energy balance. According to the US Energy Information Administration, over the last decade the proportion of total American gas production provided by shale gas has increased from 1% to over 40%—this surge in production has made the US the largest producer of natural gas in the world. Moreover, unconventional oil (shale oil, oil shale/kerogen, oil sands, and extra

---

[56]Markus K. Brunnermeier, Deciphering the Liquidity and Credit Crunch 2007–2008, Journal of Economic Perspectives, 23(1), 2009, pp.77–100

[57]Justin Yifu Lin, Policy Responses to the Global Economic Crisis, Development Outreach, World Bank Institute, Volume 11, Issue 3, December 2009, pp.29–33

[58]Historical Budget Data—August 2013, Revenues, Outlays, Deficits, Surpluses, and Debt Held by the Public since 1973, Congressional Budget Office, August 12, 2013

[59]The Daily History of the Debt Results: historical returns from 07/16/2007 through 12/31/2009, US Department of the Treasury, Bureau of the Fiscal Service

heavy oil) contributed 35% of total US crude oil production by 2012 [60]. In the same year, the International Energy Agency predicted that the United States will become almost self-sufficient in energy by 2035 [61].

During the shale revolution, the United States lost interest in Saudi Arabia as a key provider of oil for the American economy: total US imports of crude oil went down by 35%, from 10.7 million barrels/day in June 2005 to seven million in August 2013; and oil imports from Saudi Arabia to the United States dropped by around 45% in that period, from 1.566 to 0.894 million barrels/day [62]. This is partly explained by the huge domestic shale oil output enabled by high oil prices, which allowed US oil companies to produce expensive oil from American shale formations and still make a profit. The environment of high oil prices which had dominated the last decade to 2013 enabled a range of oil and gas projects with high break-even—shale development, oil sands, ultra-deep oil production, etc.—which flooded the world oil market and began to threaten the market share of Saudi Arabia and OPEC. In autumn 2014, Saudi Arabia, Kuwait and some other Sunni OPEC members responded by increasing production and lowering oil prices to retain their market share, damage the state finances of Shiite Iran—which needed $134/barrel to balance its budget [63]—and bring the US shale industry to bankruptcy, knocking out the unconventional producers [64,65,66,67,68].

Citigroup estimated that, with oil at $70/barrel, almost all US shale oil production is profitable, at $60/barrel 40% of unconventional oil production becomes uneconomical, and oil at $50/barrel would make almost 90% of it unprofitable [69]. The Saudi oil minister Ali Al-Naimi confirmed the OPEC strategy of defending their market share from the encroachments of unconventional energy producers: *"Why should Saudi Arabia cut? The US is a big producer too now. Should they cut? The market will stabilize itself eventually... Today, it is not the role of Saudi Arabia, or other certain OPEC nations, to subsidize higher cost producers by ceding market share ... If they [countries outside OPEC] want to cut production they are welcome: we are not going to cut, certainly Saudi Arabia is not going to cut... Is it reasonable for a highly efficient producer to reduce output, while the producer of poor efficiency continues to produce? If I reduce, what happens to my market share? The price will*

---

[60]The Annual Energy Outlook 2014, US Energy Information Administration, May 2014, p.ES-2

[61]Peg Mackey, US to overtake Saudi as top oil producer: IEA, Reuters, November 12, 2012

[62]US Net Imports by Country (1993–2014), US Energy Information Administration, November 2014

[63]Gaurav Sharma, Saudis Vs Oil Markets: Who's Playing Whom?, Forbes, October, 24, 2014

[64]Will Kennedy and Jillian Ward, OPEC Policy Ensures US Shale Crash, Russian Tycoon Says, Bloomberg, November 27, 2014

[65]US Shale Oil Output Growth May Stall at $60 a Barrel, Oil may fall to $50, EIA Says, Bloomberg, November 18, 2014

[66]Saudi Arabia to keep politics out of OPEC, will let market stabilize price, RT, November 26, 2014

[67]Anjli Raval, Prices, not Opec, to balance oil supply, FT, November 28, 2014

[68]Factbox - OPEC oil ministers position ahead of Thursday meeting, Reuters, November 26, 2014

[69]Andrews R., Oil Wars: Why OPEC Will Win, Oil Price. December 11, 2014, http://oilprice.com/Energy/Oil-Prices/Oil-Wars-Why-OPEC-Will-Win.html

go up and the Russians, the Brazilians, US shale oil producers will take my share... The best thing for everybody is to let the most efficient produce... The [oil] policy of the kingdom is based on a strict economic basis, nothing more, nothing less... I don't care if prices crash to $20 — we're not budging" [70,71,72,73,74,75,76,77,78]. Mohammed al-Sada, oil minister of Qatar, noted that *"We believe in the role of market fundamentals in dictating prices"* [79,80]. Abdalla Salem el-Badri, the secretary general of OPEC, commented in March 2015, *"We welcome tight [shale] oil... but this source of energy costs too much to produce. You cannot produce it at $70-$80 or $90, you need $100 plus to produce, sell it and make income out of it... OPEC cannot subsidise another source of energy... We cannot every time keep reducing our production, it (tight oil) is not a challenge for us ... we welcome it, but let the market decide now... Projects are being canceled. Investments are being revised. Costs are being squeezed... If we don't have more supply, there will be a shortage and the price will rise again... When OPEC didn't reduce its production, everything collapsed for the US shale-oil-rig market"* [81,82].

The deliberate action of Saudi Arabia and other OPEC members, in refusing to reduce production during 2014–2016 when evidence of over-supply transpired, in order to regain the largest share of the oil market, led to a dramatic fall in oil prices from $95/barrel in September 2014 to $26/barrel in January 2016 [83]. Due to plummeting oil prices, the US rig count dropped by 77%, from 1930 oil rigs in September 2014 to 443 in October 2016 [84]. Production in the United States went down by 12% from 9.6 million barrels/day in July 2015 to 8.45 million barrels/day in

[70]Saudi Arabia to keep politics out of OPEC, will let market stabilize price, RT, November 26, 2014

[71]Anjli Raval, Prices, not Opec, to balance oil supply, FT, November 28, 2014

[72]Factbox - OPEC oil ministers positions ahead of Thursday meeting, Reuters, November 26, 2014

[73]SAUDI OIL MINISTER: I Don't Care If Prices Crash To $20 — We're Not Budging, Reuters, December 22, 2014

[74]Saudi oil chief says price drop not a conspiracy, Associated Press, December 21, 2014

[75]Rania El Gamal, Maha El Dahan, Writing by William Maclean, UPDATE 1-Saudi's Naimi: Non-OPEC, speculators led to oil price slide, Reuters, December 21, 2014

[76]Rania El Gamal and Maha El Dahan, Saudi Arabia says won't cut oil output, Reuters, December 21, 2014

[77]Former Saudi oil boss says it can cope with low price, BBC, January 19, 2015

[78]Matt Egan, Saudi Arabia: Don't blame us for oil's big plunge, CNN, March 4, 2015

[79]Saudi Arabia to keep politics out of OPEC, will let market stabilize price, RT, November 26, 2014

[80]SAUDI OIL MINISTER: I Don't Care If Prices Crash To $20 — We're Not Budging, Reuters, December 22, 2014

[81]Bahrain to Use MEOS Conference as Platform for Sustainable Energy, Egypt Oil & Gas, March 9, 2015, http://www.egyptoil-gas.com/news/bahrain-to-use-meos-conference-as-platform-for-sustainable-energy/

[82]Benoit Faucon, OPEC Chief Says Cartel Is Hurting US Shale Producers, The Wall Street Journal, March 8, 2015

[83]US Energy Information Administration, 2016

[84]Baker Hughes, 2016

October 2016 [85]. Saudi Arabia also suffered severely from this strategy—even deciding in the future to sell some shares of Saudi Aramco, its pearl, to foreign investors through the New York stock exchange—and ultimately agreed to participate in a deal between OPEC and Russia to stabilize oil prices at the end of 2016. The Saudi war against American shale producers also damaged the business of higher-cost oil and gas producers in the North Sea, Western Canada, Northern Australia and the Russian Arctic. The global recovery of non-conventional energy began only in 2017, when the OPEC countries reduced their own oil output and the resulting over-supply on the market.

The response to this kind of challenge should include ongoing in-depth analysis of macroeconomic data from different countries and global supply/demand statistics, prediction of the likely movement of stock and commodity markets and of the actions of competitors and speculators, hedging of commodity prices in the event of evidence of potential volatility, and participation in cartels like the current OPEC or the erstwhile Texas Railroad Commission, in order to regulate supply and avoid overproduction, unsustainable losses for energy producers, and so on.

New Technological and Environment Trends

Suppression and Bankruptcy of the Coal Industry in the United States Due to Changes in Regulation (USA, 2013–2016) Natural gas is a clean energy compared to coal, in that it has a lower level of carbon dioxide emission during combustion: burning natural gas releases about half as much carbon dioxide per unit of energy as burning coal and a third less than oil [86,87]. As a consequence, the Obama administration favoured natural gas as the best "*bridge fuel*" during the transition from greenhouse gas-emitting fossil energy systems towards systems with zero carbon emissions. The American coal industry was dismissed as obsolete and dirty, and the administration began to suppress it with stricter regulation of coal power plants, promoting instead the American shale revolution and "*green*" energy solutions like solar, wind and hydroelectric power. Shale gas became the major source of energy for US power stations and even pushed coal down in the US energy balance: coal declined from 48% of all power generation in 2007 to 30% by 2016 [88,89].

---

[85]US Energy Information Administration, 2016

[86]Why Natural Gas Clean, America's Natural Gas Alliance: Clean & Efficient, http://anga.us/issues-and-policy/power-generation/clean-and-efficient#.U4NTklPPmVE

[87]Eli Kintisch, Plugging methane leaks in the urban maze could be key to making shale gas climate-friendly, Science, Vol 344, Issue 6191, June 27, 2014, p.1472

[88]Energy Vision 2013. Energy transitions: Past and Future Industry Vision, The Committee of World Economic Forum in partnership with IHS CERA, January 2013, p.17

[89]Short-Term Energy Outlook, US Energy Information Administration, February 2017

The share of natural gas in the domestic electricity market mainly rose because the price of natural gas fell below that of coal. The White House, through the EPA, issued new regulations restricting emissions of mercury, lead, ozone, sulphur dioxide and nitrogen dioxide from all coal-fired power plants, reducing their permitted consumption of water for cooling, raising standards on the quality of ash after coal burning, and finally setting the New Source Performance Standard for any new power stations within the United States. This standard requires all new power plants, including those that use coal, to put out no more than 1000 pounds of carbon dioxide per megawatt hour—the rate of an average gas-fired plant. According to industrial analysts, for coal plants to continue in the US the CO_2 emissions limit would have to be in the range of 1300 lbs./MWh to 1900 lbs./MWh. In other words, with the new limit of 1000 lbs./MWh, no new coal-fired plant could be built within the US: the electricity industry has been given a clear signal that they should now rely on natural gas as the main fuel for electricity production. These changes have already led to bankruptcy for one coal-powered electricity plant—the Longview plant in West Virginia, which started operation in 2011 [90].

At the same time, in order to support coal producers, and reduce lobbying efforts from the coal industry against shale gas, the US government has been promoting increased exports of American coal to Europe and Asia. Nevertheless, the suppression of coal-based power generation in the US slashed coal prices and provoked the bankruptcy of 26 coal producers by the spring of 2015—including Peabody Energy Corp, the world's biggest private-sector coal producer, which filed for bankruptcy in April 2015. Restoration of the coal industry in the United States began only after the cancelation of several Obama decrees on environmental issues by the Trump administration in 2017. The first new coal mine in the recent memory of the United States was opened in Pennsylvania in June 2017.

Worldwide regulation on CO_2 emissions Over recent years, there has been increasingly stringent worldwide regulation on CO_2 emissions from the burning of coal and other fossil fuels; we have just discussed the disastrous impact this has had on the coal industry in the United States.

During the UN Climate Change Conference in Paris in December 2015, nearly 200 nations agreed on mutual measures to reduce CO_2 emissions. Their ambitious but essential target is to ensure that global temperatures do not increase by more than 2 °C before 2100. To honor this agreement, we would need to reach zero carbon emissions by 2050—and this is only attainable by suspending the extraction of fossil fuels worldwide. According to analysts at Citibank, *"the value of unburnable reserves could amount to over $100 trillion out to 2050"* [91]. The European Investment Bank now has plans to finance the development of alternative sources

---

[90]Valerie Volcovici, US coal industry braces for EPA emissions crackdown, Reuters, September 12, 2013

[91]Energy Darwinism II, Why a Low Carbon Future Doesn't Have to Cost the Earth, Citi, August 2015, p.82, https://ir.citi.com/hsq32Jl1m4aIzicMqH8sBkPnbsqfnwy4Jgb1J2kIPYWIw5eM8yD3FY9VbGpK%2Baax

of energy, investing €1 trillion between 2020 and 2030. By 2020, further financing of projects related to fossil fuels will be suspended [92]. For the current energy giants, agreeing not to extract such huge fuel reserves of fuel represents tremendous losses, and an unprecedented revolution in their business model, as they shift towards the development of alternative sources of energy with little or no carbon emissions. The Paris Agreement was one of the Obama administration's proudest achievements; but in June 2017, Donald Trump reneged on the agreement, describing it as "...*very unfair, at the highest level, to the United States ... The Paris Accord would undermine our economy ... [and] put us at a permanent disadvantage to the other countries of the world... [B]y 2040, compliance with the commitments put into place by the previous administration would cut production for the following sectors: paper down 12 percent; cement down 23 percent; iron and steel down 38 percent; coal — and I happen to love the coal miners — down 86 percent; natural gas down 31 percent. The cost to the economy at this time would be close to $3 trillion in lost GDP and 6.5 million industrial jobs, while households would have $7,000 less income and, in many cases, much worse than that*" [93]. Trump justified US withdrawal from the Paris Agreement on similar grounds to those advanced in the late 1990s by the American Petroleum Institute—the main lobbying organization for the US oil and gas industry—to reject the Kyoto Protocol [94]. The Protocol had already been signed by the Clinton administration, but the subsequent administration under George W Bush managed to prevent it from being ratified by the US Senate.

When there is a trend towards additional government regulation over an industry, the big players can invest more time and money in lobbying—actively explaining their position on the area of proposed new regulation, and making the case for the benefits of deregulation or the threats of excessive government control.

Emergencies on Production Sites

An accident at an industrial site can become a national or even international disaster due to the scale of destruction, the number of casualties, and the threat to the environment and government tax revenues. An adequate response usually requires the large-scale involvement of governmental agencies. Accidents can affect national and global output, because the affected sites can take years to restore and government control over industry may need to be strengthened. As a rule, such industrial accidents harm local communities: families are destroyed by bereavement or injury, the taxation base of local government is wiped out, there is local or more widespread

---

92 Jerome Marin, L'Europe veut financer 1.000 milliards d'euros de projets verts en dix ans, La Tribune, September 30, 2019

93 Statement by President Trump on the Paris Climate Accord, The White House, June 01, 2017, https://www.whitehouse.gov/the-press-office/2017/06/01/statement-president-trump-paris-climate-accord

94 Benjamin Franta, Trump pulled out the oil industry playbook and players for Paris, The Guardian, July 26, 2017

environmental damage. The company involved often faces lawsuits running on for years, with the prospect of paying billions in compensation to victims and the government. In some cases, the public accuses not only the organization operating the affected site, but also the government for weak and inadequate oversight of the site, or even over an entire extraction industry. There are several key cases.

Underground Mining Accidents

Benxihu Coal Mine Disaster (China, 1942)   The exploitation of underground coal deposits is usually accompanied by the release of methane from coal belts, which can become explosive at a certain concentration in the air. Coal mining also produces dust, which again can lead to explosions. The most severe coal mining accidents have been caused by one of these factors or a combination of them. The deadliest accident in coal mining history occurred in China during the Second World War in 1942, when a mixture of methane and coal dust exploded in the Benxihu coal mine in Honkeiko. In order to extinguish the underground fire, Japanese Army officers blocked the ventilation system of the mine and halted the inflow of oxygen to the shaft. But because the entrance of the mine had been destroyed by the explosion, Chinese miners could not leave the mine and consequently 1549 of them died, mainly from carbon monoxide poisoning. There have been hundreds of similar cases all around the world, but with fewer casualties.

To reduce methane concentration in mines, modern coal producers conduct preventive degassing. This involves drilling numerous exhaust wells within coal deposits and vacuum pumping the gas from the rock, installing powerful ventilation systems that can remove gas during the operation of powerful coal cutter-loaders, and sophisticated methane detectors throughout the mine to monitor gas concentration and trigger preventive blackouts and shutdown of electric circuits if required. The main solutions for reducing coal dust in the air include humidification of coal seams, dedusting ventilation systems, rock dusting (increasing ash content by the addition of slate dust), and so on. A key element of any comprehensive safety system is a compensation model for field staff that prioritizes safety over productivity.

Coalbrook Coal Mine Disaster (South Africa, 1960) and San José Copper–Gold Mine Rescue (Chile, 2010)   900 underground pillars collapsed in an area of 3 km^2 within the Coalbrook North Mine in 1960, which led to 435 deaths of coal miners and rescue workers. It was the deadliest rock fall in the history of the industry. A contrasting example is the successful rescue of 33 miners trapped by a rock fall 700 m below the surface in the San José copper–gold mine in Chile, in an operation lasting 69 days.

To manage this risk effectively, mining engineers need to make a detailed analysis of previous accidents within similar deposits, carefully design mining sequences and support systems, develop comprehensive systems to monitor ground and roof stability, and so on.

Ground Subsidence Due to Mining Operations Berezniki and Solikamsk—industrial cities in Russia—were constructed in the 1930s over the development of underground potash mines, which were excavating at depths of just 200–300 m. Over recent decades, several large sinkholes have opened up in the cities, the largest being 100 m deep and 500 m in diameter. These pits have consumed several structures including country houses, railways, roads, and so on. The instability of the overburden has led to cracks in many other buildings. Municipal authorities are trying to relocate thousands of residents of potentially dangerous dwellings, but such movement inevitably causes protests.

Mining companies need to take great care in considering new territories for prospecting and potential extraction of resources. In addition to the safe construction and operation practices we have mentioned above, production activity in populated areas must go hand in hand with the long-term development of social infrastructure to avoid this kind of disastrous impact on local communities.

Surface Mining Accidents

Massive Landslide at Bingham Canyon Mine (USA, 2013) Landslides are a common occurrence at open pit mines. They can be caused by a number of factors, including earthquakes and artificial explosions of rock formations, instability of slopes saturated by ingress or ground water, artificial dam impoundments [95,96] and intensive vibration from trucks or heavy machinery. One of the largest landslides in the surface mining industry occurred at the Kennecott Utah Copper mine—operated by Rio Tinto and the deepest open pit mine in the world—in April 2013. The mine produced up to 25% of US copper consumption. The volume of the landslide was 55–65 million m^3—roughly the equivalent of a cube with 400 m sides. Fortunately, it was a predicted disaster: staff were evacuated in advance and nobody was killed or injured. Remarkably, the landslide triggered earthquakes of up to 5.1 on the Richter scale, and was the largest non-volcanic slide in North America's modern history [97]. It was successfully anticipated because advanced sensor systems had been installed within the mine—radar, lasers, seismic detectors, and GPS tracking of the slopes—which allowed any ground movement to be detected in good time.

Landslide mitigation solutions include advance research on the structure of deposits, careful and informed design of slopes, roads, drainage, and pumping

---

[95] A. Helmstetter, D. Sornette, J.-R. Grasso, J. V. Andersen, S. Gluzman, V. Pisarenko, Slider-Block Friction Model for Landslides: Application to Vaiont and La Clapiere Landslides, Journal of Geophysical Research, 109, 2004

[96] D. Sornette, A. Helmstetter, J.V. Andersen, S. Gluzman, J.-R. Grasso, V.F. Pisarenko, Towards Landslide Predictions: Two Case Studies, Physica A, 338, 2004, pp.605–632

[97] Kris Pankow, Jeff Moore, Mine Landslide Triggered Quakes, University of Utah Seismograph Stations, 2014

systems, appropriate exploitation of the slopes, and exhaustive monitoring informed by modeling and analysis [98].

Oil and Gas Extraction

The worst accidents during surface extraction of oil and gas are caused by wellhead blowouts leading to uncontrollable spillage or discharge of hydrocarbons onto the ground, into the water or into the air. Due to the highly explosive and flammable nature of oil and gas, another serious challenge is maintaining the integrity of all pipes used to transport the hydrocarbon mixture during drilling and processing, and detecting potential leaks in time. In the past, there were serious problems with geological misjudgments in the assessment and exploration of deposits.

Lakeview Gusher (USA, 1909) and Other Surface Oil Spills Caused by Wellhead Blowouts Such events can cause tremendous environmental damage, lead to the loss of valuable raw resources and destroy the life of local communities. The largest oil spill ever caused by a surface wellhead blowout was in Kern County, California in 1909, when nine million barrels of oil were lost over 18 months. At that time, the technology of blowout preventers was in its infancy, so the spill lasted much longer. Other serious surface oil spills took place in Libya in 1980, and in the Fergana valley in Uzbekistan in 1992.

Ixtoc I Offshore Well Blowout (Mexico, 1979) One of the largest ever marine oil spills. A blowout of drilling mud led to the release of three million barrels of oil into the Gulf of Mexico over 294 days. The water depth was just 50 m, but the technology of the time did not allow the leakage to be brought under control more quickly.

Deepwater Horizon and Macondo Well Blowout (USA, 2010) This was the largest marine oil spill in world history (4.9 million barrels). Due to the poor quality of cement work at the Macondo well in the Gulf of Mexico, and the extreme high pressure and temperature of the exploration target, there was a major blowout. For a number of technical reasons, the blowout preventer on the well failed to prevent the uncontrolled flow of oil and gas, which continued for 87 days. After the disaster, US President Barack Obama declared: *"this oil spill is the worst environmental disaster America has ever faced"*. BP was forced to cover all expenses incurred in shutting down the deepwater leak and cleaning up the American part of the Gulf of Mexico coastline—an area with 14 million inhabitants. More than 47,000 people and 7000 vessels took part in the response to the spill. BP's total losses from the accident were estimated at US$46 billion—$28 billion was spent on the accident and $18 billion on

---

[98]D. Sornette, A. Helmstetter, J.V. Andersen, S. Gluzman, J.-R. Grasso, V.F. Pisarenko, Towards Landslide Predictions: Two Case Studies, Physica A, 338, 2004, pp.605–632

additional government fines and penalties [99]—and within 3 months, the company's stock market value fell by $70 billion [100]. After the disaster, the US government suspended all deepwater activity off the American coast for 6 months.

Piper Alpha Disaster (UK, 1988) This platform in the North Sea was destroyed because of poor communication and a failure to follow for safety rules. In this case, the problem that led to the explosion arose between two repair shifts operating within the existing *"permit-to-work system"*, when the second team was not informed that the first had removed a pressure safety valve for routine maintenance. This simple lack of information about a routine maintenance process had major consequences: the unwittingly dangerous actions of the second shift caused a leakage of condensate which exploded, causing a massive fire in which 167 crew members perished. The scale of the disaster was exacerbated by the absence of authority delegated to the operators to shut down the production on the neighboring Claymore platform, which continued to pump hydrocarbons to burning Piper Alpha.

Mumbai High North Platform Fire (India, 2005) This accident occurred on the largest offshore oil and gas field in India, when a multi-purpose support vessel tried to moor to the Mumbai High North platform to evacuate an injured cook. Accidentally, the vessel crashed into the platform and broke a high-pressure pipe carrying oil and gas mixture, resulting in a fire which engulfed the whole platform and killed 22 people.

Lake Peigneur (USA, 1980) Texaco drillers developing an oil well under Lake Peigneur made a mistake in their calculations, and instead of hitting the oil field they drilled down into an underground salt mine. Water from the lake flooded the mine and the lake was dried up. Later the level of the lake returned to normal, but the water in the lake had become salty.

Modern oil and gas extraction—especially marine drilling—is as challenging as space technology in terms of the complexity of the situation and the diversity of equipment needed. The safety of the drilling operation can be analyzed in two ways: occupational safety and process safety. Occupational safety focuses on the swift and effective treatment of injuries during the operations, while process safety aims to prevent system failures by analyzing process weaknesses and responding to them in good time. Advanced technologies help to minimize the risk of geological mistakes during exploration assessment of deposits and the subsequent development drilling. Other response measures include the ongoing search for advanced technical solutions to increase the reliability of the production process, along with the implementation of international best practice safety procedures.

---

99 Bradley Olson, Margaret Cronin Fisk, Worst Case' BP Ruling to Force Billions More in Payout, Bloomberg, September 4, 2014

100 Steve Hargreaves, BP's $70 billion whipping, CNN Money, June 2, 2010

Oil and Gas Transportation Via Tankers and Pipelines

During the transportation of oil and gas, the highest risks are usually connected with the loss of integrity of the transportation system/vessels and the leakage of hydrocarbons, damaging the environment, local communities and even unfortunate passers-by.

Explosions at San Juan Ixhuatepec LPG Terminal (Mexico, 1984) More than 500 civilians died and at least 5000 were injured after explosions at an LPG terminal run by the state-owned PEMEX, which was located in a densely populated suburb of Mexico City. The disaster was caused by a gas leak from a ruptured pipe. The terminal operator noticed a sudden drop of pressure at a pipeline pumping station, but they could not identify the exact cause of the pressure drop in time because there was no gas detection system at the terminal. The hydrocarbon leakage eventually led to the explosion of several gas storage tanks and a massive fire, which totally destroyed the terminal and the nearby dwellings of local residents [101,102].

Exxon Valdez Oil Spill (USA, 1989) The oil tanker Exxon Valdez collided with a reef in Prince William Sound in Alaska. The total amount of leaked oil is estimated at between 250,000 and 260,000 barrels. It was by no means the largest oil spill ever caused by the wreck of an oil tanker, but the case became a poster child for the campaign against the oil industry: the slow and inadequate response to the spill resulted in extensive oil contamination of 2000 km of pristine coastline on the Gulf of Alaska. The clean-up operation in the summer of 1989 required 10,000 people, 1000 vessels, 38 oil skimmers and 72 aircraft. Over the 4 years following the accident, Exxon made huge efforts to clean the coastline of the Gulf of Alaska. Their total expenses for dealing with the consequences of the accident, including penalties, exceeded US$4.3 billion. The main cause of the collision was the captain's decision to deviate from the approved tanker route in order to avoid colliding with small icebergs from the nearby Columbia glacier. However, the third mate failed to properly maneuver the ship and collided with Bligh Reef [103]. At the time of the accident, the captain may have been drunk [104] and the third mate was suffering from exhaustion [105].

---

[101]PEMEX LPG Terminal, Mexico City, Mexico. 19th November 1984, UK Health and Safety Executive, http://www.hse.gov.uk/comah/sragtech/casepemex84.htm

[102]Arturson G., The tragedy of San Juanico--the most severe LPG disaster in history, Burns, Including Thermal Injury, 1987, 13(2) pp. 87–102

[103]Exxon Valdez Oil Spill Trustee Council, Questions and Answers, http://www.evostc.state.ak.us/facts/qanda.cfm

[104]Final Report "SPILL: The wreck of the Exxon Valdez", Alaska Oil Spill Commission, State of Alaska, February 1990, p.13

[105]Final Report "SPILL: The wreck of the Exxon Valdez", Alaska Oil Spill Commission, State of Alaska, February 1990, p.11

Ufa Train Disaster (USSR, 1989) The Western Siberia/Ural/Volga natural gas liquids pipeline ruptured, causing the build-up of a potentially explosive hydrocarbon-air mixture. Several hours later two Trans-Siberian passenger trains came into the zone of gas contamination, passing in opposite directions with a total of 37 railroad cars carrying 1284 passengers and 86 crew members. Apparently, a spark from a current collector on one of the electric locomotives ignited the lethal gas mixture, causing an explosion in which 575 people perished and 623 were injured. The explosion, equivalent to 300 tons of TNT, was the deadliest railway accident in the history of the Soviet Union and the Russian Federation.

There were several reasons for the accident. The pipeline had originally been designed for the transportation of oil but, due to the serious shortage of raw materials for the Soviet petrochemical industry, it was reformatted to transport natural gas liquids (NGLs): a mixture of methane, ethane, propane and butane. In making this decision, executives of the Soviet Ministry of Petroleum were ignoring the very different physical properties of the hydrocarbon mix in NGLs: at normal temperature and pressure, they are gases, so transporting them as a liquid mix involves very low temperature and high pressure. The safe transportation of the mixture required a pipe with a thick insulating wall and a diameter not exceeding 400 mm, while the oil pipeline had a thin wall and a diameter of 720 mm. The combination of the thin wall of the pipe and the temperature changes as the mixture flowed through it made the pipeline a highly dangerous structure. To make matters worse, in May 1984 executives of the Soviet Ministry of Petroleum canceled the installation of an automatic telemetry system for real-time control of possible leaks from the pipeline.

Nevertheless, regular helicopter sorties to check for possible high concentrations of methane in the atmosphere near the pipeline, and squads of trackmen with gas leakage detectors, worked effectively for the first few years after the launch of the pipeline in October 1985. Because of the tortuous landscape of the Ural mountains, and in order to reduce costs and give easy access to maintenance using nearby transport infrastructure, the pipeline was constructed dangerously close to the railway: for 273 km, the pipeline and the railway were less than 1 km apart. Moreover, the pipeline crossed the bed of the railroad, which included the high traffic Trans-Siberian railway, in 14 places.

During the construction of a bypass to a section of the pipeline in autumn 1985, a powerful excavator caused considerable mechanical damage to the pipe. The investigation after the accident concluded that a 1.7 m crack had developed in the pipeline at the exact point of the bypass construction in 1985 [106]. The pipeline had lost integrity at least 3 weeks before the disaster [107,108] but, in the absence of regular

---

[106]Minutes of session of the Supreme Court of the USSR under the chairmanship of Judge V.I. Cherkasov, December 26, 1991, pp. 8–11

[107]The torch of death, 18 years ago there was an accident in Bashkortostan, which world did not face before, MediaKorSet (Ufa), June 3, 2007

[108]Alexey Skripov, Asha explosion. Why the largest in the history of the country's rail disaster occurred, Rossiyskaya Gazeta - Week – Ural, June 11, 2009

monitoring, the leak was not identified in time. On the night of the disaster, the operators of the pipeline had increased the pressure in the system to meet increased demand. This provoked a more serious rupture of the pipe at the already weakened joint—and the massive gas release that ensued caused the explosion.

Varanus Island Gas Production Plant Fire (Australia, 2008) A natural gas leak from a corroded pipe started a fire at the Varanus Island natural gas production plant, cutting off the supply of many industrial sites in Western Australia for months. Financial losses incurred by the suspension of a third of the state's gas supply were estimated at A\$ 6.7 billion [109]. The main impact was not on residential customers but on large mining companies, metal smelting, manufacturing, electricity production, services etc. [110].

Transportation of oil and gas is clearly a high-risk operation. Managing the risks effectively will include a comprehensive assessment of possible threats to the external environment from transportation infrastructure; care in the design of reliable vessels (for example, the rejection of single-hulled tankers in favor of more expensive double-hulled tankers) and pipelines that remain safe even in extreme conditions; the installation of advanced navigation systems and sophisticated monitoring systems; ongoing training for operating staff; and having the resources permanently on standby for a quick and adequate response to any spill.

Long-Term Contamination of Surrounding Areas

Systematic pollution during surface mining by open pits, extraction of conventional oil, and the development of unconventional oil and gas, and poor conditions of pipelines, can severely damage the surrounding area, initiate a long-term struggle with local communities, create tension with the national government and invite protest action from ecological activists.

Environmental Catastrophe Over Decades in the Norilsk Industrial Zone (Russia, Since 1930s) The city of Norilsk was built by the hands of GULAG prisoners in the 1930s, to support the development of extensive mineral deposits and the smelting of non-ferrous metals in the Russian Arctic. Over more than 80 years of intensive mining and production of nickel, copper and lead, the soil, water and air within the city and nearby territory has become heavily polluted. Such a toxic environment is a serious threat to the health of workers and their families, so most of them choose to leave the city after several decades of employment. Norilsk Nickel—the mining company—supports such movement of its personnel, and is trying to reduce new emissions by shutting down the most polluting industrial plants or installing modern filtering systems. There has

---

[109]WA faces \$6.7b gas bill, Sydney Morning Herald, July 10, 2008
[110]Thomas Rid, Cyber War Will Not Take Place, US Army War College, October 5, 2011,

been similar degradation of the environment in other heavy industrial zones such as La Oroya in Peru (non-ferrous metals production), Linfen in China (coal mining), the Sukinda Valley in India (chromite mining), and Kabwe in Zambia (copper, lead and zinc mining and production)[111].

Contamination of the Jungle Around the Lago Agrio Oil Field (Ecuador, 1965–1990) For decades, oil was produced in this area using a process where backwater from the oil wells was collected in open pits instead of being re-injected underground. This resulted in catastrophic deforestation and destroyed the traditional way of life of local indigenous communities. The contamination initiated decades of legal action by the indigenous people against Texaco and later Chevron: the first suits were filed in 1993, but the case is still not resolved.

Collapse of Church Rock Uranium Mill Dam (USA, 1979) More than 1100 tons of solid uranium mining waste and 350,000 m^3 of radioactive and highly acidic tailings solution (the liquid waste product of ore processing) were spilled into the Rio Puerco in New Mexico, after the dam of the mill's tailings pond collapsed in July 1979, because the site was geologically unsound. Levels of radioactivity in the river reached 7000 times the legal limit for drinking water [112]. Radio broadcasts and signs warning about the spill were of limited help to the locals—many of whom did not speak English—including Native American farmers who relied on water from the river for drinking, irrigation and watering their livestock in this arid land. This was the largest accidental release of radioactive material in the history of the United States [113].

Complex Conflicts with Local Communities in the Niger Delta (Nigeria, Since 1970s) The Niger Delta possesses gigantic oil and gas fields, which have been the driving force for the economic development of Nigeria, accounting for up to 90% of the country's export earnings. However, more than six million Nigerians live in the area where the hydrocarbons are being produced. In 2011, the United Nations Environment Programme (UNEP) made a study of 200 locations in Ogoniland. They concluded that after 50 years of oil production in the region, which involved oil spills, oil flaring and waste discharge, the alluvial soil of the Niger Delta was no longer viable for agriculture. Obviously, this has had a serious impact on the living conditions and even the livelihood of local people, who depend on fertile land and fishing water for survival. Understandably, the locals seek to halt or restrict production and reclaim the land—and conflict is almost inevitable. Crises with locals in the Niger Delta have become a persistent headache for major oil companies and the

---

111 The World's Most Polluted Places, The Blacksmith Institute, 2006–2013

112 Bruce E. Johansen, The High Cost of Uranium in Navajoland, Akwesasne Notes New Series, 1997, Volume 2, pp. 10–12

113 W. L. Graf, Fluvial Dynamics of Thorium-230 in the Church Rock Event, Puerco River, New Mexico, Annals of the Association of American Geographers, 1990, 80, p. 327

Nigerian government, and the entrenched local opposition has made the region one of the most violent in the world for oil production.

The most prominent player in the struggle for a just distribution of wealth from the exploitation of the Niger Delta, and for stricter regulation to limit the damage to rivers and agricultural land, is the Movement for the Survival of the Ogoni People (MOSOP). Shell has been producing oil from Ogoniland, where around half a million people live, since 1958. In the 1990s, the organization declared as its aims the "*control and use of a fair proportion of Ogoni economic resources for Ogoni development, adequate and direct representation as of right in all Nigerian institutions, and the right to protect the Ogoni environment and ecology from further degradation* [114]". MOSOP issued an ultimatum to Shell: the company should pay the Ogoni people 10 billion dollars in compensation for the damage they had done to the environment of Ogoniland, or the people would launch massive protests and halt all further oil production. The ultimatum was ignored and around 300 thousand Ogoni nonviolent protestors turned out, occupying Shell facilities and significantly restricting production; one Shell worker was beaten. Shell decided to withdraw from the region. But the Ogoni continued to demonstrate and the Nigerian government, who saw their demands for self-determination as a threat to the integrity of the country and its federalization, used force to suppress the protestors. Many Ogoni villages were wiped off the face of the earth; ultimately more than 100,000 Ogoni became internal refuges and 2000 died [115]. MOSOP's leaders were arrested and framed for murder; ultimately nine were executed.

The suppression of protestors and execution of their leaders generally reduced the activity of the Ogoni people, but ignited a worldwide public relations crisis. Nigeria's membership of the Commonwealth of Nations was suspended because of the executions; supporters of the Ogoni people demonstrated outside Nigerian embassies and Shell offices; the Western media blamed Shell for doing much too little to avoid oil pollution to the region, and accused them of being implicated in the trial of the Ogoni leaders; and many private Shell customers supported a boycott of the company's petrol stations in Europe (similar to the protests against Shell's plans for the disposal of Brent Spar, a buoy used for oil storage in the North Sea, in the 1990s). In 2009, Shell reached an agreement with the families of the executed leaders and paid them US$15.5 million.

Aggressive action against the oil industry has continued in the Niger Delta. In the eyes of locals, the feeling that there has been inadequate compensation from oil companies legitimizes the stealing of oil from pipelines. This theft not only brings losses to energy producers and the government, but also further contaminates the soil and water for nearby communities when the unauthorized penetration of the pipelines leads to ongoing leaks. In some cases, this has led to disaster: for example, in 1998, a pipeline explosion occurred in the community of Jesse that took the lives of

---

[114]J.I. Dibua, Citizenship and resource control in Nigeria: the case of minority communities in the Niger Delta, Afrika Spectrum, 1(40), 2005, pp. 5–28

[115]Marino Busdachi, Environmental Situation of Ogoni, Climate Change, Bioenergy and Food Security: Challenges for the New Millennium, June 2008

1082 locals. According to the Nigerian government, the explosion was triggered by the intentional puncture of the pipeline by oil thieves. It was the deadliest explosion in Nigerian oil industry, but similar events with fewer casualties take place regularly in this region. Numerous acts of vandalism against energy infrastructure are recorded every year in the delta, and kidnappings of expatriate oil workers are frequent.

Contamination of Water Sources in Shale-Rich Regions in North America (Since 2005) The American shale revolution is based on the technology of hydraulic fracturing (*"fracking"*) for both gas and oil extraction from shale formations. A clear description of this technique can be found in a 2001 report by the National Energy Policy Development Group (NEPDG, also known as the Energy Task Force): *"This is a common procedure used by producers to complete gas wells by stimulating the well's ability to flow increased volumes of gas from the reservoir rock into the wellbore. During a fracture procedure, fluid and a propping agent (usually sand) are pumped into the reservoir rock, widening natural fractures to provide paths for the gas to migrate to the wellbore. In certain formations, it has been demonstrated that the gas flow rate may be increased as much as twenty-fold by hydraulic fracturing"* [116]. In the late 1990s and early 2000s, Halliburton—one of the leading service companies in the petroleum industry—was updating the hydraulic fracturing process. Dick Cheney was CEO of Halliburton in 1995–2000, when this work was under way. Cheney had been US Secretary of Defense during the term of George H. W. Bush from 1989 to 1993, including the first Iraqi campaign in 1990–1991; he was to be Vice-President for the two terms of George W. Bush between 2001 and 2009. During the 2000s, Dick Cheney also directed the Energy Task Force, whose description of fracking we have quoted above.

However, the authors of the report omitted an important detail: that what they simply refer to as *"fluid"* pumped into the shale formations actually contains a mixture of 500 chemicals, which are added to the water and allow it to permeate more effectively through the rock during fracturing. This mixture of water and chemicals is pumped at very high artificial pressure—up to 13,500 psi or 920 atmospheres—into a shale well. The pressurized fluid fractures the rock at depths from 400 to 5200 m depending on shale type, ensuring the release of hydrocarbons from the shale formation up to the surface. The fractures are kept open with *"proppants"*, generally sand, to ensure the continued flow of resources. Sand and water comprise 99.5% of the total volumes used [117]. The fact is that this chemical mix, invented by Halliburton, which has gone to considerable effort to keep the *"recipe"* from public

---

[116]National Energy Policy, Report of the National Energy Policy Development Group, May 2001, pp.5–6

[117]Modern Shale Gas Development in the United States: A Primer, US Department of Energy, April 2009, p.61

knowledge, is extremely harmful to the environment [118,119,120,121,122,123]. Some of the constituent chemicals are also recognized as hazardous to human health—for example, ethylene glycol, which can damage the kidneys; formaldehyde, which is known to cause cancer; and naphthalene, another possible carcinogen [124]. The authors of the NEPDG report also neglected to mention the tremendous quantities of fresh water used during hydraulic fracturing. The amount of water required for an average shale gas well exceeds that for a conventional gas well by a factor of 10–100 [125, 126]: in its lifetime, a shale well consumes an average of 15,000 tons of water [127]. However, the geological structures of shale formations vary widely, so an unconventional well in Texas requires up to 23,000 tons of water, while in California each well requires only around 1000 tons [128]. These statistics do not include subsequent re-fracking procedures: because the natural pressure within a shale formation is weak, further fracking is usually required to stimulate additional hydrocarbon release several years after the initial process.

Not surprisingly, this technology was in clear contravention of strict American standards on environmental pollution. In 2005, Dick Cheney lobbied for amendments in several acts to exclude fracking fluids and shale well air pollution from

---

[118]Glenn Miller, Review of the Revised Draft Supplemental Generic Environmental Impact Statement on the Oil, Gas and Solution Mining Regulatory Program, Toxicity and Exposure to Substances in Fracturing Fluids and in the Waste water Associated with the Hydrocarbon Bearing Shale, Consulting Environmental Toxicologist to the Natural Resources Defense Council December 29, 2009

[119]Glenn Miller, Review of the Revised Draft Supplemental Generic Environmental Impact Statement on the Oil, Gas and Solution Mining Regulatory Program Well Permit Issuance for Horizontal Drilling and High-Volume Hydraulic Fracturing to Develop the Marcellus Shale and Other Low-Permeability Gas Reservoirs, Prepared for Natural Resources Defense Council, January 6, 2012

[120]Madelon Finkel, Jake Hays, and Adam Law. The Shale Gas Boom and the Need for Rational Policy. American Journal of Public Health: July 2013, Vol. 103, No. 7, pp. 1161–1163

[121]Theo Colborna, Carol Kwiatkowskia, Kim Schultza, Mary Bachran, Natural gas operations from a public health perspective. Human & Ecological Risk Assessment, 2011, 17, pp. 1039–1056

[122]Roxana Witter, Lisa McKenzie, Meredith Towle, Kaylan Stinson, Kenneth Scott, Lee Newman, John Adgate, Health Impact Assessment for Battlement Mesa, Garfield County Colorado, Colorado School of Public Health, University of Colorado Denver, September 2010

[123]Carol Linnitt, Report "Fracking the Future - How Unconventional Gas Threatens our Water, Health and Climate", DeSmogBlog Project, 2010

[124]Ohio: Shale drillers must report chemicals locally, Associated Press, October 1, 2013

[125]Standing Committee on Natural Resources, number 040, third session, 40th parliament, Parliament of Canada, February 1, 2011

[126]Erik Stokstad Will fracking put too much fizz in your water?, Science, Vol 344, Issue 6191, June 27, 2014, p.1468

[127]Erik Stokstad Will fracking put too much fizz in your water?, Science, Vol 344, Issue 6191, June 27, 2014, p.1468

[128]Garance Burke, Colorado's Fracking Woes Show Fight Brewing In Oklahoma, Texas And Other Drought-Ridden Areas, Huffington Post, June 16, 2013

federal government regulation under the Energy Policy Act [129]. Changes were made in the provisions of most of the relevant legislation, including the Safe Drinking Water Act, the Clean Water Act, the Clean Air Act, the National Environmental Policy Act, the Comprehensive Environmental Response, Compensation, and Liability Act, the Resource Conservation and Recovery Act, and the Toxic Release Inventory under the Emergency Planning and Community Right-to-Know Act [130]. Between 2005 and 2013, more than 82,000 unconventional wells were drilled and fracked on 31 shale plays within the United States; around one billion tons of water were contaminated and more than 1400 km^2 of land damaged [131].

After the extraction of hydrocarbons from a shale formation, a significant volume of the water—around 80% [132]—remains in the well, and the remaining "*backflow water*" returns to the surface with the hydrocarbon stream, and is deposited in special wastewater disposal wells or in special reservoirs for evaporation. This backflow water also contains naturally occurring radioactive matter, totally dissolved solids, liquid hydrocarbons—including benzene, toluene, ethylbenzene, and xylene—and heavy metals. Only about 10% of this backflow water is purified because this would add to the cost of the production process. 90% of the water used during fracking at the Marcellus shale wells in Pennsylvania between 2005 and 2013 was freshwater, and just 10% of the fracked water was recovered and re-used [133].

In spite of the fact that the oil and gas industry consumes far less water than agriculture—less than 1% of the average annual water consumption of the US—the high density of fracking wells leads to a situation in some counties of Texas and other shale states where the percentage of fracking water consumption reaches double digits [134]. According to The Wall Street Journal, "*At Schlumberger, which predicts that a million new wells will be fracked around the world between now and 2035, reducing freshwater use 'is no longer just an environmental issue — it has to be an issue of strategic importance*" [135]. Thus, the main environmental

---

[129]Kevin Grandia, How Cheney's Loophole is Fracking Up America, Huffington Post, May 17, 2010, updated May 25, 2011, http://www.huffingtonpost.com/kevin-grandia/how-cheneys-loophole-is-f_b_502924.html

[130]Renee Lewis Kosnik, The Oil and Gas Industry's Exclusions and Exemptions to Major Environmental Statutes, Oil & Gas Accountability Project, a Project of Earthworks, October 2007, p.2

[131]Fracking by the Numbers, Key Impacts of Dirty Drilling at the State and National Level, Elizabeth Ridlington, John Rumpler, Environment America Research & Policy Center, October 2013, p.4

[132]Erik Stokstad Will fracking put too much fizz in your water?, Science, Vol 344, Issue 6191, June 27, 2014, p.1469

[133]Russell Gold, Energy Firm Makes Costly Fracking Bet — on Water, The Wall Street Journal, August 13, 2013

[134]Jean-Philippe Nicot and Bridget R. Scanlon, Water Use for Shale-Gas Production in Texas, US, Bureau of Economic Geology, Jackson School of Geosciences, University of Texas at Austin, March 1, 2012

[135]Alison Sider, Russell Gold, Ben Lefebvre, Drillers Begin Reusing 'Frack Water' Energy Firms Explore Recycling Options for an Industry That Consumes Water on Pace With Chicago, The Wall Street Journal, November 20, 2012

problem of hydraulic fracturing is the huge quantity of water that is withdrawn from national consumption for hundreds of years because of chemical contamination; there is also the possibility of areas surrounding the production sites being directly contaminated because of defective cementing in the shale well walls, corrosion of steel elements and deterioration of the structural integrity of the wells over the following decades.

In 1992, the US Environmental Protection Agency (EPA) reported that an estimated 1.2 million conventional oil and gas wells were abandoned in the US, of which 200,000 may not be properly plugged—around 16% of the sample [136]— while these wells used much less fluid than unconventional water-dependent wells. In 2003, Schlumberger's Oilfield Review stated that *"Since the earliest gas wells, uncontrolled migration of hydrocarbons to the surface has challenged the oil and gas industry... [It is a] significant problem affecting wells in many hydrocarbon-regions of the world"* [137]. Statistics across all operators in Pennsylvania show that about 6–7% of new shale wells drilled between 2009 and 2011 had compromised structural integrity [138]. And the US Government Accountability Office revealed: *"The fracturing process itself is unlikely to directly affect freshwater aquifers because fracturing typically takes place at a depth of 6,000 to 10,000 feet, while drinking water tables are typically less than 1,000 feet deep... [Nevertheless,] underground migration of gases and chemicals poses a risk of contamination to water quality. Underground migration can occur as a result of improper casing and cementing of the wellbore as well as the intersection of induced fractures with natural fractures, faults, or improperly plugged dry or abandoned wells. Moreover, there are concerns that induced fractures can grow over time and intersect with drinking water aquifers... [Contamination could also be caused by] inadequate cement in the annular space around the surface casing, and ineffective cement that cracks or breaks down under the stress of high pressures"* [139]. In September 2014, a group of scientists from Duke, Ohio State, Stanford, Dartmouth and the University of Rochester confirmed that *"Faulty well integrity, not hydraulic fracturing deep underground, is the primary cause of drinking water contamination from shale gas extraction in parts of Pennsylvania and Texas"* [140]. Usually, there is a 25-year

---

[136]Roberto Suro, Abandoned Oil and Gas Wells Become Pollution Portals, The New York Times, May 3, 1992

[137]Claudio Brufatto, Jamie Cochran, Lee Conn, David Power, Said Zaki Abd Alla El-Zeghaty, Bernard Fraboulet, Tom Griffin, Simon James, Trevor Munk, Frederico Justus, Joseph R. Levine, Dominic Murphy, Jochen Pfeiffer, Tiraputra Pornpoch, Lara Rishmani, From Mud to Cement— Building Gas Wells, Schlumberger's Oil Field Review, August 2003, p.63

[138]Anthony R. Ingraffea, Fluid Migration Mechanisms Due To Faulty Well Design And/Or Construction: An Overview And Recent Experiences In The Pennsylvania Marcellus Play, October 2012, Physicians, Scientists and Engineers for Healthy Energy, p.8

[139]Oil and gas: Information on Shale Resources, Development, and Environmental and Public Health Risks, US Government Accountability Office, September 2012, pp.45–47

[140]Contaminated Water in Two States Linked to Faulty Shale Gas Wells, Nicholas School of the Environment of Duke University, September 15, 2014

guarantee on the integrity of a fracking well: after the guarantee term, nobody can say what will happen with the leakage of previously pumped toxic water, backwater or methane penetration through old pipes on depleted shale energy fields.

A number of independent research projects with residents of areas where shale plays are located, and landowners of plots for shale wells, show that hydraulic fracturing technology contaminates water sources not only during deep underground injections, but also during the backflow extraction process, when methane and chemical elements are leaking into surface drinking water sources [141,142,143]. In February 2012, the US National Oceanic and Atmospheric Administration and the University of Colorado in Boulder found that up to 4% of the methane produced at shale gas fields near Denver was escaping into the atmosphere. The American Geophysical Union reported even higher rates of methane leakage at the Uinta Basin in Utah—up to 9% of the total production. By comparison, the EPA suggested that 2.4% of total natural gas production was lost to leakage in 2009 [144,145]. Some researchers suggest that the impact of methane emission on global climate change is up to 72 times greater than that of carbon dioxide emission [146,147]; and this too seems to be worse with fracking than with conventional extraction: *"[D]uring the life cycle of an average shale-gas well, 3.6 to 7.9% of the total production of the well is emitted to the atmosphere as methane. This is at least 30% more and perhaps more than twice as great as the life-cycle methane emissions we estimate for conventional gas, 1.7% to 6%"* [148]. In 2013, Robert Jackson of Duke University and colleagues analyzed water from 141 wells in Pennsylvania's Marcellus region. Methane was detected in 82% of drinking water samples (115 out of 141), with average concentrations six times higher for homes located within a kilometer of natural gas wells than for homes located farther away from gas wells. Ethane was found in 30% cases (in 40 out of 133 drinking water samples); ethane concentrations were 23 times higher on average for homes located within 1 km from a gas well in

---

[141]Documentary "Gasland", 2010, Director Josh Fox

[142]Letter to United States Environmental Protection Agency, Region IX about "Underground Injection Control Drinking Water Source Evaluation", The State Water Resources Control Board of California, September 15, 2014

[143]Stephen Stock, Liza Meak, Mark Villarreal and Scott Pham, Waste Water from Oil Fracking Injected into Clean Aquifers, NBC Bay Area, November 14, 2014

[144]Jeff Tollefson, Methane leaks erode green credentials of natural gas. Nature, 493, 12, January 2, 2013

[145]Methane leaks from oil and gas fields, detected from space, American Geophysical Union, Eos, Transactions American Geophysical Union, Vol. 95, No. 46, 18 November 2014, pp.427–428

[146]Overview of Greenhouse Gases, US Environmental Protection Agency, http://epa.gov/climatechange/ghgemissions/gases/ch4.html

[147]Eli Kintisch, Plugging methane leaks in the urban maze could be key to making shale gas climate-friendly, Science, Vol 344, Issue 6191, June 27, 2014, p.1472

[148]Robert W. Howarth, Renee Santoro, Anthony Ingraffea, Methane and the greenhouse-gas footprint of natural gas from shale formations, Climatic Change, Volume 106, Issue 4, June 2011, pp.679–690

comparison with remote homes [149]. The US Government Accountability Office stated that *"According to EPA analysis, natural gas well completions involving hydraulic fracturing vent approximately 230 times more natural gas and volatile organic compounds than natural gas well completions that do not involve hydraulic fracturing"* [150]. Eventually, US nationwide media coverage of the environmental impact of fracking provoked accusations of inaction by state and federal regulators. In several states, public pressure achieved a ban on the exploitation of shale formations by fracking until the environmental safety of the technology had been scientifically proven.

Pascua Lama Copper Project (Chile-Argentina, 2000–2010s) and Conga Gold and Copper Mining Project (Peru, 2011)   Barrick Gold, based in Canada and the largest gold mining company in the world, proposed an $8.5 billion project to extract copper high in the Andes on the border between Chile and Argentina. The project met with entrenched local opposition because it threatened to contaminate rivers that are the main sources of water for farmers in the valleys below. When it became clear that the company intended to break up and move several glaciers in order to reach deposits beneath them, this caused public outcry. Local, national and international protest against the project finally persuaded the Chilean government and financial institutions not to proceed with it. Similar public concern and protest about the contamination of water sources led to the cancellation of the $5 billion Conga gold and copper mining project in Peru in 2011, proposed by the US-based Newmont Mining Corporation.

Protests Against Development to Extract Bauxite in the Niyamgiri Hills (India, 2010s)   Vedanta Resources had plans to extract bauxite from the Niyamgiri Hills, an area of pristine forest and the homeland of the Dongria Kondh indigenous people. Because of local resistance, the project never went ahead, and the company faced a shortage of bauxite for its nearby Lanjigarh aluminum refinery.

Collapse of the Fundão Tailings Dam at Samarco Iron Ore Mining Site (Brazil, 2015)   The Samarco mining complex is a 50–50 joint venture of two global mining giants—the Australian BHP Billiton and the Brazilian Vale. The complex consists of two sites: the Germano-Alegria mining unit, located in the Southeastern Brazilian state of Minas Gerais, where iron ore is extracted and initial ore-dressing treatment occurs; and the Ubu production and port unit, located in the state of Espirito Santo, where pelletizing and coating take place in preparation for the shipment of iron ore pellets and fines to customers abroad. The sites are connected by 400 km long

---

[149]Robert Jackson, Avner Vengosh, Thomas Darrah, Nathaniel Warner, Adrian Down, Robert Poreda, Stephen Osborn, Kaiguang Zhao, Jonathan Karra, Increased stray gas abundance in a subset of drinking water wells near Marcellus shale gas extraction, Proceedings of the National Academy of Sciences of the United States of America, July 9, 2013, 110(28), pp. 11250–11255, https://www.ncbi.nlm.nih.gov/pmc/articles/PMC3710833/

[150]Oil and gas: Information on Shale Resources, Development, and Environmental and Public Health Risks, US Government Accountability Office, September 2012, p.35

pipelines to transport the iron ore slurry from the mining site to the Atlantic Ocean. Launched in 1977, the complex had become by 2014 one of the largest in the world, producing more than 25 million tons of iron ore per annum.

In 2015, a disaster suspended production indefinitely. As part of the ore-dressing treatment at the Germano-Alegria unit, several tailings dams [151] had been constructed to store waste products from the crushing and milling process. One of these, the Fundão dam, collapsed on November 5 2015. A tidal wave of up to 62 million m^3 of toxic waste mud—from both Fundao and the Santarem dam located directly below it—flooded down the Gualaxo do Norte river and into the Doce river downstream, traveling more than 600 km to the Atlantic Ocean within 17 days. The toxic mudflow killed 19 locals or site workers, destroyed villages and left hundreds homeless; its impact on the river ecosystem downstream was also catastrophic. In ecological terms, the accident was considered by Brazilian officials to be the largest man-made environmental disaster in the country's history. 850 km of river bank were contaminated by heavy metals in high concentration; 200 towns, numbering hundreds of thousands of people, depend on the river water for home, agricultural and fishing needs and there was a rich riverine ecosystem [152].

An independent study of the technical causes of the collapse, mainly conducted by North American scientists and engineers, revealed the existence of drainage flaws during the construction and operation of the dam years before the disaster. In addition, there was an interaction over time between sands and "*slimes*" in the tailings, leading to a greater proportion of the waste products contained in the dam becoming saturated with water. According to the investigation, there had also been an incident with the drainage system in 2009, which required the redesign of the dam and, in August 2014, cracks were found in the body of the dam [153]. Furthermore, Brazilian officials point out that Samarco increased ore production in 2014 and 2015 compared to previous years, in response to rising commodity prices. To cope with the sharply rising discharge of mining waste, the height and weight of the dam were increased to expand its capacity [154]. Court documents filed against Samarco by one of the company's investors include this account: "*Subsequent investigations revealed that Samarco had for years disregarded safety concerns raised with respect to the Fundão dam... In 2011, Pimenta de Avila, the Fundão tailings dam's designer, authored a technical report for Samarco that warned of serious problems*

---

[151] A tailings dam is typically an earth-fill embankment dam used to store byproducts of mining operations after separating the ore from the gangue. Tailings dams rank among the largest engineered structures on Earth.

[152] An open letter to the International Community & Australian Parliament on the ongoing crisis in Brazil, February 18, 2016, http://www.aph.gov.au/~/media/Committees/fadt_ctte/estimates/add_1516/Openletter.pdf

[153] Report on the Immediate Causes of the Failure of the Fundão Dam, Fundão Tailings Dam Review Panel (Norbert R. Morgenstern, Steven G. Vick, Cássio B. Viotti, Bryan D. Watts), August 25, 2016, http://fundaoinvestigation.com/wp-content/uploads/general/PR/en/FinalReport.pdf

[154] Catastrophic Failure: The biggest environmental disaster in Brazil's mining history, ABC Australia/Journeyman Pictures, 2016, https://www.youtube.com/watch?v=KF3Clm6T_kI

if the dam was expanded. In October 2013, the Instituto Pristino, a Brazilian not-for-profit organization, prepared a report — subsequently confirmed to have been in Samarco's possession in 2013 — that warned of design deficiencies associated with the... dam and its planned expansion... Nevertheless, Samarco disregarded these concerns and ramped up production at its Germano facilities to offset falling ore prices, thereby increasing the waste volume to be contained by the Fundão tailings dam. However, as Brazilian prosecutors discovered, '[i]nstead of planning a new dam, with a new structure, [Samarco] looked for a patchwork solution'. In 2014, working as a consultant for Samarco, Avila identified several cracks in the dam during an inspection. Believing that the cracks were the beginning of a rupture, Avila made several recommendations, including installation of piezometers to allow for daily monitoring of water pressure. After the dam burst, Avila testified that he 'never received any feedback or request for clarification about his reports' and that an after-the-fact review of consulting reports prepared for Samarco gave no indication that the recommended piezometers had been installed or their readings monitored" [155]. There are official allegations against 26 Samarco executives and personnel that they were informed about the possible collapse of the dam many months before it happened, but the company rejects these claims and insists that it received no such information [156,157]. Samarco was also blamed for a slow and ineffective response in warning locals downstream of the dams within the first minutes and hours after they burst. With such poor warning transmission, if the collapse had happened during the night the number of casualties could have been many times higher [158]. In March 2016 Samarco, with the full support of BHP Billiton and Vale, made a settlement with the Brazilian government to pay around US$2.3 billion in compensation, including payments to those directly affected and support for the recovery of the local environment and social infrastructure damaged by the accident [159]. In January 2019, the Brazilian mining multinational Vale experienced another dam collapse at its Córrego do Feijão iron ore mine, killing at least 248 people, most of whom were mine workers.

---

[155]BANCO SAFRA S.A. – CAYMAN ISLANDS BRANCH, Individually and on Behalf of All Others Similarly Situated, Plaintiff, v. SAMARCO MINERAÇÃO S.A. and RICARDO VESCOVI DE ARAGÃO, Defendants. United States District Court, Southern District Of New York, November 14, 2016, http://securities.stanford.edu/filings-documents/1059/SMS00_01/20161114_f01c_16CV08800.pdf

[156]Paul Kiernan, Samarco Warned of Problems at Dam, Engineer Says, The Wall Street Journal, January 17, 2016

[157]Rob Davies, BHP Billiton employees face criminal charges on Brazil dam disaster, The Guardian, October 21, 2016

[158]Catastrophic Failure: The biggest environmental disaster in Brazil's mining history, ABC Australia/Journeyman Pictures, 2016, https://www.youtube.com/watch?v=KF3Clm6T_kI

[159]Mariaan Webb, Samarco unlikely to reopen this year, job cuts planned – BHP Billiton, Mining Weekly, July 14, 2016, http://www.miningweekly.com/article/samarco-unlikely-to-reopen-this-year-bhp-billiton-2016-07-14

It is clear that the development of raw resources in highly populated areas may provoke conflicts with locals: this can be seen in the cases we have described in Nigeria, Ecuador and the USA. In order to avoid decades of court battles—probably leading to multi-billion compensation payments—extraction companies have to develop deposits strictly within the national legal framework, buying land near deposits ahead of development and re-settling locals in other regions, applying the highest available ecological standards and using the most advanced technologies to minimize any adverse effect on the local environment and communities. And after the development of a deposit, the companies who have profited from that development need to invest some of those profits in restoring the territories they have impacted. Heavy industries may wish to study and implement some of the design principles set out by Elinor Olstrom, who shared the 2009 Nobel Memorial Prize in Economic Sciences for her work on fostering stable, local, common-pool resource management. The design principles are [160]: (I) set clearly defined boundaries (to effectively exclude external un-entitled parties); (II) establish rules regarding the appropriation and provision of common resources that are adapted to local conditions; (III) develop collective-choice arrangements that allow most resource appropriators to participate in the decision-making process; (IV) ensure effective monitoring by monitors who are part of or accountable to the appropriators; (V) implement a scale of graduated sanctions for resource appropriators who violate community rules; (VI) develop mechanisms of conflict resolution that are cheap and of easy access; (VII) allow for self-determination of the community recognized by higher-level authorities; (VIII) to manage larger common-pool resources, there will need to be organization in the form of multiple layers of nested enterprises, with small local CPRs (common-pool resources) at the base level.

Other Crises

Strikes Major strikes have brought coal mining to a standstill in several countries. In the USA in 1977–1978, industrial action for better wages, working conditions, safety, health protection and pension guarantees lasted 111 days; in Great Britain in the 1980s, strikers across the country fought, but ultimately failed, to halt colliery closures; in the USSR at the end of the 1980s, miners initially sought a better money allowance, but later began to raise political questions and played their part in challenging the authority of the Communist Party before the collapse of the Soviet Union. In the oil industry, strikes on the ARAMCO oil fields in Saudi Arabia in 1952–1956 forced the management to increase workers' wages. And a strike by 37,000 oil workers in Iran brought oil production there down from six million barrels/day to around 1.5 million barrels; it was the oil workers action, along with wider social upheaval, which ultimately led to the Islamic revolution in February

[160]Ostrom, Elinor, Beyond markets and states: polycentric governance of complex economic systems. American Economic Review, June 2010, 100 (3), pp. 641–672.

1979 [161]. There was a 64-day oil strike in Venezuela in 2002–2003 after a failed military coup against the democratically-elected Hugo Chavez. The Venezuelan opposition, and executives and some middle managers of the state-owed PDVSA, sought to undermine Chavez' popular support and force him to call new elections by halting Venezuelan oil and gasoline exports to bring on an economic crisis. Oil workers in Zhanaozen, part of the Ozen oil field in Western Kazakhstan, were on strike for months during 2011 for better pay and working conditions. The strike was finally suppressed by the Kazakh government, which blamed the protestors for violence during celebrations for the 20th anniversary of Kazakh independence, in which at least 15 people died. And in 2013, after the Libyan uprisings, a labor strike at the Gialo oil field cut production by 120,000 barrels/day. In response, the Libyan government threatened strikers with military action. In Tunisia—where the economic situation had not improved after the Arab Spring—disaffected young people began to picket phosphate and oil production facilities, in order to motivate the national government to create new jobs and share out revenues from the export of mineral resources. The blockades of phosphate industrial sites alone cost up to US$2 billion [162]. After several years, the government was forced to deploy army units to defend production activity from protesters.

Stealing Oil-Related Products from Pipelines Theft of gasoline from pipelines is a common problem in many oil-rich but poor countries. However, in Mexico, this activity has reached an unprecedented level, costing US$3 billion annually by 2019. Criminals tap into pipelines near the refineries of the state-owned PEMEX, steal fuel from the refineries themselves with the help of refinery workers and hijack fuel trucks. Some PEMEX franchisees have been involved in the illegal retail sale of the stolen fuel. To tackle this illegal activity, the Mexican government deployed up to 4000 troops and marines at six PEMEX refineries and 20 marine oil import terminals.

Natural Disasters as a Trigger for Industrial Disasters Earthquakes can provoke slope collapses in open pit mines. This is a challenge for mining in earthquake-prone regions like Chile and Peru, China, New Zealand, and the USA (Alaska and California). Storms and hurricanes are a serious threat to floating infrastructure for the extraction of oil and gas. For instance, in 1980 in the Norwegian part of the North Sea, the Alexander L. Kielland sea platform capsized. The platform was a living module for offshore drilling staff. It capsized accidentally when one of the vertical legs supporting the platform was torn off during a storm; 123 people died. In 1982 in Canada, 84 people were killed when the Ocean Ranger oil rig platform sank in a severe storm: a large rogue wave flooded the ballast control panel and later the platform lost buoyancy. In 1983, the drillship Glomar Java Sea capsized and sank in the South China Sea during the Lex tropical storm; 81 people died. In 1989, the drillship Seacrest capsized during Typhoon Gay in Thailand, killing 91 people. In

---

[161]IRAN: Another Crisis for the Shah, The Time, November 13, 1978

[162]Tunisia orders army to protect oil and gas fields, Al Jazeera, May 10, 2017

2005, Hurricane Katrina slashed oil production in the Gulf of Mexico by 95%, damaging more than 110 oil platforms and damaging more than 450 pipelines. In 2011, the Kol'skaya oil rig capsized and sank in the Sea of Okhotsk during transportation in stormy winter weather; 53 perished. During a storm in the Caspian Sea in 2015, 30 people died when a natural gas pipeline was ruptured and the Guneshli oil platform caught fire. In August 2017, during Hurricane Harvey, 105 out of 737 oil and gas production platforms in the Mexican Gulf were evacuated. Oil production dropped by 22%, natural gas by 26%, and ten refinery plants and eleven ports that received their oil from the area were shut down [163,164]. Some of the largest tanker oil spills also occurred during storms: the Odyssey (North Atlantic, 1971), the Amoco Cadiz (France, 1978), the Atlantic Empress (Trinidad and Tobago, 1979) and others.

Cyber Attacks The first known instance of government-sponsored cyber attack occurred in 1982, when a US agency introduced malware into the software controlling Soviet oil pipeline pumps in Siberia. The resulting explosion is considered the largest non-nuclear explosion on record [165]. In August 2012, the Saudi Aramco oil company was attacked by the Shamoon virus. Data from more than 2000 servers and 30,000 desktop computers was wiped. But Shamoon was far simpler than the infamous American-Israeli Stuxnet virus, which destroyed 984 of around 5000 uranium enrichment centrifuges at the Natanz underground nuclear site in Iran [166,167]. Experts concluded that the Shamoon virus had been written by private "*hacktivists*", rather than nation-state developers with unlimited resources like the creators of Stuxnet [168,169,170]. Nobody claimed responsibility for the attack, and no clear evidence came to light to implicate any state or organization; but Iran seemed to

[163] Jill Disis, Matt Egan, Chris Isidore, 10 refineries close as Harvey drenches Texas energy hub, CNNMoney, August 28, 2017

[164] BSEE Tropical Storm Harvey Activity Statistics: August 27, 2017, https://www.bsee.gov/news room/latest-news/statements-and-releases/press-releases/bsee-tropical-storm-harvey-activity-0

[165] Richard A. Clarke, Robert Knake, Cyber War: The Next Threat to National Security and What to Do About It, Harper Collins, 2012

[166] David E. Sanger, Confront and Conceal: Obama's Secret Wars and Surprising Use of American Power, Crown Publ. Group, 2012, pp. 188–209

[167] David E. Sanger, Obama Order Sped Up Wave of Cyberattacks Against Iran, The New York Times, June 1, 2012

[168] Kim Zetter, The NSA Acknowledges What We All Feared: Iran Learns From US Cyberattacks, WIRED, February 10, 2015, http://www.wired.com/2015/02/nsa-acknowledges-feared-iran-learns-us-cyberattacks/

[169] Dmitry Tarakanov, Shamoon The Wiper: Further Details (Part II), Kaspersky Labs' Global Research & Analysis Team, September 11, 2012,

[170] Norman Johnson, Shamoon Wiper: The most damaging corporate attack - Did you miss it?, http://www.cyber50.org/blog/shamoon-wiper

have most to gain from an attack on their main rivals in the region [171,172]. Clearly, the attack was intended to disrupt Saudi oil production [173]. However, the Shamoon virus, like Stuxnet and its subsequent developments—exploited bugs in Microsoft software to attack computers using Windows operating systems; computers using the supervisory control and data acquisition (SCADA) system were unaffected [174,175]. RasGas, a joint venture between Qatar Petroleum and ExxonMobil, also had its network disabled by the Shamoon virus a few weeks later [176].

Key Risk Mitigation Measures in Mining, Oil and Gas Extraction

- Deep understanding of geopolitical changes and trends, which have a serious influence on the business of extraction companies.
- Maintaining good and stable relations with the authorities in countries and regions where prospecting for mineral resources is taking place—in order to obtain and control low-cost and resource-rich deposits.
- Developing strong relations with local authorities, local community leaders, NGOs and environmental advocates and the local population, so that mineral deposits can be developed over the long term with broad local support.
- Monitoring changes in the legislation and regulation of extraction companies and associated businesses, along with lobbying efforts to push for a good compromise between the interests of business and those of the wider society.
- Predicting or even conducting market-making of prices on commodities, the majority of which are determined by quotation.
- Careful selection and retention of highly qualified staff with the skill and discipline to operate highly sophisticated industrial objects safely.
- Focusing on the reliability and safety of production sites in order to reduce workplace incidents, and minimize the chance of larger industrial

(continued)

---

[171] Kim Zetter, The NSA Acknowledges What We All Feared: Iran Learns From US Cyberattacks, WIRED, February 10, 2015, http://www.wired.com/2015/02/nsa-acknowledges-feared-iran-learns-us-cyberattacks/

[172] Sean Lawson, Anonymous Sources Provide No Evidence of Iran Cyber Attacks, Forbes, October 31, 2012

[173] Saudi Arabia says cyber attack aimed to disrupt oil, gas flow, Reuters, December 9, 2012

[174] Thomas Rid, Cyber War Will Not Take Place. Oxford University Press, 2013, p.66

[175] Heather MacKenzie, Shamoon Malware and SCADA Security – What are the Impacts, Tofino Security, October 25, 2012, https://www.tofinosecurity.com/blog/shamoon-malware-and-scada-security-%E2%80%93-what-are-impacts

[176] Chris Bronk, Eneken Tikk-Ringas, Hack or Attack? Shamoon and the Evolution of Cyber Conflict, Baker Institute for Public Policy, February 1, 2013

accidents—which could become nationwide disasters, cause widespread contamination and incur tremendous expenses for recovery.
- Closely cooperating with national and regional law enforcement to ensure physical and cyber protection of production sites and logistics from any action by third parties.
- Securing long-term and low-cost investment resources to provide for capital-intensive development even in the event of global economic obstacles, rapid changes in demand for mineral resources, etc.

2.2   Manufacturing

This subsector includes the following industries according to ISIC Rev. 4:

- 10—Manufacture of food products
- 11—Manufacture of beverages
- 12—Manufacture of tobacco products
- 13—Manufacture of textiles
- 14—Manufacture of wearing apparel
- 15—Manufacture of leather and related products
- 16—Manufacture of wood and of products of wood and cork, except furniture; manufacture of articles of straw and plaiting materials
- 17—Manufacture of paper and paper products
- 18—Printing and reproduction of recorded media
- 19—Manufacture of coke and refined petroleum products
- 20—Manufacture of chemicals and chemical products
- 21—Manufacture of basic pharmaceutical products and pharmaceutical preparations
- 22—Manufacture of rubber and plastics products
- 23—Manufacture of other non-metallic mineral products
- 24—Manufacture of basic metals
- 25—Manufacture of fabricated metal products, except machinery and equipment
- 26—Manufacture of computer, electronic and optical products
- 27—Manufacture of electrical equipment
- 28—Manufacture of machinery and equipment not classified elsewhere
- 29—Manufacture of motor vehicles, trailers and semi-trailers
- 30—Manufacture of other transport equipment
- 31—Manufacture of furniture
- 32—Other manufacturing
- 33—Repair and installation of machinery and equipment

General Description and Key Features of this Subsector and Incorporated Industries

We will classify industries within the manufacturing subsector according to the kind of customers for whom companies manufacture their goods: retail business (sometimes referred to as business-to-customer or B2C) and corporate business (business-to-business or B2B). B2C companies from this subsector (those involved in the production of food, beverages, tobacco, wearing apparel, furniture, cars, pharmaceutical products, computers and other electronic devices, etc) generally focus on the development of innovative products, mass production and very fast distribution to numerous clients worldwide. Retail manufacturers' most valuable assets are not production lines, but well-known product brands respected by customers, R&D staff, and extensive logistical networks that can distribute goods from outsourced production plants located in one part of the world to loyal customers on other continents. The commitment to relentless innovation in a highly competitive environment means that each new product will soon be superseded.

Because the manufacturing process of many products can easily be duplicated at plants all over the world, there is fierce global competition for customers in this subsector. Globalization and the opening of borders for international trade have had a significant influence on the production process, facilitating the search for cost-effective suppliers of raw materials and enabling companies to transfer assembly to countries with cheaper labor. Producers are continuously benchmarking products and prices in a given industry, and rapidly implementing improvements in their production using cutting-edge technologies. Advances in technology allow producers to make the production process ever more automated to meet the challenge of reducing defect rates: there is constant pressure to cut back on manual labor in order to minimise payroll costs, human error and the potential disruptive influence of unionized assembly workers.

Heavy manufacturing industries process raw materials (oil, gas, ore, wood) and/or synthesize chemicals mainly for corporate customers, who in turn manufacture specific products for the retail customer. Most of the industries in this group face similar types of risks: regulation-related risks, occupational safety challenges, and the potential threat posed to the environment by hazardous technological processes. Such heavy industries are generally recognized as part of the national infrastructure and make a huge contribution to the GDP and tax revenue of a government. Therefore, their actions are strictly regulated by government and regulators, in particular regarding the discharge of hazardous substances during production, national security concerns (national or state control over the industry and its supply and distribution), waste management solutions, and approval for massive production projects that could completely change the lives of nearby local communities. Occupational safety is a challenge for every industry in this group because the production processes involved are potentially highly dangerous—for example the production of pesticides, smelting of metal, and even the loading and transport of dangerous materials or the presence of workers at hazardous production sites. The production process could get out of control, or be poorly designed, in ways that damage the natural environment through the leakage of highly hazardous substances: discharge of chemicals, emission of harmful gases, metals and dust during metal smelting and so on. The surrounding environment can be laid to waste after the construction of a major industrial facility. Corporate customers will often depend on the supply from a plant, and there may be no alternative producers, so large numbers of employees may depend on the plant's activity. Moreover, there is likely to be high insurance coverage, and dealing with a major disaster may require government support.

More advanced or sophisticated manufacturing for corporate customers—the production of machinery, equipment, electronic and optical products, and so on—resembles retail manufacturing in many respects. There are high expenses on R&D, product quality is critical, there is fierce competition, branding is very important, distribution requires a global logistics network, and production lines have usually been relocated to places with lower production and labour costs. From our point of view, sophisticated manufacturing is generally devoid of serious risks to the environment, because its manufacturing processes use high value added materials, already produced at the more hazardous plants we have mentioned. In some cases, sophisticated manufacturing for corporate customers may be seen as a national security matter requiring government oversight—for example, companies would be expected to seek government approval to export sophisticated machinery or technology that could be adapted for military use.

Critical Success Factors for an Organization Within the Subsector

Product quality is one of the top critical success factors in this subsector, for both retail and corporate customers. In order to maintain consistent quality of goods, a producer will focus on innovative design and on new solutions to enable customers to get more out of its product. Manufacturers for retail, and of more sophisticated products for corporate customers, also try to minimize the percentage of defects throughout the production and distribution processes. This involves sourcing high-quality raw materials and spares from reliable suppliers, continuously improving the production process, and monitoring the work and output of assemblers; even once the product leaves the factory, the producer will seek to maintain control over the distribution of merchandise to the customer, storage conditions, customer care and fast repair or replacement in case of product fault or breakdown—or the urgent recall of a product if high quality has not been maintained in a particular batch, or if the design of the product turns out to be flawed. Producers may outsource or relocate production to other countries, but will expect to do so without adversely affecting product quality and reliable distribution to their main markets. Heavy manufacturing focuses on ensuring that its products comply with industrial standards and meet the requirements of specific customers.

This brings us to the second critical success factor that determines the profitability and sustainability of any manufacturer, and gives a product a competitive advantage on the market: low production cost. Producers try to keep their expenses down by continuously developing product design, searching for new suppliers, optimizing production lines and labor costs, transferring some production processes to areas with cheaper labor, and streamlining distribution networks. New strategies for "*lean management*" are especially in demand in this subsector. The constant benchmarking of products and prices leads to a market in which many rival manufacturers are producing quite similar goods at more or less the same pricing level. This tendency places great importance on the branding and marketing of goods: if you are turning out essentially the same product as your competitors, something must be done to differentiate your product in the eyes of customers and keep customers loyal to your brand. And because distributors are at the frontline of interaction with consumers, it is their level of service that will determine customer satisfaction throughout the life cycle of the product. Thus, companies in this subsector need to keep a tight rein on their distributors, to regulate customer care regarding the sale and repair of goods through the distribution network.

There are certain areas where customers are less sensitive to price but set great store by the quality and perceived "*exclusivity*" of a product, and the opposite cost strategy can work: the production of luxury cars, executive watches, yachts, etc., referred to in economics as "Veblen" goods [177]. In these cases, rich customers are ready to pay what most people would consider to be exorbitant prices in order to emphasize their ability to consume such luxury goods and confirm their status among a narrow circle of exceptional people. Such limited niche markets are the exception: the general orientation of this subsector is towards the reduction of production costs to survive fierce competition on pricing.

A third critical success factor, which becomes more and more important with technology development and global competition, is flexibility. Flexibility in manufacturing includes the ability to adapt to changing parts, to allow modifications in parts assembly and in process sequence, to convert to different product design and to quickly adjust production volume to needs [178]. Flexibility should also include adaptation both to predicted and unexpected changes.

The following success factors are critical in heavy manufacturing and chemicals production: maintaining good relations with the authorities, particular regarding the discharge of hazardous substances during production and national security concerns, operating production sites in an environmentally responsible way as the basis for long-term sustainable development and low accident rate.

---

[177]Veblen, T. B. (1899). The Theory of the Leisure Class. An Economic Study of Institutions. London: Macmillan Publishers.

[178]T. Tolin, Design of Flexible Production Systems – Methodologies and Tools. Berlin: Springer, 2009.

Stakeholders in this Subsector [179]

- Customers—50%
- Regulators—10%
- Suppliers—10%
- Distributors—10%
- Employees—10%
- Local communities—5%
- Other—5%

Typology of Common Risks, Main Features of Major Accidents and Risk Mitigation Measures Within the Subsector

Mistakes in Product Design

The main threat to the stability of retail manufacturing and sophisticated manufacturing for corporate customers comes from product-related crises induced by mistakes in product design and poor quality of production. When it *"goes public"* that there have been large numbers of customer complaints about a defective product, there is widespread awareness and concern among consumers. Such crises sometimes lead to a reduction in the market share of an organization when competitors move into the gap left by the defective product. If mistakes in design are revealed, an organization usually prefers to recall the affected merchandise, compensate customers, and suspend the whole production line until the design flaw has been eliminated.

Phocomelia and Peripheral Neuritis Outbreaks Caused by Thalidomide Consumption (Germany and Worldwide, 1950s–1960s) Thalidomide was an effective sedative drug, but it was not adequately tested for possible side effects during the pre-sale phase. After several years of distribution mainly in Europe, the United States and Japan, it became clear that consumption of the drug by pregnant women had led to birth defects in which infants were born without hands and feet, with severely deformed limbs (phocomelia) or with peripheral neuropathy. The total number of children affected was unclear, but thought to be about 10,000 babies worldwide. The drug was recalled, and its producer and distributors paid hundreds of millions of dollars over decades to compensate the damage. In later clinical tests, it was proven that thalidomide is an effective drug for leprosy, and in the treatment of some specific cancers.

---

[179]Our informed appraisal of the influence of each audience on a typical organization within the subsector (100% = combined influence of all audiences)

Unilever and Planta Red Margarine (The Netherlands, 1960) In the 1950s, Unilever was the first FMCG (fast-moving consumer goods) producer to start selling plant-based margarine, under the brand name Planta. During further research, the company discovered that the addition of an emulsifier called ME-18 made the margarine more easily spreadable and prevented splashing. The new improved margarine was released on the Dutch market under the brand Planta Red in 1960. Tests before its release were conducted only on animals; Unilever's staff tasted it in limited volumes. Shortly after the launch of Planta Red, at least 40,000 customers complained about the blister disease, a type of skin rash, which turned out to be caused by consumption of the new margarine; four people died and hundreds were hospitalized. Unilever completely recalled the new product—and some margarines under other brands, because traces of the margarine mixture with the ME-18 emulsifier were detected in tanks on the Unilever production site. The Planta brand was never sold on the Dutch market again, although margarine continues to sell under this name in Belgium and Portugal [180,181,182].

Ford Pinto and Ford Automatic Transmissions (USA, 1960s–1980s) The Ford Pinto was designed in record time, taking only 2 years from concept to production compared with an industry average of 3.5 years. At the final stage of the design process, during crash tests, engineers found that even a minor rear-end collision of the model could puncture the fuel tank and set the vehicle on fire. A full redesign was rejected: the additional costs could be significant, the engineers were determined to keep the weight of the model below 900 kg, and they were racing to bring out the model ahead of European and Japanese competitors [183]. Finally, the Pinto was released on the market in September 1970. This decision led to the deaths of almost 900 people [184]. In 1973 Ford published a cost-benefit analysis, as part of a wider objection to stricter regulations by the US National Highway Traffic Safety Administration (NHTSA). The new rules were demanding stronger fuel systems to withstand rollovers, rear-end and other collisions; they did not specifically mention the Pinto model. The company maintained that the implementation of enhanced fuel systems within the American car industry could increase the price to customers by up to $137 million—around $11/vehicle—while the benefit from such changes was estimated at $49.5 million and 180 lives saved. For this analysis, the value of an adult's life was appraised by Ford at $200,000, while a child's life was considered to

---

[180]Marcel Metze, The Planta-affair, 1960, Nofota History Project, http://www.nofota.info/the-planta-affair/

[181]Mirjam Gulmans, Femke Veltman, De Planta-affaire, Dodelijke slachtoffers door onschuldig pakje margarine, http://anderetijden.nl/aflevering/124/De-Planta-affaire

[182]Peter Anthonissen, Crisis Communication: Practical PR Strategies for Reputation Management & Company Survival, Kogan Page Publishers, 2008, p.19

[183]Dennis A. Gioia, Pinto Fires. In Trevino, L. & Nelson, K. Managing Business Ethics: Straight Talk, About How To Do It Right, Wiley, 1995, pp.80–84

[184]Peter Wyden, The unknown Iacocca, Morrow, 1987, p.238

be worth $100,000 and an injury $67,000 [185]. Media revelations about the fuel system shortcomings of the model led to an NHTSA investigation and the "*voluntary*" recall of 1.5 million Pintos and 30,000 Mercury Bobcats for the installation of protective devices within the fuel system.

Following the media furor over the concealment of the defect before the release, and the disreputable business philosophy Ford had demonstrated in their cost-benefit analysis, some of the victims of Pinto crashes took Ford to court: the company faced 117 lawsuits [186]. After a jury decision on one case, they were required to pay record damages of $127.8 million, although this was later reduced by an appeal judge to $six million. Ford was widely condemned: it appeared from their actions that the bottom line was more important to them than the lives of their customers. Of course, it stands to reason from an engineering point of view that carmakers will never be able to produce absolutely safe cars: engineers always try to reduce risk to an acceptable level. There is an unbridgeable gap between customers' understandable wish for "*absolute*" safety and the "*acceptable*" risks of real-world engineering. But with tough competition on the car market and a lifelong demand for cars, public confirmation of such a rational—but apparently cynical—engineering approach, which makes a subjective judgment about what constitutes an acceptable risk, can destroy customer loyalty in the long term. When it comes to buying a new vehicle, even a loyal and devoted user of Ford cars will wonder if Ford is the right choice when the company seems to have such a cavalier attitude towards the safety of its customers. It is worth noting that, in this subsector, risk management is first and foremost about marketing and public perception, and only then about engineering.

Several years after the Pinto episode, Ford was again facing a design-induced crisis. By 1980, the NHTSA had received tens of thousands of complaints from drivers of Ford vehicles with automatic transmission about park-to-reverse slippage in the gear boxes, which sometimes caused "parked" vehicles to move. General Motors had a similar issue with their transmissions, but Ford received 12 times the number of complaints: more than 23 million Ford vehicles had such defective gear boxes. Instead of recalling, to eliminate the mistake by correcting the tension in the automatic gear shift mechanism, Ford reached a deal with the NHTSA: they would send the affected customers a label warning them of the possibility of sudden vehicle movement, with short instructions on how to avoid it. The label would be stuck onto the instrument panel to warn the driver. This solution was clearly intended to transfer responsibility for any possible accident from the manufacturer to the customer. Remarkably, the estimated expenses for Ford to do a recall would have been about US$250 million—whereas their decision to avoid the recall ultimately cost them $1.7 billion in damages, when the families of more than 300 victims of crashes caused by the faulty transmission took the company to court [187].

---

[185]E.S. Grush and C.S. Saundy, Fatalities Associated With Crash Induced Fuel Leakage and Fires, Ford Motor Company, Environmental and Safety Engineering, 1973

[186]Peter Wyden, The unknown Iacocca, Morrow, 1987, p.238

[187]Ford Transmissions Failure to Hold in Park, Center for Auto Safety, 2009, https://www.autosafety.org/ford-transmissions-failure-hold-park/

Problems with the Rear Cargo Door of McDonnell Douglas DC-10 (Worldwide, 1970s) Difficulties with the latching of the cargo door of the DC-10 were first noticed in 1970—3 months before the aircraft went into commercial service—during tests of the air conditioning system. The DC-10's cargo door was designed to open outward to allow more storage space in the hold, so any problem with the latching system could pose a serious danger. However, McDonnell Douglas was in a race with Boeing and Lockheed to release the new plane as soon as possible, so the discovery of the rear cargo door problem was ignored. Over the following few months, reports came back to the manufacturers that the handlers were having problems closing the door electronically and were often forced to close manually. Then on June 12, 1972, a DC-10 lost its cargo door in flight near Windsor, Ontario, and the salon of the plane lost pressure. Fortunately, the pilots were able to land the damaged aircraft without serious injury to passengers or crew. After the incident, the manufacturers tried to solve the problem and partially followed the recommendations of air crash investigators, but the implementation of the changes they recommended was left to the airlines' discretion.

On March 3, 1974, a Turkish Airlines flight from Paris to London crashed, and this time it cost the lives of 346 passengers and crew. The decompression of the salon and failure of the hydraulic system had again been caused by the aft cargo door of the aircraft opening during flight. An investigation concluded that technical specialists at Turkish Airlines had weakened the latching mechanism because of problems with the proper closure of the door. As a result, the door was even less able to withstand internal pressure in flight. On the day of the flight, the handler at Paris Orly airport believed he had properly closed the door, but did not check the latch indicator in a special window: the sign detailing the proper latching procedure was in English and Turkish, but he was an Algerian immigrant who spoke Arabic and French and could not read the sign. The flight engineer of the crashed aircraft was also required to check the correct latching of the aft bulk door but, due to rush on that day, the check was skipped. Finally, the pilots too mistakenly believed the door was properly closed because they had a signal to that effect from an electronic micro-switch. In reality, the plane took off with an unlocked door.

The reputation of McDonnell Douglas was so severely damaged by this door issue, and subsequent crashes due to other causes, that the DC-10 was nicknamed the *"Death Cruiser-10"*. The company merged with Boeing at the end of 1996.

Another example of a major aircraft design flaw came to light in 1954 after two similar crashes involving the Comet 1, one of the first commercial jetliners in the world and produced by the British manufacturer de Havilland. As a result of errors in the design of the fuselage and the production process of the aircraft, there were microscopic cracks in the fuselage. Over time, this caused cumulative metal fatigue, and the fuselages disintegrated in mid-flight.

In October 2018, Indonesian Lion Air Flight 610 and, in March 2019, Ethiopian Airlines Flight 302, crashed claiming the lives of 346 people. Both were newly released Boeing 737 MAX 8 planes. Aviation authorities around the world grounded all planes of this model on suspicion of errors within the Maneuvering Characteristics Augmentation System (MCAS), which is designed to avoid the plane's nose pitching down. Indonesian investigators considered that "*the design and certification of [MCAS] was inadequate. The aircraft flight manual and flight crew training did not include information about MCAS*" [188]. In March 2019, Boeing was forced to recall all 737 MAX 8/9 planes worldwide to update the MCAS software and make special recommendations for pilot trainings. Before the recall, more than 5000 planes of these two models had been ordered by 107 customers.

Procter&Gamble and Rely Tampons (USA, 1978–1980) In order to offer a unique product on a very competitive market, researchers at Procter&Gamble proposed to use polyester foam cubes and chips of carboxymethylcellulose for the production of tampons instead of cotton and rayon. The innovation allowed a far greater absorption of menstrual flow than common tampons could provide; the advertising promised that the revolutionary new tampon could "*even absorb the worry*". However, after the new tampons were released under the brand name Rely, some customers were hospitalized with toxic shock syndrome caused by the bacterium Staphylococcus Aureus. It was hypothesized that the Rely tampon reduced the natural humidity of the vagina by absorbing so much menstrual fluid, which created a favorable environment for the bacteria to reproduce. About 20% of the general population carries these bacteria on the skin, and they are part of the normal ecology of some women's bodies, usually suppressed by the immune system. However, when the new super-absorptive tampon collected a tremendous amount of menstrual liquid under the right conditions of pH, chemical composition and viscosity, it triggered the proliferation of the bacteria in the vagina. If this continued unchecked, the toxins produced by the bacteria could cause potentially lethal systemic poisoning. A dispute followed between Procter&Gamble and the US Centers of Disease Control and Prevention: the manufacturer initially claimed that the usage of Rely tampons could not be scientifically proven to have caused toxic shock syndrome [189]. But ultimately, the company was forced to conduct a recall and eliminate the brand. The recall helped to mitigate the negative publicity over the company and its products.

Intel Pentium FDIV Bug Crisis (USA, 1994) Intel brought out a new Pentium processor, having already discovered that some units had a bug that affected complex division calculations. The bug was very rare, and the vast majority of customers would not be affected by it during common operations. Nevertheless, Intel was heavily criticized by some users and the wider tech community: not only had they concealed the problem, but they had made it difficult for customers to replace the defective processors, requiring them to prove that they had used Intel Pentium for complex calculations and even demanding that they show mistakes in their

---

[188]KNKT.18.10.35.04, Aircraft Accident Investigation Report, Komite Nasional Keselamatan Transportasi, http://knkt.dephub.go.id/knkt/ntsc_aviation/baru/2018%20-%20035%20-%20PK-LQP%20Final%20Report.pdf

[189]Sharra L. Vostral, Rely and Toxic Shock Syndrome: A Technological Health Crisis, Yale Journal of Biology and Medicine, 84(4), 2011, pp. 447–459

calculations. After pressure from customers and the media, Intel agreed to give replacements on request with no such conditions. Later, the CEO admitted that the manufacturer had failed to understand the psychology of the market: *"we crossed over the line to where there are millions of consumers out there who think they are better able to judge quality than we are and we were insensitive to that"* [190].

Bayer and Baycol (Worldwide, 1997–2001)   Statins are a class of drugs intended to reduce cholesterol levels in human blood. Bayer developed a statin under the brand name Baycol, which was far more powerful than the brands of competitors. Statins rarely have adverse side effects, but in some cases rhabdomyolysis can occur. This condition involves muscle pain and weakness combined with decreased renal function, and without timely diagnosis it can be fatal. Several years after the launch of Baycol in 1997, regulators and the manufacturer observed increased incidence of rhabdomyolysis among users, especially those on higher doses and those who also took gemfibrozil, another lipid-lowering drug, concurrently with Baycol. Though Bayer warned customers about the incompatibility of Baycol and gemfibrozil in the enclosed product information sheet, some patients were still taking both drugs concurrently. Over 300 became ill and 31 died in the United States alone; and while some competing statins caused lower incidences of rhabdomyolysis, the concurrent use of gemfibrozil with the other statins did not seem to have such a strong causal link with the condition. Finally, in August 2001, the manufacturer made a voluntary recall of the drug and ceased production. The recall had a big financial impact on the company: projected sales of Baycol for 2001 were around €1 billion and the drug was Bayer's third-largest pharmaceutical product. The company settled more than 3100 Baycol-related cases worldwide without acknowledging liability, paying out more than US$1.1 billion [191].

Ford-Firestone Tire Controversy (USA, 2000)   Poor coordination between Ford and Firestone in the engineering of tires for the new Ford Explorer, Mercury Mountaineer and Mazda Navajo models led to large numbers of these SUVs rolling over. The crisis forced the recall of tires and led to a public dispute between Ford and Firestone, ending nearly a century of partnership between the producers.

Merck and Vioxx (USA, 2000–2004)   In the 1990s, the American pharmaceutical giants ushered in a new generation of painkillers which inhibit the action of COX-2, an enzyme involved in inflammation and pain. Merck & Co marketed one such drug, rofecoxib, under the brand name Vioxx. Unlike older painkillers, COX-2 inhibitors relieved pain without the adverse effects of ulcers and other gastrointestinal diseases: thousands of people had died from internal bleeding caused by the use of painkillers in the past. With this huge advantage, and in the absence of clinical trials showing any serious negative side effects, Vioxx was approved by the US Food and Drug

---

[190]Michael Regester, Judy Larkin, Risk issues and crisis management: a casebook of best practice, 2005, third ed., p.63

[191]Reinhard Angelmar, The rise and fall of Baycol/Lipobay, Journal of Medical Marketing, 2007, Vol. 7, pp. 77–88

Administration (FDA, the American federal regulator for the food and pharmaceutical industries) for market release in 1999. Merck was in a rush to bring out the new drug because the patents on many of their popular drugs were due to expire imminently: the company needed a new best-selling product to rescue their falling profits, and had no time to insist on more exhaustive trials to look for possible side effects. In addition, competitors were moving fast to promote similar painkillers—Pfizer, for example, already had a COX-2 inhibitor out under the brand name Celebrex.

But after the drug's release, Merck and the FDA began to hear clinical evidence suggesting that the use of Vioxx could cause heart attacks. In March 2000, one such study confirmed a reduced risk of internal bleeding compared to other painkillers, but four times the risk of heart problems. Of course, the FDA were also aware of these statistics, but did not consider that Vioxx should be withdrawn because the drug had such a clear positive impact on the treatment of patients with ulcers. Further studies followed, and all came to similar conclusions; and there were internal discussions within Merck about the side effects revealed. The manufacturer decided that it would be unethical to conduct a new cardiovascular study, preferring to monitor already active clinical trials with a specific focus on any cardiovascular problems that may emerge. They also proposed to record and monitor statistics on possible heart problems for the millions of existing users of Vioxx. The FDA agreed with these proposals [192].

However, there is ample evidence of *"conflicts of interest pervading the FDA, including the fact that many members of FDA advisory committees are paid consultants for drug companies"* [193]. According to the New York Times, around a third of FDA drug advisers in this period had done consultancy work in recent years for the drug companies [194]. And while Merck's preference not to undertake a specific cardiovascular trial may have been partly based on ethical grounds, it is noteworthy that the company's marketing executives were also against it, and their researchers discussed the reformulation of Vioxx to reduce the cardiovascular side effects in 2000. Later it became known that *"Merck instructed sales people to show physicians a pamphlet showing Vioxx might be 8 to 11 times safer than other anti-inflammatory drugs. Another memo told sales representatives not to bring up the drug's heart risks with doctors"* [195].

In 2004—after nearly 5 years on the market with annual sales of US$2.5 billion and up to 20 million customers, and several studies which confirmed the link to cardiovascular problems—Merck made a voluntary recall of Vioxx, presenting the

[192]Alex Berenson, Gardiner Harris, Barry Meier And Andrew Pollack, Despite Warnings, Drug Giant Took Long Path to Vioxx Recall, November 14, 2004

[193]Marcia Angell, Your Dangerous Drugstore, New York Review of Books, 2006, 53(10), pp. 38–40

[194]Gardiner Harris, Alex Berenson, 10 Voters on Panel Backing Pain Pills Had Industry Ties, The New York Times, February 25, 2005

[195]Christine Lagorio, Merck Knew Vioxx Dangers In 2000, The Associated Press, June 22, 2005

decision as a response to side-effects having been discovered and confirmed, but failing to acknowledge how long they had been aware of them. Dr. David Graham, Associate Director for Science at the FDA's Office of Drug Safety from 1999 to 2003, estimated that Vioxx may have caused 27,785 heart attacks, with a mortality rate of 27% [196]—in other words, around 7500 Vioxx users may have died. The company paid $4.8 billion to settle lawsuits from the families of the thousands of patients killed or injured, who blamed Merck for neglecting to warn doctors and the public about known and proven cardiovascular risks. In 2005, Pfizer also voluntarily recalled Bextra, its COX-2 inhibitor painkiller, because it had the same adverse effects; Celebrex, a weaker version of Vioxx, was allowed to remain but with special warning sign on the box and a lower recommended dose [197].

Bromate in Dasani Purified Tap Water (UK, 2004)   Only 1 month after launch, Coca-Cola had to recall 500,000 bottles of its purified water in the UK when high levels of the carcinogen bromate were found in the water. The manufacturer used tap water in the production of Dasani, purifying it through three separate filters in a process similar to that used by NASA to purify drinking fluids for astronauts. However, the resulting pure water then needed to be salted artificially: the filtering was so effective that it almost eliminated the natural mineralization of the original tap water. To try and recreate a *"natural taste"* calcium chloride, which contains bromide, was added. But in the final purification step of ozonizing the water, this bromide was oxidized into bromate. Coca Cola never reintroduced the Dasani brand to the UK market.

Sony Batteries for Laptops (Worldwide, 2006)   In 2006, Sony and its vendors (Dell, HP, Fujitsu, Lenovo, Apple, and Toshiba) recalled 9.6 million lithium-ion batteries for notebooks, after six instances since December 2005 in which notebooks had overheated or caught fire [198]. Experts had known for years that such batteries could catch fire in extremely rare circumstances, but the recent cases involving notebook computers caused a wave of negative publicity over their safety. Explaining the voluntary recall, a Sony spokesman concluded: *"This is not a safety issue. This is about addressing people's concerns which have become a social problem, and we made the managerial decision that the recall was necessary"* [199]. Sony promised to make improvements in production, design and inspection in order to prevent laptop overheating problems in future. In the previous book on

---

[196]David J. Graham, Risk of acute myocardial infarction and sudden cardiac death in patients treated with COX-2 selective and non-selective NSAIDs, FDA internal memorandum, September 30, 2004, http://www.fda.gov/downloads/drugs/drugsafety/postmarketdrugsafetyinformationforpatientsandproviders/ucm106880.pdf
[197]Questions and Answers FDA Regulatory Actions for the COX-2 Selective and Non-Selective Non-Steroidal Anti-inflammatory drugs (NSAIDs), US Food and Drug Administration, April 2005
[198]Damon Darlin, Dell Recalls Batteries Because of Fire Threat, The New York Times, August 14, 2006
[199]Yuri Kageyama, Sony Apologizes for Battery Recall The Associated Press, October 24, 2006

risk information concealment by the present authors, this case is discussed as a positive response by management to a crisis [200].

Apple iPhone 4 Antenna (USA, 2010) After the release of a new version of Apple's popular smart phone, customers detected problems with signal reception. Initial Apple CEO Steve Jobs dismissed enquiries about possible antenna problems in the iPhone 4. However, after several weeks of media and customer criticism, he was forced to admit the shortcoming [201]. Moreover, it emerged that Jobs had been aware of weak signal reception in this model before the release, but had ordered the immediate launch of the product without proper tests. Eventually, Apple offered free special care for Apple iPhone 4 customers, and updated the corresponding software to increase signal strength.

Toyota Pedal Crisis (USA-Japan, 2009–2011) The largest car producer in the world was forced to make the largest product recall in the history of the industry— involving more than nine million cars of different models. Drivers had experienced unintended acceleration in several Toyota vehicles because the floor mat had slipped on top of the accelerator pedal; the recall aimed to redesign both components. There were also allegations that crashes had occurred because the company had installed an electronic drive-by-wire throttle system—but on investigation, the US authorities did not find electronic defects in Toyota vehicles and placed responsibility on drivers in these cases. At the beginning of the crisis, Toyota decided to admit all allegations, even though some were later disproved: it was more important to demonstrate to millions of current and potential customers that the company's overriding concern was for the safety of drivers [202].

GM Defective Ignition Switch (USA, 2014) It was revealed that General Motors had been hiding a defect with a tiny part in the ignition switch for 10 years. For the sake of a part worth just 57 cents, the concealment caused 13 traffic deaths, and 2.6 million cars had to be recalled.

Volkswagen Diesel Engine Emissions Scandal (USA-Worldwide, 2000s– 2010s) Tough new regulations on diesel engine emissions in the United States required car producers to find innovative engine solutions. But instead of making real innovations to achieve emission levels that seemed beyond the reach of engineers, VW decided to install a defeat device into its diesel vehicles, which reduced noxious gas emissions to approved levels only during emission tests: in real-world driving conditions, the same models released exhaust gases at 35–40 times the new legally permitted level. The company reached a $14.7 billion settlement with US

---

[200]Dmitry Chernov and Didier Sornette, Man-made catastrophes and risk information concealment (25 case studies of major disasters and human fallibility), Springer, 2016

[201]Miguel Helft, Apple Acknowledges Flaw in iPhone Signal Meter, The New York Times, July 2, 2010

[202]This case is studied in detail in chap. 2.4.2 of Dmitry Chernov and Didier Sornette, Man-made catastrophes and risk information concealment (25 case studies of major disasters and human fallibility), Springer, 2016

federal and Californian regulators, and the 560,000 American owners of the cars affected. This was the largest civil penalty imposed on any carmaker in US history. The Volkswagen scandal raised awareness over the high levels of pollution being emitted by diesel vehicles built by a wide range of carmakers, including Volvo, Renault, Jeep, Hyundai, Citroen, BMW, Mercedes, Mazda, Fiat, Ford and Peugeot [203,204,205].

IKEA Shelves (Worldwide, 2015–2016) The retailer IKEA produces and sells furniture under its own name. The company made a voluntary recall of more than 27 million dressers after three children were killed when the shelves toppled and fell on them [206,207]. Improvements in the design of the dressers stipulated the provision of special fasteners to bolt the furniture to the walls of a room.

Samsung Galaxy Note 7 (2016) Following the launch of Samsung's new smart phone, hundreds of customers found that their phones caught fire during recharge (the problem affected *"less than 0.1 per cent of the entire volume sold"* [208]). The company recalled the first batch of defective devices and released a new updated version. However, the new batch had similar problems with overheating batteries. To protect the brand and stop negative publicity about Samsung products, the producer made a total recall of the model and cancelled further production. This cost Samsung more than US$3 billion in lost profit, but for the company the loyalty of their customers was paramount [209].

Manufacturers can take a range of actions to reduce the probability of mistakes in design. These could include making serious investments in R&D, setting up competition between teams of researchers to develop the best possible solutions, refraining from putting extreme time pressure on developers in recognition of the dire consequences such pressure can bring, implementing proper testing of prototypes in different conditions, and inviting the involvement of loyal customers in the testing and evaluation of products. They should also make preparations in

---

[203]David Connett, Volkswagen emissions scandal: More carmakers implicated as tests reveal pollution levels of popular diesels, Independent, September 30, 2015

[204]Damian Carrington, Wide range of cars emit more pollution in realistic driving tests, data shows, The Guardian, September 30, 2015

[205]Kartikay Mehrotra, Dodge Truck Owners Accuse Chrysler of VW-Like Cheating, Bloomberg, November 14, 2016

[206]Lydia Willgress, IKEA issues urgent warning over popular range of drawers after THIRD child is crushed to death by a toppling unit, Daily Mail, April 20, 2016

[207]Ikea recalls 29 million dressers after 6 kids killed, Associated Press, June 28, 2016

[208]Amelia Heathman, We finally know why Samsung's Galaxy Note 7 s 'exploded', Wired, January 24, 2017

[209]Samsung Electronics Estimates Mid-3 Trillion Won Negative Impact for Q4 2016 and Q1 2017 Due to Galaxy Note7 Discontinuation, Samsung, October 14, 2016, https://news.samsung.com/global/samsung-electronics-estimates-mid-3-trillion-won-negative-impact-for-q4-2016-and-q1-2017-due-to-galaxy-note7-discontinuation

advance for a fast recall in the event of a serious shortcoming being revealed, and for compensating any customers affected.

Crises Induced by Low-Quality Production Process

If a low-quality production incident comes to light, retail manufacturers and those producing more sophisticated items for corporate customers usually prefer to recall the specific batch that included defective goods, and convince customers that it was a limited case and that all remaining products still in distribution are safe for consumption.

Perrier (Worldwide, 1990) In 1990, the French mineral water producer Perrier conducted a voluntary global recall of 230 million bottles of its water when it emerged that it was contaminated by carcinogenic benzene.

Before the discovery, Perrier was the most popular imported carbonated water in the United States with US$150 million in annual sales [210]. Because the brand was associated with purity as well as style, Perrier water was a hundred or even a thousand times more expensive than common tap water. Their advertising boldly claimed: *"It's Perfect. It's Perrier"*. The owner of a French restaurant in New York observed: *"People think it's prestigious [to drink Perrier]; it's an 'in' thing. We sell a lot of it. To me, I think it's the biggest hype since the Beatles"* [211].

At the beginning of 1990, a laboratory in North Carolina was testing domestic mineral waters, and selected Perrier as a benchmark of purity for matching. Surprisingly, the scientists found excessive levels of benzene in Perrier's samples—from 12.3 to 19.9 parts per billion (ppb). The US Environmental Protection Agency (EPA) sets the maximum allowed contamination of benzene for public drinking water supplies at a level of 5 ppb [212], while the World Health Organization (WHO) currently stipulates 10 ppb and the European Union standard is 1 ppb. The levels of benzene contamination recorded in Perrier bottles in North Carolina did not threaten customers with cancer. A representative of the US Food and Drug Administration (FDA) declared: *"At these levels, there is no immediate hazard. The hazard would be that over many years, if you consumed about 16 fluid ounces a day, your lifetime risk of cancer might increase by one in a million, which we consider a negligible risk. You don't have to be concerned if you just had a bottle of Perrier"* [213]. So the FDA did not ask the producer to make a recall of Perrier bottles, but the management team

---

[210]Tom Furlong, Perrier Water Scare Goes Flat in Southland, The Las Angeles Times, February 12, 1990

[211]George James, Perrier Recalls Its Water in US After Benzene Is Found in Bottles, The New Times, February 10, 1990

[212]Ibid

[213]Ibid

of the American branch of Perrier decided to make a voluntary recall of 70 million bottles from shelves throughout the United States.

It is important to emphasize that the recall was implemented not to avert a direct health threat to customers, but to support the brand in the long term: Perrier's positioning was based on associations with absolute purity, extremely high environmental standards, perfection and luxury. If the company had left benzene-contaminated bottles of Perrier in stores, exclusive restaurants and luxury hotels in America, all of those associations could be shattered, potentially leading to a total collapse of the brand among high income customers. At the beginning of the crisis, the management team of the American branch of Perrier had no exact information about the cause of the contamination. This led to premature statements: they suggested that it must be a staff mistake during the packaging and distribution process of bottles for the American market: *"Perrier… [blamed] a cleaner's improper use of a cleaning solvent on machinery filling bottles bound for the USA for the contamination"* [214]. The team considered that it would only take 2 or 3 months to produce enough water from the company's source in Vergèze (southern France) to bring completely pure Perrier water back onto American shelves. But when the crisis erupted in America, laboratories in Holland and Denmark began to test samples of Perrier from domestic stores and once again confirmed levels of contamination similar to those found in North Carolina. There were suppositions that the Vergèze source itself must be contaminated. Media and customers worldwide blamed Perrier not only for the benzene contamination, but for withholding information about the contamination risk by limiting the recall only to the United States. Finally, the French parent company had to make a global voluntary recall, taking back a tremendous number of bottles—160 million, bringing the total to 230 million bottles including the American recall. To put this in perspective, annual pre-1990 production of Perrier was 1 billion bottles: 400 million were sold in France and 600 million were exported.

During the investigation, it was established that the cause of contamination was a month-long failure of the filtering system at the manufacturing plant in Vergèze, which had to purge away benzene and other noxious elements from the natural carbon dioxide gas produced from the depths. The clogging of the filters was undiscovered until the crisis. The water source itself was not contaminated. Before the crisis, Perrier water was positioned as *"naturally carbonated"*, but after the benzene contamination discovery, the company admitted that mineral water was pumped separately from carbon dioxide because of harmful substances in the pumped-out gas. After filtering, the natural carbon dioxide was used to carbonate the water artificially during the production process. When Perrier had fixed the filter problem and the benzene content had been brought back down to permitted levels, the FDA required them to eliminate the wording *"naturally carbonated water"* from bottle labels and from all of their advertising. The full restoration of Perrier to the marketplace took 5 months. But the brand never returned to its US pre-1990 market

---

[214]Andrew Caesar-Gordon, Lessons to Learn from a Product Recall, PR week, October 28, 2015

volume of 44% of imported mineral water, yielding significant market share to the French Evian and the Italian San Pellegrino; American producers like Saratoga and Coors also gained from the crisis. In 1992, the Perrier brand, the Vergèze water source and the manufacturing plant were sold to Nestlé, the world's leading bottled water company.

Poly Implant Prothese (PIP) Fraud (France, 1993–2010) For 17 years, this French producer manufactured thousands of breast implants containing a significant percentage of industrial gel in addition to medical gel, in order to reduce production costs. Due to shortcomings in the deregulated legislation of the cosmetics industry in France and the European Union, it had become possible to hide changes in the gel formula for decades. Using the new gel formula allowed PIP to save up to US$1.6 million/year. More than 300,000 women still carry these implants [215].

Nutricia and Olvarit Baby Food (the Netherlands, 1993) Nutricia recalled two million jars of its baby food when excessive levels of disinfectant were recorded. The company was so committed to hygiene and sterility in its cleaning process that it went too far, leaving traces of disinfectant on the jars [216].

Odwalla Fresh Juice E. coli Outbreak (USA, 1996) Odwalla produced fresh pasteurized juices, and wanted to maintain the intense natural flavor of their products. So they rejected the idea of chlorinating the water used to wash their fruit, maintaining that existing sanitization procedures were adequate. But an outbreak of E. coli bacteria was traced to a batch of Odwalla apple juice, resulting in one death and making 66 people ill. The producer made a total recall of any apple juice presented. The crisis reduced sales by 90%. The company struggled to recover for several years, and was finally bought out by Coca Cola.

Staphylococcus Outbreak in Snow Brand Dairy Products (Japan, 2000) This Japanese dairy producer concealed a staphylococcus aureus outbreak in milk products produced at its Osaka plant. Workers at the plant washed equipment valves poorly and failed to sterilize them properly. There was also falsification of documentation. The outbreak affected 14,000 people, and ruined Snow Brand's reputation as the leading producer on the Japanese dairy market [217,218,219].

---

[215]This case is studied in detail in chap. 2.4.3 of Dmitry Chernov and Didier Sornette, Man-made catastrophes and risk information concealment (25 case studies of major disasters and human fallibility), Springer, 2016

[216]Peter Anthonissen, Crisis Communication: Practical PR Strategies for Reputation Management, Kogan Page, 2008

[217]Sydney Finkelstein, Why Smart Executives Fail: And What You Can Learn from Their Mistakes, Penguin, 2004

[218]Matt Haig, Brand Failures: The Truth about the 100 Biggest Branding Mistakes of All Time, Kogan Page Publishers, 2005, pp. 134–136

[219]Staphylococcus food poisoning in Japan, Infectious Disease Surveillance Center, 2001, 22 (8), No.258, pp. 185–186

Failure of US National Oceanic and Atmospheric Administration N-Prime Satellite (USA, 2003) The satellite fell to the factory floor while being turned over during the production process at Lockheed Martin Space Systems. The fall occurred because 24 bolts connecting the satellite to the turn-over cart had not been put in place. The investigation stated that this oversight stemmed from "*the operations team's lack of discipline in following procedures evolved from complacent attitudes toward routine spacecraft handling, poor communication and coordination among operations team, and poorly written or modified procedures*" [220]. The repair required the replacement of 75% of the components of the craft and cost more than US$217 million [221]. The satellite was successfully sent into space in 2009.

Salmonella Outbreak at Cadbury Marlbrook Factory (UK, 2006) An outbreak of salmonella affecting up to 40 people was traced to Cadbury's chocolate, after the existence of the bacteria at Marlbrook factory had been concealed for 6 months [222].

Chinese Milk Scandals (China, 2004 & 2008) Chinese companies producing infant milk formula were discovered to have reduced its nutritional value. Sales of watered-down "fake milk powder" caused 13 baby deaths from malnutrition in 2004. Four years later, some producers were found to have been adding melamine to their products to increase their apparent protein content during protein tests. This time 54,000 babies were hospitalized for kidney problems and six died. In total, the deception affected more than 300,000 babies.

Manufacturing Deficiencies at GlaxoSmithKline's Puerto Rico Plant (USA, 2010) SB Pharmco Puerto Rico, a subsidiary of GlaxoSmithKline in Cidra, Puerto Rico, produced adulterated drugs for years, undetected by customers: some batches were more or less potent than they were supposed to be, some pills had a low-quality appearance, and there were even mistakes in packing the correct drug in the correct bottle. Fortunately, no customers were harmed. The plant was closed and GlaxoSmithKline agreed to pay US$750 million to settle the case [223].

Worldwide Recall of Lactalis Baby Milk Products Due to Salmonella Outbreak (France and Worldwide, 2018) Lactalis is one of the largest dairy holdings in the world; its most famous brand is Président butter and cheese. In 2018, 12 million boxes of Lactalis baby milk products were recalled in 83 countries after 37 children in Europe suffered from salmonella. The bacteria were traced back to a factory at

---

[220]NOAA N-PRIME Mishap Investigation. Final Report, NASA, September 13, 2004, https://www.nasa.gov/pdf/65776main_noaa_np_mishap.pdf

[221]Stephen Clark, Back from the brink: Broken satellite now fixed and ready, Spaceflight Now, February 1, 2009, https://www.spaceflightnow.com/delta/d338/090201preview.html

[222]Jeevan Vasagar, Chocolate may have poisoned more than 40, The Guardian, June 24, 2006

[223]GlaxoSmithKline to Plead Guilty & Pay $750 Million to Resolve Criminal and Civil Liability Regarding Manufacturing Deficiencies at Puerto Rico Plant, The United States Department of Justice, October 26, 2010

Craon in the département of Mayenne in north-western France. Remarkably, this factory had been the source of a previous salmonella outbreak in 2005, which affected 149 infants. According to executives of the holding, Lactalis baby milk products *"could have been contaminated with salmonella for thirteen years [since 2005]"* [224]. In 2017, several bacteria tests at the factory had confirmed the existence of salmonella in a broom and on a tile near drying equipment, but the holding had not reacted to these findings *"because we had no element showing our products were affected"* [225]. A private testing laboratory had also failed to confirm the existence of salmonella within baby milk powder. After the outbreak and ensuing public attention, the factory was closed for anti-bacteria treatment. The holding faced a huge challenge to immediately withdraw 12 million boxes of milk powder from shelves worldwide; even the French finance minister Bruno Le Maire had to admit: *"I cannot guarantee that right now there isn't a single tin of baby milk left on a shelf in a giant warehouse or in a pharmacy ... I think this [a further recall in January 2018 following an initial recall in December 2017] is the strongest guarantee we can give"* [226].

Looking for the best response to the discovery of defects during production was the basis of the lean management approach developed by Toyota. Lean management takes the principles of continuous improvement in the quality of production, and constant vigilance in minimizing waste of resources, which are employed in the Toyota Production System. Other important actions could include: improving the testing of goods before shipment; developing claim management through better detection and systematization of customer complaints; enhancing coordination with distributors and retailers, to ensure that defective merchandise is immediately removed from the shelves and that customers are well looked after at the beginning of recalls; rapid elimination of shortcomings, by preventive voluntary recalls of defective batches and immediate improvements to the production process; a simple and effective compensation system for customers affected by the recall; and better crisis communication during recalls for proper communication with victims, customers, regulators and investors, so that the producer's actions are clearly explained and the brand is protected.

---

[224]Ben Chapman, Lactalis baby milk could have been infected with salmonella for 13 years, CEO says, Reuters, February 2, 2018

[225]Kim Willsher, Lactalis to withdraw 12 m boxes of baby milk in salmonella scandal, The Guardian, January 14, 2018

[226]Gus Trompiz, Richard Lough, France's Lactalis forced into new recall in baby milk scare, Reuters, January 12, 2018

Sabotage and Unauthorized Production

Generally, when a manufacturer faces a sabotage or an unauthorized forgery, the company is not responsible for the crisis—and is as much a victim of the illegal action as are the affected customers. Therefore, the manufacturer can ask state regulators and the law to find and prosecute the fraudsters and help the company recover. If the company is seen to deal with the crisis skillfully and with integrity, the free publicity brought by such actions can attract the attention of customers, ultimately increasing sales.

Johnson & Johnson and Cyanide in Tylenol (USA, 1982) In 1982, an unidentified person or group of people put cyanide in bottles of Tylenol, at the time the most popular painkiller in the country with around 37% of the US market. Seven people in Chicago were killed. 75 poisoned capsules were detected in this city alone during a nationwide inspection of eight million pills; the FDA confirmed 36 cases of *"true tampering"*. The company immediately coordinated with the Chicago Police Department, FBI and FDA to prove that the poisoning had been sabotage, and offered a reward for information leading to the arrest of the saboteur. To allay panic among customers, and show the public that customer safety was paramount to them, Johnson & Johnson conducted an immediate total recall of 31 million bottles of the drug in the United States—at a cost of more than 100 million dollars. Within a month, the company had developed special triple-sealed packaging to avoid tampering, and put the new bottles out with a large price reduction to bring its market share back to pre-crisis levels. Due to the positive perception among customers of the company's reaction to the crisis, the product recovered in less than a year. This became a textbook case in crisis management because of the speed of Johnson & Johnson's reaction to the challenge, the commitment to safety the company had shown through their immediate warnings and total recall, their openness to any media or customer request and their intensive collaboration with regulators and the law to investigate the crisis and contain the threat [227]. As a result of the case, product tampering became a federal crime in the United States.

Pepsi and "Syringes in the Cans" (USA, 1993) On June 10, 1993, in Washington state, there was an allegation that a syringe had been discovered in a can of Pepsi. The allegation attracted national media attention, which in its turn provoked copycat allegations from customers from other states. Pepsi was clear that it would be impossible for a foreign object to get into the cans during the production process: *"A can is the most tamper-proof packaging in the food supply. We are 99.99% certain that this didn't happen in Pepsi plants"* [228]. The company called in the FDA to inspect the production plant where the first allegedly tampered can was manufactured. The regulator found nothing, recommended not to conduct a recall

---

[227]Judith Rehak, Tylenol made a hero of Johnson & Johnson: The recall that started them all, The New York Times, March 23, 2002

[228]Colin Doeg, Crisis Management in the Food and Drinks Industry: A Practical Approach, Springer Science & Business Media, 2006, p.180

and warned customers that the penalty for product tampering was $250,000 with a maximum of 5 years in prison. Nevertheless, new allegations appeared and soon there were 65 cases. Several days after the eruption of the crisis, Pepsi received a videotape from a surveillance camera at a Colorado store, showing a woman putting a syringe into a newly-opened Pepsi can. Pepsi informed the FDA Criminal Investigation about this evidence, the tamperer was arrested and the videotape was passed to the hysterical media. After the broadcast of the recording, the crisis was soon resolved in favor of Pepsi, though the company lost $30 million in sales over the crisis period. However, the media buzz around the hoax attracted the attention of the American public, who were impressed by Pepsi's behaviour in the crisis, and the company did very well during the summer of 1993.

Pfizer's Counterfeit Viagra (Worldwide) According to the World Health Organization, from 100,000 to 1 million people die annually from the consumption of counterfeit drugs. The majority are from developing countries where control over drug production and distribution is weak—for instance in Pakistan, where 40–50% of all drugs are counterfeited. The annual turnover of the industry worldwide was US $431 billion in 2012 [229]. In the United States, pharmaceutical companies, the healthcare sector and distribution networks generally do not allow counterfeit products to reach customers. However, people who do not have medical insurance or hesitate to ask doctors about some intimate issues can buy necessary medicines through a myriad of specialised web sites, which offer drugs for reasonable prices from abroad via express mail services. The complexity of detecting fraudulent trading through this distribution channel attracts counterfeiters, who can earn up to $200–500 from one dollar invested [230]. Drugs for the treatment of erectile dysfunction are the most popular among falsified medicines, and sildenafil (under the brand of Viagra) is the most commonly forged of all. The official wholesale price of Viagra from its legitimate manufacturer Pfizer in the US is $18/tablet, while on the foreign online pharmacies, *"generic"* Viagra can cost as little as $1/tablet (around 50% of adulterated medicines are sold online). Pharmaceuticals plants in India and China produce the generics, and illegally transport them to Dubai and on to the UK and the Bahamas; from here, the drugs are mailed to North America. In 2000, the WHO estimated that 32.1% of counterfeit drugs investigated (46 cases from 20 countries) had no active ingredients whatsoever; 20.2% had incorrect quantities of active ingredients; 21.4% included some of the wrong ingredients; 15.6% had the correct ingredients, but fake packaging; 8.5% had high levels of impurities; and 1% were copies of an original product [231]. Generally, the counterfeiters do not want to kill consumers with their drugs and attract the attention of criminal investigators from different countries; but the customers could still suffer or die because the counterfeits have no positive effect, or through the side effects of adulterated medicines.

---

[229] Surge in illegal sales of drugs as gangs exploit 'phenomenal market' online, The Guardian, December 28, 2014

[230] Gena Somra, Deadly fake Viagra: Online pharmacies suspected of selling counterfeit drugs, CNN, August 31, 2015

[231] Paul Toscano, The Dangerous World of Counterfeit Prescription Drugs, CNBC, October 4, 2011

Unauthorized production and sophisticated distribution is also a headache for producers of perfume, apparel, shoes, watches, electronics and toys, but the adulteration of medicine, food, alcohol and cigarettes poses a real threat to human health.

Manufacturers could improve their ties with lawmakers, regulators and law enforcement agencies to restrict illicit activity towards their merchandise. Improving packaging to resist tampering is also an important response. Raising awareness among customers reduces the probability of them unwittingly buying adulterated or counterfeit goods, limits the manufacturer's liability in the event of an incident and demonstrates the producer's commitment to customer safety.

Crises Induced by Negligence in the Safe Operation of Manufacturing Sites

The destruction of a manufacturing site through fire, explosion, collapse of buildings and so on generally has a limited influence on an organization's long-term output: there is an abundance of free manufacturing capacity around the world, and manufacturing can easily be relocated. Nevertheless, disasters in heavy industries—fires or leaks at chemical plants, ammonia leaks from the refrigerating systems of retail food processing plants, and so on—can pose a deadly threat to both plant workers and local communities.

Regular Minor Injuries and Isolated Deaths of Workers Due to Violation of Health and Safety Rules on Production Sites

Regular Dust Explosions on Grain Elevators, Mills and Sugar Plants Due to the microscopic size of particles of flour and sugar, the particles become airborne and can be explosive. This aspect of the production of milled goods has claimed the lives of thousands of shop floor workers in global food industry over the last 200 years.

Great Molasses Flood (USA, 1919) A giant storage tank—15 m tall and 27 m wide—collapsed due to several design flaws on Boston's waterfront at the Purity Distilling Co., releasing more than 8.5 million litres of molasses (the sweet substance commonly found in cookies) into Boston's North End. The resulting stream of molasses, flowing at more than 55 km/h, flooded the streets, killing 21 people and injuring 150. The accident was exacerbated by the cold temperature, which made the molasses four times more viscous: those who perished were unable to escape from the viscous mass [232].

---

[232]Gretel Kauffman, Scientists finally decode the Great Molasses Flood of 1919, The Christian Science Monitor, November 29, 2016

Explosion at Flixborough Chemical Plant (UK, 1974) The plant produced caprolactam, which is used in the manufacture of nylon. A temporary pipe had been constructed to bypass one of the six reactors involved in the production process, because a crack had been detected in the reactor. The pipe ruptured, causing an explosion and setting fire to the plant. 28 people lost their lives, more than 50 were severely injured, the plant was completely destroyed and around 2000 residential buildings, shops and factories nearby were damaged. If the accident had not occurred over the weekend but on a normal working day—with up to 500 workers on site—the number of casualties could have been many times larger. The disaster triggered a widespread public outcry over process plant safety and is often quoted in justification of a more systematic approach to process safety in UK process industries. In conjunction with the Seveso disaster (see below), it led to explicit UK government regulation of facilities which process and store large inventories of hazardous materials.

Seveso Chemical Plant Dioxin Leak (ICMESA, Italy, 1976) On July 10 1976, at the ICMESA chemical plant (owned by the Swiss multinational pharmaceutical Hoffmann-La Roche) in the Milan suburb of Seveso in Northern Italy, a cloud of 2,3,7,8-tetrachlorodibenzodioxin gas (better known as dioxin or TCDD) was released into the atmosphere. Fortunately, none of the residents of Seveso or neighbouring cities died from dioxin poisoning. However, the land and vegetation were contaminated and more than 2000 people had to be treated for dioxin poisoning. The event was a turning point in changing European regulation on the prevention of industrial disasters in the chemical industry. The new legislation, called the Seveso Directive, includes risk ranking and assessment, the exchange of information on near-misses and incidents within the industry, and a procedure for the obligatory disclosure of risk information that is implemented on more than ten thousand facilities in Europe where dangerous substances are located.

Bhopal Disaster (Union Carbide, India, 1984) In 1969, the Union Carbide Corporation, an American company, opened a pesticide plant in Bhopal (Madhya Pradesh state, India). In the early years, the plant produced pesticides extracted from US-imported concentrate. In 1979, Union Carbide India Ltd launched production of an insecticide called carbaryl under the trademark SEVIN, using locally produced methyl isocyanide (MIC). MIC is a catalyst in pesticide production processes and has an extremely toxic impact on human health. Because safety measures had been neglected and even dropped to save money, more than 40 tons of MIC and other gases leaked into the atmosphere in December 1984. Over the days following the accident, between 3000 and 10,000 citizens of Bhopal died, 100,000 suffered irreversible damage to their health and more than 500,000 were exposed to toxic gases [233], out of a total population of around 850,000 residents. In terms of casualty numbers, this makes Bhopal the second largest industrial accident in world history;

---

[233]Ingrid Eckerman, The Bhopal Saga: Causes and Consequences of the World's Largest Industrial Disaster, Universities Press, 2005

the largest was the breach of the Banqiao and Shimantan Dams in Central China in 1975 due to Typhoon Nina, when around 230,000 people perished in the following months from disease and famine [234,235].

Jilin Chemical Plant Explosion (China, 2005) An accident at the plant provoked the evacuation of more than ten thousand people, in order to avoid massive casualties in the event of a further explosion. Benzene and nitrobenzene pollution of the River Songhua, and contamination of the River Amur in Russia, threatened the provincial capital city of Harbin and large cities in the far east of Russia.

Qinghe Special Steel Corporation Disaster (China, 2007) Because of inadequate safety precautions, 32 workers were burnt to death when a ladle holding molten steel overturned at a foundry in Liaoning Province.

Mattress and Furniture Factory Fire (Morocco, 2008) 55 workers died in a fire at a mattress and furniture factory near Casablanca. A lot of highly flammable raw materials were stored within the building, and there was a policy of keeping doors and windows locked during the production process to prevent materials from being stolen by factory workers. When the fire broke out, the combination of these factors made it impossible for many employees to leave the factory. A quite similar event had occurred 15 years earlier in Thailand at the Kader Doll Factory, when 190 workers perished and 500 were injured during a fire at the plant [236].

Savar Building Collapse (Bangladesh, 2013) On 24 April 2013, in Savar in the Greater Dhaka Area of Bangladesh, 1127 workers at garment factories died when the Rana Plaza building collapsed on them. There are more than 5000 competing garment factories in Bangladesh, which provide cheap labor for the tailoring of many world-famous brands. The Rana Plaza was originally designed as a six-story building for shops and offices, but the owner of the plaza illegally constructed three additional floors using low-quality materials—and sited five garment factories there, deploying heavy machinery, which generated excessive vibrations. The day before the collapse, local authorities discovered cracks in the building and issued an order to evacuate the whole building. The personnel on the lower floors with shops and a bank were not permitted into their workplaces until inspectors had confirmed the safety of the building; but managers of the garment factories insisted that their staff should go to work, threatening them with the loss of a whole month's salary if they refused [237]. Moreover, they misled the sewers by telling them that the building had

---

[234]Typhoon Nina–Banqiao dam failure, Encyclopedia Britannica

[235]David Longshore, Encyclopedia of Hurricanes, Typhoons, and Cyclones, Infobase Publishing, 2009, p.124

[236]Zakia Abdennebi, Morocco mattress factory fire kills 55, Reuters, April 27, 2008

[237]Arun Devnath, Mehul Srivastava, 'Suddenly the Floor Wasn't There,' Factory Survivor Says. Bloomberg, 25 April 2013

been inspected and declared safe [238]. The motives of the managers were simple: if operations were shut down, they would be fined by their customers—world-famous high street clothing brands—for delays with shipping, and could lose contracts in a highly competitive market. Two years earlier in 2011, Walmart and GAP had refused to sign a new industry agreement to pay Bangladeshi factories a higher price, so the garment industry could not afford safety upgrades on their sewing factories [239]. Devastating fires at garment factories had already occurred in the United States in 1911 (146 casualties at Triangle garment factory in New York), in Bangladesh in 2006 and 2012 and in Pakistan in 2012, costing the lives of more than 450 workers.

Ecological Crises Caused by Long-Term Contamination of Land Near Production Sites In 1956, in Minamata city in Japan, a strange epilepsy-like neurological disease was discovered among locals, as well as in their cats and dogs. They called the disease "*dancing cat fever*". Initially, scientists thought that it was an infectious disease but, when they tested marine creatures on the coast nearby, they discovered extremely high levels of mercury contamination, which was traced to industrial wastewater discharge from the adjacent Chisso Corporation chemical factory. The factory used mercury sulfate as a catalyst in the production of acetaldehyde, and had been discharging the compound into Minamata Bay for 25 years. Seafood from the bay had been the main diet of local residents and their domestic animals for decades. Ultimately, 2955 victims were officially recorded, more than seventeen hundred of whom died from the poisoning. They eventually received financial compensation from the company, which paid out a total of US$130 million [240].

Measures to improve safety on production sites include fostering a sharper focus on occupational safety throughout the workforce to prevent incidents and injuries among employees, and the analysis and improvement of production processes to make manufacturing more reliable. Where production involves the use of harmful substances, plant operators must have a clearly understandable and well-drilled contingency plan in the event of a leak; this needs to be set up in coordination with local authorities and residents to enable a timely and safe evacuation if necessary.

---

238 Syed Zain Al-Mahmood, Rebecca Smithers, Matalan supplier among manufacturers in Bangladesh building collapse, The Guardian, April 24, 2013

239 Arun Devnath, Mehul Srivastava, 'Suddenly the Floor Wasn't There,' Factory Survivor Says. Bloomberg, April 25, 2013

240 Minamata Disease The History and Measures, Environmental health department, Ministry of the environment, Government of Japan, 2002, http://www.env.go.jp/en/chemi/hs/minamata2002/index.html

Supplier Errors and Failures in Supply Chains

The increasingly globalized operation of modern business is very clearly demonstrated in modern manufacturing: to assemble a typical piece of hi-tech gadgetry, a producer will purchase thousands of components from tens of different countries and hundreds of suppliers. Sometimes a breakdown with, or illegal action from, one of these suppliers could pose a threat to the business of the manufacturers and provoke the dissatisfaction of customers.

Heineken Recall (The Netherlands and Other Countries, 1993) The beer producers recalled 3.12 million bottles of their Export strong lager after glass shards were detected in 32,000 bottles, caused by chipping of the mouth of the bottles during packaging or opening. Heineken blamed United Glass Manufacturers, the bottle suppliers, for delivering bottles with weak necks.

Nokia and Ericsson Supply Chain Crisis (Worldwide, 2000) In March 2000, there was a minor fire at the Philips semiconductor factory in Albuquerque in the United States, which was extinguished by plant workers. However, the local fire brigade was also called, and firefighters soaked some critical machinery at the plant and damaged two out of four cleanrooms for the production of chips for cell phones. Philips informed Nokia and Ericsson, its main clients, about the accident and forecasted about 1 week delay in supplies. On receiving this warning, Nokia's supply managers worked with Philips staff to make sure they had a realistic estimate of the delay and started looking for alternative suppliers; Ericsson staff, on the other hand, accepted the rather optimistic estimate from Philips and took no further action because they had reserve stock to cover such delays. Subsequently, Philips gave a revised estimate of several months to get the Albuquerque plant back to full production. Nokia, who had started to find alternative supply channels for the chips, had booked the entire global production output of their new suppliers. Ericsson were left unable to find the necessary components for their cellphones, which threatened the revenue of their mobile phone division and almost led to the company's complete withdrawal from this market (Ericsson's phone division was eventually merged with Sony). Meanwhile, Nokia increased its share on the handset market from 27 to 30% [241]. Similar supply chain challenges took place after the Tohoku seaquake and tsunami in Japan in March 2011 (the global car and electronics industries had previously relied on sophisticated components made in Japan) and after the flooding in Thailand in 2011 (around 45% of the global output of computer hard drives came from Thailand at that time). If a full-blown military conflict develops in the Korean peninsula—by no means inconceivable in the current geopolitical climate—the global electronics industry could be paralyzed, because many critical components for hi-tech equipment are only produced in South Korea [242].

---

[241]Yossi Sheffi, The Resilient Enterprise: Overcoming Vulnerability for Competitive Advantage, MIT Press, February 23, 2007, pp.3–12

[242]Bruce Einhorn, Kanga Kong, Kyunghee Park, A Korean War Could Cut Pipeline of Vital Technologies to the World "If Korea is hit by a missile, all electronics production will stop", Bloomberg, April 27, 2017

Lead Paint in Mattel Toys (Worldwide, 2007) The American toy manufacturer Mattel—most famous for the Barbie doll, and a leading producer in its sector—faced a worldwide crisis when a European retailer discovered lead in the paint used on some of their toys. Mattel immediately apologized, accusing one of their Chinese contractors of using lead-based paint to reduce production costs, and promised to establish better control over the whole of their supply chain. The company decided to recall not only the toys with lead paint, but also some other badly-designed models containing small magnets that could be a danger to children; 150 different models were recalled, amounting to a total of 20 million toys, only three million of which were lead-painted. Though it was clearly Mattel and not the Chinese who were responsible for mistakes in the design of the toys with dangerous magnets, the coverage of the story in the global media made the Chinese manufacturing industry the scapegoat for the entire recall. Mattel's CEO was forced to formally apologize to Chinese officials [243]: by not correcting the press story, Mattel had allowed the safety of Chinese manufacturing in general to be called into question. Meanwhile in China, the consequences went far beyond the embarrassment of a public climb-down. In response to Mattel's apology, the Chinese government reviewed the entire domestic toy industry and revoked the export licenses of 300 producers; the review led to the layoff of 1.5 million Chinese workers [244]. The owner of the company which supplied the lead-painted toys to Mattel, and had been a Mattel supplier for 15 years, committed suicide. And the former head of the Chinese equivalent of the FDA—which had certified the paint for the Mattel toys as *"lead free"*—was executed for corruption.

Baxter and Contaminated Heparin (USA, 2008) The American pharmaceutical firm Baxter produced heparin, a blood thinner, from ingredients supplied from China. In 2008, while analyzing statistics of patient complaints, doctors at Baxter noticed excessive numbers of what appeared to be allergic reactions to heparin. Further tests revealed that the drug was contaminated by over-sulfated chondroitin sulfate (OSCS). But *"the contaminant had been chemically modified and was therefore so heparin-like in nature that it wasn't detected"* through the routine tests conducted by Baxter on every batch. According to the manufacturer *"OSCS was intentionally introduced into heparin made by Baxter during the early points in the product's supply chain [from China]"* [245]. Because of this deliberate but undetected adulteration, 81 patients died and more than 700 were injured. The company recalled the contaminated drug and ceased all commercial activity in this field, selling off the division that produced heparin and some other generic medications. A quite similar case occurred in Panama in 2007, when a manufacturer of

---

[243]Jyoti Thottam, Why Mattel Apologized to China, Time, Friday, Sept. 21, 2007, http://content.time.com/time/business/article/0,8599,1,664,428,00.html

[244]W. Timothy Coombs, Sherry J. Holladay, The Handbook of Crisis Communication, John Wiley & Sons, 2012, p. 468

[245]Heparin Sodium Injection, Baxter, http://www.baxter.com/products-expertise/product-safety-information/heparin-background-information.page

cough syrup used toxic diethylene glycol (antifreeze) mistakenly believing they were using glycerine. The mistake cost at least 365 lives. Investigation concluded that a Chinese supplier had delivered counterfeit glycerine, which had gone undetected because the producers had not tested the "*glycerine*" before using it in the production process. Similar instances of mistaken usage of diethylene glycol during drug manufacture have occurred in Haiti, Bangladesh, Argentina, Nigeria and India [246].

To avoid crises like those above, producers need to increase their control over the procurement of raw materials by suppliers, ensure that the production processes employed by those suppliers comply with their own requirements, maintain the comprehensive testing of shipped components, and back up their supplies and logistics networks.

Governmental Regulation and Geopolitical Conflicts

Some manufacturers produce goods that are damaging to customer health, or have a production process that harms the environment and local communities. In such cases, there is public pressure on governments at all levels to impose additional regulation over the activity of these manufacturers. Some products can even be recognized by governments as a threat to national security, or banned within the broader context of a geopolitical struggle.

Additional Regulation and Taxation over Non-healthy Products The most prominent example of such regulatory pressure is in the tobacco industry, the activity of which is causing up to six million deaths per year from smoking worldwide, and generating up to US$1.4 trillion in economic losses [247]. Even the industry itself was eventually forced to admit that its product poses serious health risks to customers. It took billions of dollars in lost lawsuits, and the revelation of their deliberate and disgraceful misinformation of regulators and the public, to reach this point. But after decades of denial, a more constructive strategy has emerged from leading tobacco firm Philip Morris International in response to the pressure of tightening regulation: they have invested in the development of products that could deliver nicotine to customers without the harmful health impact of cigarette smoke (e.g. IQOS technology [248]). As well as its own vapor e-cigarettes, the company is proposing a range of other smokeless tobacco solutions, all of which are 90–95%

246Walt Bogdanich and Jake Hooker, From China to Panama, a Trail of Poisoned Medicine, The New York Times, May 6, 2007

247Tobacco, The World Health Organization, July 2013

248Some have suggested that IQOS stands for "*I quit ordinary smoking*", but one can read on the *official website of Philip Morris International: "Although many have tried to come up with explanations of IQOS, none of these creative ideas are correct. IQOS is not an acronym but a brand name created to denote innovation and game-changing technology coupled with advanced science, and that we could protect and use in many markets."*

less toxic than classic cigarettes [249]. The new approach of Philip Morris International has the support of regulators, customers and investors—who approve of the company's more protective and sustainable strategy in an environment of increasing pressure from the WHO, governments, social activists, and customers whose health has been irreparably damaged by smoking. Nevertheless, it is important to mention that according to research by the Blauen Kreuzes Bern-Solothurn-Freiburg laboratory in Switzerland, the heating of IQOS polymer filters up to +100 °C releases isocyanates, which are dangerous toxins [250].

Soft drinks, which contain significant amounts of sugar, are also under attack from regulators because their consumption increases the likelihood of type II diabetes, heart diseases and obesity. Just as with tobacco products, some countries have taken the approach of imposing taxes over soft drinks in order to motivate customers to reduce their consumption. In the United States, the first such tax was imposed in Berkeley, California in 2014. In the first 3 years of taxation, consumption of sugar-sweetened beverages dropped by 52% in the city while sales of bottled drinking water rose [251].

International Crises and Conflicts Affecting Trade Almost any international conflict will impact producers in one or all of the countries involved. Mexicans boycotted American products to protest against Trump's "*wall tax*"; Turkish fruit and vegetables were banned for import into Russia after a Russian bomber was shot down in Syria by a Turkish fighter jet; the conflict between Russia and the West over Ukraine led to Western economic sanctions on Russia, followed by reactive Russian sanctions against food imported from the EU, Australia and Canada; Chinese resentment of Japan for its militaristic history resulted in the widespread boycott of Japanese products there; conversely, China's oppression of Tibet has prompted long-standing calls to avoid products made in China; the invasion of Iraq in 2003 led to calls for a boycott of American merchandise in both Muslim and European countries; Americans boycotted French products due to French opposition to the Iraqi invasion; the Danish producers lost hundreds of millions of dollars in 2005, after Jyllands-Posten—the Danish broadsheet newspaper—published several cartoons associating the Prophet Muhammad with terrorism; Pepsi was boycotted in 1996 for continuing to operate in Burma, despite evidence from human rights advocates of the Burmese regime's use of torture; South African produce suffered from sanctions during the decades of apartheid. The Israeli settlement construction program in Palestinian territory provoked boycotts not only of Israeli food, but also of machinery produced by Caterpillar, a leading American manufacturer of construction and mining equipment, for supplying Israel with the bulldozers they used to destroy Palestinian homes.

---

[249]Martinne Geller, Philip Morris looks beyond cigarettes with alternative products, Evening Standard, November 30, 2016

[250]Mischa Aebi, Gift in Filtern der E-Zigaretten von Philip Morris, Tages-Anzeiger, April 7, 2019

[251]Dan Charles, U.S. Soda Taxes Work, Studies Suggest — But Maybe Not As Well As Hoped, NPR, February 21, 2019

U.S. Government Opposition to Huawei 5G Networks   This Chinese company is the leading global producer of telecommunications equipment for mobile networks, and the second largest smartphone producer in the world after Samsung. Over several decades of hard work in the telecommunications field, continuous investment in R&D and loans from Chinese state banks, the small company from Shenzhen became a global pioneer in solutions for 5G networks, which could revolutionize the implementation of artificial intelligence, driverless cars, and the internet of things in the lives of billions of people. However, the U.S. government opposes the construction of 5G networks based on Huawei equipment and software—not only in the United States, but also in allied countries—justifying its opposition on grounds of national security. The main goal of this intervention is to prevent the Chinese company from becoming a global leader in a development of global telecommunications that could have major strategic significance. The Foreign Policy magazine assessed the situation as follows: *"Washington fears that Beijing will replace it as the world's premier intelligence power and perhaps even deny it access to the networks that make global commerce and the projection of military power possible. For decades, U.S. intelligence agencies have capitalized on the central role of U.S. companies in global telecommunications networks to spy on adversaries and gather crucial intelligence. And now, whether by design or luck or some combination of the two, the Chinese Communist Party may have the means to overturn that disadvantage—especially because the United States itself, despite the prominence of Silicon Valley, doesn't have a national champion of its own developing 5G. ... As a result of these fears, Washington has essentially banned Huawei equipment inside the United States. ... While a few U.S. allies, such as Australia and Japan, have followed Washington's lead and already banned Huawei technology, many others are still considering it... In December 2018, the Justice Department also ordered the arrest of Meng Wanzhou, Huawei's chief financial officer [and daughter of Huawei's founder], on charges of trying to steal American technology and lying about the company's business in Iran. She is currently fighting extradition to the United States from Canada... Washington has never publicly presented evidence backing up its assertions that Huawei equipment plays a role in Chinese espionage operations... [M]any experts say security fears singling out Huawei equipment are overblown, as nearly all the big telecom equipment makers use Chinese factories to churn out their components..."* [252]. In May 2019, the Trump administration imposed a ban on any collaboration between American business and Huawei: no products, services or technology covered by the U.S. Export Administration Regulations can be sold to the company. The ban has affected Huawei especially by blocking the import of American semiconductors from companies like Qualcomm, Intel and Broadcom, and of software for smartphones (Google's Android) and laptops (Microsoft's Windows). In 2018, Huawei sold a record 200 million smartphones and its annual revenue was around US$107 billion. According to Huawei executives, the US

---

[252]Keith Johnson, Elias Groll, The Improbable Rise of Huawei, Foreign Policy, April 3, 2019, https://foreignpolicy.com/2019/04/03/the-improbable-rise-of-huawei-5g-global-network-china/

blacklisting could wipe out $30 billion of revenue, and overseas smartphone sales could decrease by 40–60 million units annually because Huawei cannot now install a full-fledged version of Google's Android [253]. China was set to invest more than $400 billion in its domestic 5G networks in the period from 2020 to 2030 [254]. Obviously the American ban aims to slow down Huawei's development, and especially the deployment of global 5G networks using the company's technology. Between 2015 and 2017, ZTE Corp, another Chinese producer of telecommunication equipment, faced quite similar restrictions from the US government over its collaboration with Iran and North Korea. Ultimately, ZTE *"acknowledge[d] the mistakes it [had] made"* [255], agreed to pay a $1.2 billion fine, replaced its top management and allowed a special compliance team to monitor the company's decisions and actions in order to avoid any violations of US sanctions.

Influence of Political Events on the Production Process A remarkable example of this is the connection between two major events in India, one political and one industrial: on October 31 1984, Indira Gandhi, the 3rd Prime Minister of India, was assassinated by two of her Sikh bodyguards. There was public outrage and widespread rioting. Amid the upheaval, the chemical plant at Bhopal could not safely continue producing SEVIN pesticide from its tremendous stocks of methyl isocyanate. In November, the Indian government announced nearly 2 weeks of national mourning and brought in a curfew to stop communal and religious violence in the country. Consequently, workers on the second and third shifts at the Bhopal plant had trouble fulfilling their duties and production of SEVIN from existing MIC stocks was slow [256]. As a result, the plant accumulated a large amount of MIC in its tanks: stocks of this lethal chemical reached 62 tons, with only 3–4 tons required daily for the production of SEVIN. This was in contravention of common practice in the chemical industry, which is to *"always keep only a strict minimum of dangerous materials on site"* [257]. The tragic accident that followed on the night of 2–3 December 1984—vastly exacerbated by the huge quantities of MIC in storage—shows why this principle is generally followed. Another case arose during the NATO operation against Yugoslavia in 1999. In addition to military targets, NATO air forces bombed the Pančevo chemical complex, which included petrochemical, fertilizer and refinery plants. Chlorine, mercury, hydrocarbons, ammonia, nitrogen and sulfur oxides, phosphorus compounds, hydrogen halides, and other

---

[253]Li Tao, Huawei CEO says he did not expect such a ferocious, large-scale US attack on the Chinese telecoms giant, South China Morning Post, Jun 17, 2019

[254]Bien Perez, Why China is set to spend US$411 billion on 5G mobile networks, South China Morning Post, June 19, 2017

[255]Bien Perez, ZTE to pay record US$1.2 billion fine for violating Iran, North Korea sanctions, March 8, 2017

[256]M.J. Peterson, Case Study: Bhopal Plant Disaster (with appendixes), University of Massachusetts – Amherst, 2009

[257]Ingrid Eckerman, The Bhopal Saga: Causes and Consequences of the World's Largest Industrial Disaster, Universities Press, 2005, p. 25

chemicals were released, causing widespread contamination of surrounding land, air and water [258].

Responding to these risks requires manufacturing activity to comply strictly with national regulations—which may or may not be served by active lobbying for the interests of manufacturers to the political establishment of countries where production plants are located or goods are sold. Where there is involvement in a geopolitical crisis, a manufacturer should be out of any discussion concerning politics. This will reassure foreign governments and customers that its main goal is the production of safe and useful goods at a reasonable price for its valued customers, without political preference for any side in the conflict. In case of war or social upheavals, it is advised to relocate the production facility and staff in a safe place or close whole production lines and ensure proper storage or removal of dangerous substances used in the production process (like ammonia for industrial freezers on dairy factories, hazardous chemicals in pesticide plants, etc.), while notifying the two sides of the conflict about existing dangers at the production site.

Other Crises

Conflicts Between Manufacturers and their Distribution Networks or Retailers Because of competition between manufacturers for advantageous shelving of their products, distribution networks and retailers have a powerful influence. They can therefore dictate unfavourable conditions to producers, which sometimes causes disputes.

Irresponsible Procurement of Raw Materials, or Employment of Unscrupulous Suppliers can invite harsh criticism from customers and the general public in developed countries. Thus, Nestle and Unilever were blamed for massive bulk buying of palm oil, produced by farms which had cut down tropical jungles in Malaysia and Indonesia. Deforestation leads to the extinction of many species including orangutans. IKEA was criticized for harvesting timber from ancient Russian forests to produce its furniture. Kimberly-Clark, producer of hygiene disposables for children and adults (their brands include Kleenex, Scott and Cottonelle), was under similar pressure to halt the destruction of ancient forests like the Boreal Forest. Many of these companies eventually declared new environmental policies to conserve forests and support sustainable forestry.

Impact of External Disasters on Production Process and Quality of Production In May 2011, researchers at the University of California, Berkeley, detected iodine 131, cesium 134 and cesium 137 in milk samples provided by a dairy in Sonoma County, California where the cows were fed by local grass. The main

---

[258]Michael Heylin, NATO bombs take out chemical complex, Chemical & Engineering News, Volume 77, Number 19, May 10, 1999

cause of the radioactive contamination—which fortunately was no threat to human health—was the fallout from the Fukushima Daiichi Nuclear Power Plant after the meltdown of three reactors in March 2011, caused in turn by the tsunami created by the Tohoku seaquake of March 11, 2011. Moreover, the Tohoku quake caused disaster not only at Fukushima Daiichi NPP, but also at the Cosmo Oil Refinery, where several LNG tanks leaked and exploded, creating enormous fireballs and burning uncontrollably for nearly 10 days. The Izmit earthquake in Turkey in 1999 also caused serious damage to the Turpas oil refinery. Power outages and the flooding of backup generators caused by Hurricane Harvey led to explosions at the Arkema chemical plant in Texas, when refrigerators for cooling highly inflammable substances used in pesticide production were temporarily shut down.

Illegal Employment of Children Globalization allows manufacturers to transfer the production process to countries with extremely low wages to keep their production costs down. Nike, Levi Strauss, Gap and many others were condemned for outsourcing sewing work to garment factories in developing countries, which ignored International Labour Organization rules by employing children and young people. Unilever, Nestle, Kellogg and Procter&Gamble faced similar reproofs when it became apparent that children of 8 years old were working on palm plantations in Indonesia, which supplied the transnational corporations with palm oil. Some of Apple's suppliers also hired children, but the companies were more sternly criticized for collaborating with Chinese assemblers like FoxConn, whose use of cheap local labor amounted to modern corporate slavery: their sweatshop conditions and violation of labor rights pushed some Chinese workers to suicide.

Strikes are also a common challenge for manufacturers, especially car makers from developed countries where labor rights are heavily protected. Assemblers have not only put tools down in the fight for better wages and pensions, but also in protest against plans to move production lines to countries with lower salary levels and unrestricted labor legislation.

Cyber Risks Disruptions of business and production processes through unauthorized penetration of the IT systems of manufacturers can lead to serious losses. Thus, the French building supplier Saint-Gobain lost US$230 million in sales from the cyber attack of the NotPetya virus. Some fast-moving consumer goods producers have also admitted losses as a result of the virus: Reckitt Benckiser lost $129 million and Mondelez International suffered a 3% drop in their second quarter sales in 2017.

Irresponsible/Outrageous Behavior of Workers at Manufacturing Plants We will not give the names of the producers affected, but we are aware of outrageous behaviour from some plant workers: swimming in a tank of milk for cheese, peeing into a tank of wort at a brewery plant, mixing human excrement into tobacco raw materials before the assembly of the cigarettes, and so on.

Testing of Products on Animals provokes protest from animal rights activists— their targets have included Procter&Gamble, L'Oreal, Unilever and many others.

Adidas and Nike were censured for using kangaroo skin in their merchandise. VW tested the effect of diesel car exhaust fumes on monkeys as well as human volunteers.

LGBT groups were outraged by **anti-gay hiring practices** at the Coors Brewing Company in the 1970s—1990s. There was outrage against the 2014 Winter Olympics in Sochi (Russia) when the Russian government banned what it described as LGBT *"propaganda aimed at young people"*. Activists called for the boycott of vodka under the Stolichnaya brand (ironically, a brand produced in Latvia) and the Russian Standard brand (made in Russia).

The Baby Milk Action campaign organised the boycott of Nescafé coffee for years because of Nestlé's **marketing of infant formula products in developing countries at high prices and in an environment of poor access to safe water**. Worse still, the promotion led some mothers to reject breastfeeding in favour of the perceived benefits of infant formula products—despite the fact that the WHO has concluded that breastfeeding could prevent about 800,000 deaths per year among children under five if all children under 23 months old were optimally breastfed.

Key Risk Mitigation Measures in Manufacturing

- Ensuring stable product quality and low production cost in every batch produced at any plant.
- Protecting brand reputation by fast crisis management, and prompt and appropriate response to complaints from customers, authorities or third parties.
- Supporting the continuous development of innovative solutions to enable the release of cutting-edge products ahead of competitors in the market.
- Closely cooperating with national and international law enforcement to prevent or minimize the impact of unauthorized production, illegal ("grey") importing of goods and acts of sabotage.
- Developing a reliable and safe production process in order to reduce workplace incidents and industrial disasters, especially at the manufacturing stage. In hazardous industries, it is also advisable to develop good relations with local communities, inform them of any risks in the production process and implement testing of contingency plans and actions.
- Performing careful selection of raw materials suppliers—not only with regard to quality of supplies, but also to their ethical standards (avoiding the use of child labor, hazardous environmental practices, etc.).
- Developing fast and reliable distribution chains to retail networks and directly to customers.
- Strict compliance to government regulation in countries where production sites are located or where goods are sold.

2.3   Utilities

This subsector includes the following industries according to ISIC Rev. 4:

- 35—Electricity, gas, steam and air conditioning supply
- 36—Water collection, treatment and supply
- 37—Sewerage
- 38—Waste collection, treatment and disposal activities; materials recovery
- 39—Remediation activities and other waste management services

General Description and Key Features of this Subsector and Incorporated Industries

This subsector focuses on the generation, distribution and supply of electricity, hot and cold water, and natural gas to retail and corporate customers, and also on water and waste management, sewerage and remediation activities. The generally accepted principle is that utility companies have to provide non-stop supply of their product 24–7-365 for customers. Therefore, any disruptions—due to industrial accidents, natural disasters or even violence and sabotage—set off a "*domino effect*" where disconnected customers are also unable to perform their duties adequately because of the blackout. Production activity in this sector is highly localized in a limited area, while distribution and supply activities usually occur on a nationwide scale. The high importance of utilities to the national security and economy makes the subsector highly dependent on regulation by the authorities. Utilities prefer to operate large-scale complexes in order to make their production cost-effective for large-scale consumption. Utilities are highly dependent on suppliers because the cost of raw materials determines the profitability of the utility and the sale price to customers. Global changes in energy matters, environment legislation and technological progress have a high influence on utilities, because new developments could lead to additional expenses to satisfy stricter requirements, and may even force the decommissioning of some facilities. Despite some liberalization in developed countries, the subsector on a global scale remains seller-oriented. This is because customers depend on large-scale infrastructure having been constructed, which has required massive long-term investment—billions of dollars over decades—and only a limited number of players can make such an investment, usually with government support, participation or funding. Therefore this is a capital-intensive activity, which also requires constant access to long-term and low-cost investment resources (up to several billion dollars). In some cases, most notoriously nuclear energy, an accident at a production site could cause massive contamination of territory near the plant, or even radioactive pollution of whole regions and areas of ocean. It can also lead to a change in the public perception and induce policy changes, as occurred in Japan and Germany in reaction to the Fukushima-Daiichi nuclear disaster in March 2011.

Critical Success Factors for an Organization Within the Subsector

The main critical factor in this subsector is the ability to provide an uninterrupted supply to customers 24-7-365, or in the event of an interruption, to ensure a faster emergency response than competitors. In other words, the success of a utility provider depends on the reliability of its production, distribution and supply. A utility company should also maintain good relations with the authorities, allowing it to participate in decision-making on national energy strategy, environmental legislation and liberalization of the subsector, thus protects its long-term investment in infrastructure. It is critical that the company can react rapidly to changes in the global energy market, environmental legislation and technology. Finally, there is a constant search for cost-effective and reliable supplies to keep production costs down.

Stakeholders in this Subsector [259]

- Government—30%
- Customers—25%
- Employees—20%
- Suppliers—10%
- Investors—5%
- Local communities—5%
- Other—5%

---

[259]Our informed appraisal of the influence of each audience on a typical organization within the subsector (100% = combined influence of all audiences)

Typology of Common Risks, Main Features of Major Accidents and Risk Mitigation Measures Within the Subsector

Technical Failures on Utility Infrastructure

Shutdowns/Accidents at Generation Facilities

The destruction of a production site through fire, explosion, collapse of buildings, and so on has a critical influence on an organization's long-term output, and can also affect wider national economic activity over months or even years. This is because most national utilities have limited spare capacity to fall back on during an emergency, and the reconstruction of large-scale utility plants takes a long time and requires massive investment.

Three Mile Island Nuclear Accident (USA, 1979) This nuclear power plant (NPP) is situated on an island in the Susquehanna River in Pennsylvania. The largest civil nuclear accident the world had ever seen until that time happened here in March 1979, triggered by a problem with the routine servicing of the feedwater system on one of the two reactor units. The main feedwater pumps—which would normally send re-heated water to the steam generators to help cool the reactor core of Unit 2—shut down. As a result, the turbine and the reactor automatically tripped; these were duly activated, but discharge valves of the feedwater system had also been closed earlier for routine maintenance, so the system was unable to pump any water to cool the steam generators and remove the decay heat . The temperature and therefore the pressure within the system went up, automatically opening a relief valve at the top of the pressurizer tank called the pilot-operated relief valve (PORV). This reduced pressure by releasing steam and water from the reactor core into a tank on the floor of the reactor unit. However, instead of closing almost immediately once the pressure had fallen to a normal level, the PORV remained stuck open. To make matters worse, the PORV monitor light in the control room of TMI-2 indicated only whether a signal to close had been sent to the valve, rather than whether it was open or closed; since the valve had indeed been instructed to close, operators assumed that it would be closed. Another design flaw in the control system was that none of the instruments showed how much water was covering the core, so staff were unaware that cooling water was pouring out of the open PORV valve—and since the pressurized water level was still high, they assumed that the core was properly covered. For over 2 h, the valve remained open undetected, until the next shift noticed the coolant leakage and closed it. At this point, for a few minutes after the leakage was detected, there were fears that the system was overfilled due to the pressurizer level signal —so the automatic emergency cooling system was turned off, reducing water flow into the reactor to a tenth of the designed level. All of these design shortcomings and mistakes combined causing the reactor to overheat, severely damaging the nuclear fuel. The deterioration of the fuel led to a radioactivity surge, but this was confined in the containment building of TMI-2 and therefore posed little threat to the outside environment [260,261,262].

Cleaning up the TMI-2 site took 14 years and an estimated US$1 billion at 1993 prices [263]. Meanwhile, the nearby TMI-1 reactor, despite the contamination of TMI-2, has continued working over the decades since the accident, and received an extension of its operating license for a further 20 years from the NRC in 2009 [264]. The Three Mile Island accident was rated at Level 5 out of a maximum of 7 on the International Nuclear and Radiological Event Scale, and although there were no casualties it did have wider consequences. Nearly 150,000 people were evacuated from houses nearby [265], although this turned out to be unnecessary. A wave of public resistance to civil nuclear energy followed, and on May 6, 1979, 65,000 antinuclear demonstrators gathered in Washington. Ultimately all construction of new nuclear power stations within the United States was suspended for decades.

An investigation concluded that there was a serious communication problem in the industry: decision-makers had only a fragmented picture of the risks they were facing. It was not standard practice for information about operating experience— even about dangerous incidents—to be reliably shared between the US Nuclear Regulatory Commission, the designers and manufacturers of nuclear reactor systems and plants, the utility companies operating them, and the contractors and suppliers of critical components. For example, it was not the first time that the emergency cooling system at a PWR reactor is turned off by mistake: the suppliers of the nuclear steam system were well aware of it and it had occurred several times before, but operators of the plants in question had not informed their peers at other plants [266]. And the design of the monitoring systems in reactor control rooms was known across the industry to be overly complicated, and liable to overload operators with unstructured information in emergencies and therefore make it harder for them to grasp the condition of a plant and make the right decisions quickly. This had even been recognized at the design phase, but until the TMI case nothing had been done about it [267].

---

[260]Report of the President's Commission on the Accident at Three Mile Island: The Need for Change: The Legacy of TMI, October 1979, pp. 12, 90–111

[261]Backgrounder on the Three Mile Island Accident, United States Nuclear Regulatory Commission, February 11, 2013

[262]Three Mile Island: Report to the Commissioners and to the Public, M. Rogovin and G. Frampton, US Nuclear Regulatory Commission, January 1980, Vol. I, pp. 3–4

[263]14-Year Cleanup at Three Mile Island Concludes, The New York Times, August 15, 1993

[264]NRC Issues Final Safety Evaluation Report For Three Mile Island Nuclear Plant License Renewal Application, US Nuclear Regulatory Commission Press Release-09-119, June 30, 2009

[265]Robert A. Stallings, Evacuation behavior at Three Mile Island, International Journal of Mass Emergencies and Disasters, №2, 1984, p.12

[266]Report of the President's Commission on the Accident at Three Mile Island: The Need for Change: The Legacy of TMI, October 1979, pp. 10, 43, 93

[267]Report of the President's Commission on the Accident at Three Mile Island: The Need for Change: The Legacy of TMI, October 1979, pp. 29–30

Tacoa Power Plant Explosions (Venezuela, 1982) In December 1982, one of the oil tanks at the Tacoa thermal power plant exploded during the transfer of fuel between tanks. Firefighters had nearly extinguished the fire, and local journalists had already arrived at the site to report on the event, when a new explosion blew up a second tank with 16,000 tons of oil, killing 40 firefighters and 8 journalists. Because the tanks were located on a hillside, the wave of burning oil from the explosion destroyed buildings below the tank and damaged two of the plant's generation units. More than 160 people perished, more than 500 were injured and 40 thousand were evacuated. The accident completely shut the plant down, causing blackouts in Caracas that disrupted the city's water supplies, sewage system and medical services [268,269].

Chernobyl Nuclear Accident (USSR, 1986) In the spring of 1986, an experiment was under way with the emergency power supply system at the Chernobyl nuclear power plant. At 1:23 in the morning on 26 April, there was a power surge in Reactor #4, which started to burn uncontrollably. A huge area of Belarus, Russia and Ukraine was left contaminated [270]; and from traces of some isotopes discovered later, the contamination spread as far as Northern and Western Europe. This was and remains the largest accident in the history of the civil nuclear industry. In 2005, the UN estimated that up to 4000 people could eventually die of radiation from the Chernobyl NPP accident [271]. The Soviet Union spent 18 billion rubles [272], approximately US$27 billion [273], on dealing with the consequences of the disaster. But according to Academician Valery Legasov, a key member of the government investigation committee into the Chernobyl disaster, the total damage caused by this disaster was around 300 billion rubles in pre-1990 prices [274], or approximately US$450 billion at 1990 prices.

   Although the first industrial civil nuclear reactor in the world was commissioned in 1954 at Obninsk, 80 km from Moscow, the Soviet Union only began full-scale development of its nuclear power at the end of the 1960s. Until then, the powerful Soviet Planning Commission had made the mistake of relying on coal-fired plants to supply electricity to the western part of the Union. In the 1960s and 1970s, there

---

[268] Incidents That Define Process Safety, Center for Chemical Process Safety, John Wiley & Sons, 2013

[269] VENEZUELA - Power Plant Fire, USAID, http://pdf.usaid.gov/pdf_docs/PBAAB295.pdf

[270] Frequently Asked Chernobyl Questions, IAEA, http://www.iaea.org/newscenter/features/cherno byl-15/cherno-faq.shtml

[271] Chernobyl: the true scale of the accident, World Health Organization, 2009, http://www.who.int/ mediacentre/news/releases/2005/pr38/en/

[272] Interview with Mikhail Gorbachev, documentary "The Battle of Chernobyl", Director: Thomas Johnson, 2006

[273] Official exchange rate of State Bank of the USSR by the end of 1986: Soviet ruble/US dollar – 0,6783, Archive of Bank of Russia, http://cbr.ru/currency_base/OldDataFiles/USD.xls

[274] Valery Legasov, Problems of Safe Development of the Technosphere, Communist Journal, #8, 1987, pp. 92–101

were energy shortages in the region, due to double-digit industrial growth and the large-scale construction of civil infrastructure. Driven by this deficiency, the Soviet civil nuclear program was launched with urgency.

The easiest and fastest way to start producing electricity from atomic fission was to use the experience gained from the military nuclear program. Senior Soviet nuclear weapons scientists considered adapting the Obninsk reactor—a water-cooled uranium-graphite channel reactor, originally used to produce plutonium—into a high-power channel reactor or RBMK. The design of RBMK nuclear power plants (NPPs) did not stipulate the construction of a containment building to prevent the spread of radioactive elements in case of a reactor accident; this cut construction costs by 25–30% and reduced the commissioning time for this type of NPP. The construction of the new reactor also allowed the possibility of changing nuclear fuel without shutting down the reactor—unlike shell-type reactors, which require a compulsory shutdown—which made the RBMK very cost-effective in comparison with competing reactor types.

The time and cost limits imposed by the Politburo required the construction of a dozen reactors in the western part of the Union simultaneously. Thus, when a pilot reactor was commissioned near Leningrad (now Saint Petersburg) in 1973, several other reactors were already under construction in the Union according to the still untested pilot design. Even as many design solutions for the new reactor were being operationally tested in the pilot model, the construction of the first series of RBMKs was well under way: for instance, the foundations of Chernobyl NPP were already laid out in 1970.

The first serious imperfections of the RBMK design came to light in 1975, when an accident occurred during testing on the Leningrad NPP. One of these was a phenomenon known as the "*positive SCRAM effect*": during an emergency shutdown or SCRAM, control rods are lowered into the reactor core to stop the fission, but the immediate result of introducing the rods seemed to be that the lower part of the core initially became *more* reactive [275]. In 1976, investigators concluded that parts of the reactor core would have to be redesigned: they needed to stop the reactivity increase of the system shooting up when steam bubbles form (positive void coefficient), improve the performance of the control rods and increase the speed of the SCRAM system [276,277]. However, the investigative commission did not insist on fundamentally redeveloping the original RBMK design for the other units under construction—these units were not even adapted in line with the improvements recommended after the Leningrad NPP accident [278]. In 1983, first at the Ignalina

---

[275]The SCRAM system refers to the control rods that are inserted into a nuclear reactor core to suppress nuclear fission. A "positive SCRAM effect" is a localized increase of activity in the bottom of the core of a reactor during emergency shutdown with low power range: introducing graphite rods leads to decreased absorption of neutrons by the xenon in the core ("xenon poisoning") and accelerates the nuclear reaction.

[276]The Chernobyl Accident: Updating of INSAG-1, INSAG-7, IAEA Publications, Vienna, 1992, pp.47–48

[277]Anatoly Dyatlov, Chernobyl. How it was, Nauchtekhlitizdat, Moscow, 2003, p.153

[278]The Chernobyl Accident: Updating of INSAG-1, INSAG-7, IAEA Publications, Vienna, 1992, p.87

NPP and then during the launch of Reactor #4 at Chernobyl NPP, the positive SCRAM effect was observed again [279,280]. In correspondence with colleagues, the Chief Design Engineer for the program discussed the problem [281] and assured them that the RBMK design would be changed to correct it. But instead of adapting the design, he settled for recommending that new procedural measures should be included in the plant operating instructions; in fact, even these recommendations were not adopted [282]. It was not until 10 years later, after the Chernobyl disaster, that design changes were implemented.

Although the positive SCRAM effect had happened three times, the RBMK developers assumed it would only occur in rare cases—so instead of making technical changes to the reactor design, they tried to ensure the safe operation of the reactor through clearer instructions, staff training and other organizational measures [283]. They were confident that the training and discipline of the ex-military reactor operators would compensate for any technical disadvantages of the RBMK when it became operational. However, the Politburo subsequently made a strategic decision to transfer the construction of RBMK reactors from the military to the civil Ministry of Energy and Electrification of the USSR, where executives were less qualified and operative personnel less disciplined than the nuclear military.

The scientists developing the RBMK reactor were highly respected figures in Soviet society, and the Soviet senior executive accepted their assurances that the RBMK reactor was safe and that Soviet nuclear technology was infallible [284,285,286]. The developers' overconfidence persuaded Politburo members and executives at the Ministry of Energy and Electrification that it was safe to hand over the operation of nuclear power plants to personnel who had experience of running thermal power stations, but no education in nuclear science. They saw no need to inform the Politburo or the management and personnel of the Ministry of Energy about the shortcomings of the reactor design—including the existence of the

---

[279]The Chernobyl Accident: Updating of INSAG-1, INSAG-7, IAEA Publications, Vienna, 1992, p.43

[280]Nikolaii Karpan, Vengeance of peaceful atom, Dnepropetrovsk, 2006, p. 290

[281]Anatoly Dyatlov, Chernobyl. How it was, Nauchtekhlitizdat, Moscow, 2003, pp. 136

[282]The Chernobyl Accident: Updating of INSAG-1, INSAG-7, IAEA Publications, Vienna, 1992, pp.44–45

[283]Nikolaii Karpan, Vengeance of peaceful atom, Dnepropetrovsk, 2006, pp. 294–296

[284]Anatoly Dyachenko, Experience of liquidation of Chernobyl disaster, Federal State Unitary Enterprise "Institute of Strategic Stability" of Rosatom, Moscow, 2004, http://www.iss-atom.ru/book-7/glav-2-3.htm

[285]Interview with Mikhail Gorbachev, documentary "The Battle of Chernobyl", Director: Thomas Johnson, 2006

[286]Nikolaii Karpan, Vengeance of peaceful atom, Dnepropetrovsk, 2006, p. 401

positive SCRAM effect in rare cases. They assumed that NPP personnel would accurately follow instructions for the safe operation of the plant [287,288,289].

Because the only way to improve the RBMK design was through real world testing, the developers of the reactor stipulated that an experiment should be conducted on newly launched reactors with the emergency power supply system. This experiment was necessary because the emergency operating modes had not been properly tested prior to the RBMK being rolled out across the USSR [290]. The detailed steps of each experiment were left to the personnel of each plant. Thus in the case of Chernobyl Reactor #4, the plan for the experiment was not submitted to the developers because there was no requirement to obtain their approval for experiments [291]. The Chief Engineer there had no experience of running nuclear plants and little understanding of the risks involved in such a test being performed on an RBMK [292]. He decided to conduct it as he would have conducted a routine electrotechnical test within a turbine-generator system, during a regular service break of Reactor #4 in April 1986, as part of compulsory measures stipulated by the reactor project [293,294]. The experiment plan drawn up at Chernobyl NPP violated applicable operating instructions in 12 sections [295]. Unaware of the flaws in the plan, operators at Reactor #4 duly switched off many protection systems.

The shortcomings of the reactor were quickly revealed—in particular, the insertion of control rods caused a surge of reactivity—and the accident took place. The Director of Chernobyl NPP arrived at the plant a few hours later to find that an explosion had blown off the roof of Reactor #4 and that the reactor was still burning uncontrollably. Incredibly, the operators who had been running the reactor during the accident assured the director that it was not damaged. Unable to verify this statement during the first few hours [296], the director in his turn reassured his superiors in Kiev and Moscow: "*The*

---

[287] Anatoly Dyatlov, Chernobyl. How it was, Nauchtekhlitizdat, Moscow, 2003, p.102

[288] Unapprehended atom. Interview with Victor Bryukhanov, Profile, Moscow, № 29(477), April 24, 2006

[289] Vladimir Shunevich, Victor Bryuhanov: I was expelled from the party directly at a meeting of the Politburo, Fakty newspaper, Kiev, July 72,012

[290] The Chernobyl Accident: Updating of INSAG-1, INSAG-7, IAEA Publications, Vienna, 1992, p.51

[291] The Chernobyl Accident: Updating of INSAG-1, IAEA Publications, Vienna, 1992, p.52

[292] "*The causes of the accident lie not in the programme [of the experiment] as such, but in the ignorance on the part of the programme developers of the characteristics of the behavior of the RBMK-1000 reactor under the planned operating conditions*", The Chernobyl Accident: Updating of INSAG-1, IAEA Publications, Vienna, 1992, p.52

[293] Grigori Medvedev, Chernobyl Notebook, New World Magazine, №6, 1989

[294] Nikolaii Karpan, Vengeance of peaceful atom, Dnepropetrovsk, 2006, p. 446, 451

[295] Anatoly Dyatlov, Chernobyl. How it was, Nauchtekhlitizdat, Moscow, 2003, p.134

[296] Seven hours after the accident (at 10:00 a.m. on 26 April 1986), one of the engineers of Chernobyl NPP explored the reactor room and found out that the reactor was demolished, but the Director of the plant did not believe his statement. It took a helicopter ride 12 hours after the accident (around 3:00 p.m. on 26 April 1986) to establish the fact that Reactor #4 was destroyed and was throwing out radioactive material into the atmosphere (Alexandr Borovoy, Evgeny Velihov. Experience of Chernobyl, National Research Center "Kurchatovsky Institute", Moscow, 2012, p.11).

reactor is intact, continuing to pump water into the reactor, the radiation level is within the normal range" [297]. This misinformation of the authorities delayed by more than 36 h the evacuation of the residents of Pripyat, a town of 47,000 inhabitants located near the plant [298]. Michael Gorbachev made no official broadcast to the wider Russian public until 18 days later [299]. Even then, wishing to avoid nationwide panic, he minimized the scale of the accident. The international media were already reporting heightened levels of radiation in Scandinavia despite the absence of news from the Soviet Union itself. Inevitably, rumors of the accident began to reach the Soviet people from abroad. Feeling misled and betrayed by Gorbachev and the communist regime, the public had little confidence that their leaders could deal with the crisis. Gorbachev would later identify the Chernobyl disaster as a major contributor to the ultimate collapse of the Soviet Union [300]. By November 1986, a massive concrete cover was in place over Reactor #4, and most of the radioactive material released in the explosion had been collected and buried in deep landfills or stored inside the reactors. More than 600,000 people—many of them showing heroic self-sacrifice—worked together to achieve this in only 7 months [301].

The concrete sarcophagus has been progressively damaged by the high levels of radiation. It has been absorbing 10,000 röntgens/hour; a lethal dose is 500 röntgens over 5 h. In 1996, a decision was taken to replace the sarcophagus with the New Safe Confinement (NSC) (completed in 2018). The NSC is intended to contain the radioactive remains of the No. 4 unit and to last for the next 100 years, preventing the reactor complex from leaking radioactive material into the environment. The total cost of the Shelter Implementation Plan, of which the New Safe Confinement is the most prominent element, was €2.1 bln [302].

Sayano-Shushenskaya Hydropower Station Disaster (Russia, 2009) This station is the largest hydroelectric power plant in Russia and, with an installed capacity of 6400 MW, it is the country's largest power producing facility. Situated on the Yenisei river in central Siberia, the station produces 2% of Russia's electricity and 15% of its hydroelectricity. In 2009, only five hydroelectric plants in the world generated more power—the stations at Three Gorges in China, Itaipu in Brazil/Paraguay, Guri in Venezuela, Tucuruí in Brazil, and Churchill Falls in Canada.

---

[297] Alexandr Borovoy, Evgeny Velihov. Experience of Chernobyl, National Research Center "Kurchatovsky Institute", Moscow, 2012, p.11; Grigori Medvedev, Chernobyl Notebook, New World Magazine, №6, 1989

[298] The accident occurred at 1:24 a.m. 26 April 1986, but evacuation was started only at 2:00 p.m. 27 April. Central government officials arrived in Chernobyl by the end of 26 April, and recognized that the real picture differed completely from the one in reports.

[299] Documentary "Chernobyl. Chronicle of silence", Director: Irina Larina, 2006

[300] Mikhail Gorbachev, Turning Point at Chernobyl, Project Syndicate, April 14, 2006, http://www.project-syndicate.org/commentary/turning-point-at-chernobyl

[301] Occupational Radiation Protection in Severe Accident Management. Interim Report, Organisation for Economic Co-operation and Development, Nuclear Energy Agency, January 7, 2014, pp. 65–66

[302] Mu Xuequan, Chernobyl's New Safe Confinement goes into operation, Xinhua, July 11, 2019, http://www.xinhuanet.com/english/2019-07/11/c_138215894.htm

On August 17 that year, the rotor of Turbine 2 at Sayano-Shushenskaya (SSHPS) shot out. The turbine hall of the station and other rooms below its level were soon flooded, damaging eight of the remaining nine turbines and killing 75 station workers. The accident caused blackouts to several gigantic aluminum plants, and power shortages across Siberia. An investigation after the accident found that technical staff at the station were unaware of the fact that the stud-bolts on turbine caps in high-pressure hydropower stations gradually deteriorate—even though the same phenomenon had caused a similar incident in 1983, at Nurek hydropower station in what was then the Soviet Socialist Republic of Tajikistan. After that incident, the Nurek station management had given a detailed account of the phenomenon to the Soviet Ministry of Energy and Electrification, but the Ministry had not passed this critical information on to the management and technical staff at other Soviet HPSs. By 2009, the management of SSHPS had been operating the station without knowing that the turbine cap stud-bolts were especially prone to deterioration for more than 25 years.

On March 16, 2009—less than 6 months before the accident—Turbine 2 had resumed operation after a period of closure for repairs and no abnormal vibrations were detected over the following 35 days. But by late April, when the spring flood brought the highest reservoir levels of the year, vibration in all the turbines had increased [303,304]. From April 21, one of the sensors inside Turbine 2 began to register abnormally high vibration levels. But station managers decided not to order an emergency stoppage to investigate and eliminate this technical failure during the months before the disaster. This inaction may seem incredible in retrospect—but the managers still had no idea of the risks they were dealing with, and the operators did not even consider the possibility of a serious accident: engineers had been recording minor vibrations in the station turbines for decades.

When the station was launched in 1979, Turbine 2 had been equipped with a poor-quality interim rotor and had therefore been especially prone to vibration [305,306,307,308]; despite this, there had been no serious turbine accidents. The

---

[303]Dissenting opinion of R.M. Haziahmetov (member of investigation commission of Rostechnadzor) regarding the Act of technical investigation of the accident at the Sayano-Shushenskaya HPP, Destruction of Tubine 2 of Sayano-Shushenskaya Hydropower Station: causes and lessons, Volume III, Hydrotechnical Construction, Moscow, 2013, p.276

[304]N. Baykov, Analysis of the circumstances of the accident at the Sayano-Shushenskaya HPP, Destruction of Tubine 2 of Sayano-Shushenskaya Hydropower Station: causes and lessons, Volume I, Hydrotechnical Construction, Moscow, 2013, p.158

[305]Valentine Bryzgalov, Monograph "From the experience of establishment and development of Krasnoyarsk and Sayano-Shushenskaya HPSs", Krasnoyarsk, Surikov Publisher, 1999, p. 541

[306]Vladimir Demchenko, Andrew Krassikov, Sergey Teplyakov, Irina Tumakova. Was Turbine #2 on SSHPS shaking during 10 years? Izvestia, September 14, 2009

[307]F. Kogan, Abnormal operating conditions and reliability of modern hydro turbines, Destruction of Tubine 2 of Sayano-Shushenskaya Hydropower Station: causes and lessons, Volume I, Hydrotechnical Construction, Moscow, 2013, p.49

[308]N. Baykov, Analysis of the circumstances of the accident at the Sayano-Shushenskaya HPP, Destruction of Tubine 2 of Sayano-Shushenskaya Hydropower Station: causes and lessons, Volume I, Hydrotechnical Construction, Moscow, 2013, p.153

condition of the turbine cap stud-bolts was checked during routine maintenance, but visually, without ultrasonic scanning. As well as the turbine blowout at Nurek HPS in 1983, there had been another at Grand Rapids in 1992, but the managers and engineers at SSHPS knew little or nothing about these incidents—so nobody there realized that excessive or unusual vibration could have catastrophic consequences [309]. Laboratory tests after the disaster showed that the stud-bolts had about 60–65% metal fatigue, and that most of them were cracked [310].

The Minister of Emergency Situations for the Russian Federation later evaluated this accident as "*the biggest man-made emergency situation [in Russia] in the past 25 years [i.e. since Chernobyl]—for its scale of destruction, for the scale of losses it entails for our energy industry and our economy*" [311]. Reconstruction of the station took more than 5 years, and cost around US$1.5 billion.

Frequent Minor Incidents at Power Plants (Worldwide)   Electricity generation at thermal power plants involves boiling water and heating the steam to temperatures of up to +540 °C, before passing it through turbines to generate electricity. There have been many accidents involving the depressurization of stream pipes, often causing the death of personnel. Another common cause of injury or death through poor occupational safety is electrical arc flash.

Severe accidents at nuclear power plants or hydropower stations pose a threat to millions of people, and can become national or even international catastrophes. A range of measures are needed in response: strict government oversight over the running of utilities along with a genuine readiness from the industry to comply with such regulation; careful analysis of plant design throughout the life-cycle of a unit to ensure the latest safety solutions are being implemented; and ongoing international information exchange about near-miss cases, experience from actual disasters, safety solutions, etc. Due to the complexity of modern utility units, operators should collaborate intensively with plant designers and hardware and software producers to monitor the condition of equipment, make timely modernisations and updates and exchange information about the behavior of the equipment in different situations. There is a critical need to train utility personnel for the operation of sophisticated equipment, both in regular and emergency mode. Because any misjudgement in handling an emergency in nuclear- or hydro-electroenergetics could lead to radiation exposure or flooding to nearby communities, utilities need to work with state emergency services to conduct regular training and make well-tested contingency plans.

---

[309]Rostehnadzor: the accident at the Sayano-Shushenskaya HPP is not unique, in 1983 was a similar situation at Nurek HPP, Interfax, October 3, 2009 and review about accidents and other disturbances on power stations and electric networks of USSR energy system for 1983, Soyuztechenergo, Moscow, 1984

[310]B. Skorobogatykh, N.Shepilov, S.Kunavin, V.Ushakov, Investigation of the metal and the nature of damage studs of turbine cover of Turbine 2 of Sayano-Shushenskaya Hydropower Station, Destruction of Tubine 2 of Sayano-Shushenskaya Hydropower Station: causes and lessons, Volume I, Hydrotechnical Construction, Moscow, 2013, p.373

[311]Joe P. Hasler, Investigating Russia's Biggest Dam Explosion: What Went Wrong, Popular Mechanics, February 2, 2010

Blackouts on Electricity Transmission and Communal Disasters

The destruction of supply hardware and distribution is usually limited to a specific area and allows one to organize rapid responses to repair the damage (from several hours to days).

North American Blackouts (USA and Canada, 1965 and 2003) In the evening of November 9, 1965, nearly 30 million people experienced a blackout for up to 13 h. The blackout was caused by the incorrectly installed protective relay on a transmission line in the Ontario province in Canada, which tripped the line below its capacity after a small surge and resulted in a cascade of further "trippings" when lines were overloaded in several states in the northeastern US and eastern Canada. More than 800 thousand residents of New York were trapped in the subway and many others were stuck in elevators or delayed in getting home by commuter transport stoppages [312]. Fortunately, it was a clear and moonlit night, so people were not left in total darkness and many could walk home. Lyndon Johnson, the US president, concluded that *"[this] failure is a dramatic reminder of the importance of the uninterrupted flow of power to the health, safety, and well-being of our citizens and the defense of our country"* [313]. 37 years later, on August 14, 2003, a blackout affected up to 50 million people for a period of 2–4 days. This blackout was caused by a fault in alarm software, and the failure of dispatchers at the Ohio-based utility FirstEnergy Corporation to respond adequately, when a transmission line was tripped by contact with overgrown trees. Again, the initial trip caused a cascade effect through other lines in the grid, and 61,800 MW of generating capacity was taken offline in seven US Midwest and Northeast states and the Canadian province of Ontario. Estimates of the cost of damage from these blackouts range from US\$4 to \$10 billion [314].

Indian Blackouts (2001, 2012) Power outages affected more than 220 million people from Northern India and caused losses of around \$107 million on January 2, 2001, when the failure of a distribution substation in Uttar Pradesh triggered a blackout. The blackouts of July 30 and 31, 2012 took place during a very hot summer when both domestic and agricultural demand for electricity were exceptionally high. The first of the two, on July 30, was initiated by circuit breakers on an overloaded transmission line in Northern India, which initiated a cascade effect throughout the grid and the disconnection of 300 million people in nine states including Delhi, where seven water treatment stations lost power; the duration of the blackout was around 13 h. The second blackout the following day was caused by a relay problem on another overloaded transmission line in Northern India, and

---

[312]Peter Kihss, Power Failure Snarls Northeast; 800,000 Are Caught in Subways Here; Autos Tied Up; City Gropes In Dark, The New York Times, November 9, 1965

[313]Lyndon B. Johnson, Memorandum Concerning the Power Failure in the Northeastern United States, November 9, 1965. Gerhard Peters and John T. Woolley, The American Presidency Project, http://www.presidency.ucsb.edu/ws/?pid=27361

[314]The Economic Impacts of the August 2003 Blackout, Electric Consumer Research Council (ELCON), February 2, 2004

affected up to 620 million people in 22 states for between 2 and 7 h. This was the largest blackout in human history in terms of disconnected consumers.

Untreated Water Supply The largest waterborne bacterial outbreak in the history of the United States occurred in 1993 in Milwaukee, and was caused by Cryptosporidium bacteria. The subsequent investigation found that chorine processing at one of the water processing plants in the city did not kill the bacteria in turbid water pumped from Lake Michigan. Necessary improvements included the installation of additional filters to reduce the microbial contamination of the water.

In 2000, a severe E. coli outbreak occurred in Walkerton, in Ontario, Canada. The utility company involved was aware of the contamination of its water sources months before the outbreak, but concealed this fact and continued to supply residents of the city. The outbreak killed seven people and 2300, around half of the town's population, became ill. Later it was found that surface water ran into one of the underground wells from which the utility pumped its water. Moreover, the company had failed to provide adequate chlorine processing of the water from the contaminated well [315].

In 2014, tap water in Lanzhou, a Chinese city of 3.6 million residents, was found to be contaminated with benzene. The level was 20 times the maximum benzene level permitted in the country. The cause of contamination was a culvert carrying raw water from a sedimentation plant to the water treatment plant, which took benzene from oil previously leaked from local chemical plants [316]. According to the Chinese Ministry of Environmental Protection, there were 280 million people in the country without access to safe drinking water in 2014 [317].

In the same year, the Flint water crisis emerged in Michigan, United States, where incorrect water treatment allowed lead deposited from old water pipes into the water system, exposing residents to this neurotoxic heavy metal [318].

Untreated Sewage The American Society of Civil Engineers "*estimates an annual [American] national discharge of 900 billion gallons of untreated sewage through leaks, broken pipes and other [technical] mishaps*" [319]. This is a common problem affecting many countries of the world. There have been several serious accidents. For instance, in 2006 in Hawaii, 48 million gallons of sewage spilled from a ruptured pipe just near the famous Waikiki Beach, which had to be closed for swimmers. In 2007 in Gaza, Palestine, a 2-m wave of sewage from a ruptured sewage pool killed four people, injured 20, destroyed 20 homes and damaged up to 250 buildings. In the same year in Scotland, 100 million liters of sewage was spilt when a pump failed at a sewage

---

[315]Inside Walkerton: Canada's worst-ever E. coli contamination, CBC News, May 10, 2010

[316]Stian Reklev, Kathy Chen, Ben Blanchard, Michael Martina, Geert De Clercq, Andrew Roche, Chinese rush for bottled drinks after benzene pollutes tapwater, Reuters, April 11, 2014

[317]Catherine Wong Tsoi-lai, Nation wages war on water contamination, Global Times, April 17, 2015

[318]Maggie Fox, Flint Water Crisis: Feds Expand Programs to Help Kids Affected by Lead, NBC News, March 2, 2016

[319]David Fleshler, Dana Williams, Sewage overflow incidents on rise as aging pipes break, Sun Sentinel, February 18, 2012

processing plant; the flood threatened fish in waterways and nearby wildlife. In 2011 in Mexico, up to 200 homes were flooded by a ruptured sewage pipe after heavy rains [320].

Guadalajara Sewer Explosions (Mexico, 1992) On April 22 1992, a series of massive explosions in downtown Guadalajara took the lives of at least 206 inhabitants, with 500–600 missing and more than 1800 injured [321]. The explosions destroyed 13 km of streets and more than 1000 houses and buildings. An investigation concluded that there were gasoline gases within the sewer system, which had exploded with such devastating effect. In 1973, a steel pipeline had been built to carry gasoline from the Petroleos Mexicanos (PEMEX) refinery in Salamanca to a depot in Guadalajara. Many years later, municipal services constructed a water pipeline too close to the gasoline pipeline. Because this pipeline was made from zinc-coated iron, underground humidity initiated an electrolytic reaction with the steel gasoline pipe resulting in a hole of just 1 cm in diameter in the gasoline pipe. But this pipe carried gasoline at high pressure, so gasoline gases gradually began to penetrate the sewer system in one district of the city. Later, during the construction of a subway line, sewer pipes were erroneously redesigned and most of the dangerous gas which had penetrated the system collected in a limited area. Even though locals had informed municipal services about the smell of natural gas days before the explosion, and checks had already been carried out revealing abnormal gas levels, the city authorities did not evacuate residents living near the contaminated sewers. Losses from the accident were estimated at between US$0.3 and $1 billion.

Natural Gas Explosions (Worldwide Problem) On a global scale, there are thousands of deaths from household accidents in buildings supplied with natural gas for heating and cooking, due to improper use of gas appliances, jumps in natural gas pressure, and so on.

Utilities could increase the speed of their emergency response to distribution and supply breakdowns. This would include receiving and adequately responding to customer complaints, maintaining and improving the equipment of repair staff, ongoing training of repair staff to increase the speed and quality of repairs, and improving the skills of customer support staff to respond to a crisis. They could also develop a wide network of agreements with other players within the industry to ensure backup supplies in case of emergency, and use mathematical optimization models to avoid blackouts [322]. In the event of a severe accident, it is critically important to inform customers and the authorities of the real scale of a disaster, give a realistic forecast about the restoration of the situation, and request external assistance without delay to get the recovery process under way.

---

[320]Katherine Butler, When waste attacks: 5 big sewage disasters, Mother Nature Network, January 9, 2013

[321]Guadalajara Mexico, Ministry in charge of the environment of France, December 2007, https://www.aria.developpement-durable.gouv.fr/wp-content/files_mf/FD_3543_guadalajara_1992_ang.pdf

[322]Chao Zhai, Hehong Zhang, Gaoxi Xiao and Tso-Chien Pan, Modeling and Identification of Worst-Case Cascading Failures in Power Systems, arXiv:1703.05232v1, March 15, 2017

Shutdowns, Accidents and Blackouts Due to External Natural Events

Natural disasters can pose a serious threat to utility infrastructure, and the destruction of facilities can be as devastating as that seen in industrial accidents caused by technical mistakes.

Failure of Dams and Shutdown/Accidents of Power Plants

Vajont Dam Disaster (Italy, 1963) The Vajont hydropower station generated electricity from a dam and reservoir at the top of the Piave valley in the Dolomite region of the Italian Alps. In October 1963, 260 million m^3 of rock on one side of the reservoir slid into the water, triggering a 150–250 m high wave that overtopped the dam and cascaded down the valley, wiping out several villages in its path and killing at least 1921 people [323]. Fundamentally, the disaster happened because the geological composition and instability of the slopes surrounding the planned reservoir—which consisted of sand, limestone and clay—had not been fully researched and understood before the dam was built [324].

When the reservoir was beginning to fill up after the completion of the dam, power company surveyors noticed abnormal ground movement on the south side of the reservoir. They invited German and Italian geologists to look into the composition of the rock around the reservoir; their study confirmed that the southern slope was unstable, and estimated that, if the company tried to fill the reservoir completely, more than 200 million m^3 of rock could slide, because the foundation was undercut by an ancient landslide. This was unwelcome news for the power company; they commissioned new research, which reassured them that the slope was relatively stable, and prone only to much smaller landslides. The new research concluded that the maximum likely wave from a landslide up to a volume of 40 million m^3 would be no higher than 25 m. There is no information that the power company passed on the results of either study to the Italian government. Reassured by the new study, power company managers continued to push the station towards full capacity. By the autumn of 1963, the reservoir was filled to around the maximum limit and the total cumulative ground movement of the southern slope exceeded 3 m [325]. From September 1963, in spite of continuous water drainage, the ground movement

[323]F. Guzzetti, G. Lollino, Book Review of "The Story of Vaiont Told by the Geologist Who Discovered the Landslide", Natural Hazards and Earth System Sciences, 11, 2011, pp.485–486

[324]Mountain Tsunami, documentary of "Seconds from Disaster" serious, National Geographic Channel, 2012

[325]Mountain Tsunami, documentary of "Seconds from Disaster" serious, National Geographic Channel, 2012

velocity reached 20 cm/day [326,327] and residents of villages located above the reservoir were starting to see cracks in their houses. On October 9, 1963 at 10:29 pm, the huge landslide that the geologists had predicted duly occurred, and a wall of water hundreds of meters high killed dam and station personnel, and over 1900 locals from the villages downstream.

Collapse of the Malpasset Dam (France, 1959) This double-curvature arch dam was constructed on the Reyran River for irrigation and water supply in 1954. During heavy rains in 1959, the dam collapsed and the ensuing flood killed more than 400 people from the town located further down the river. The causes of the collapse were the tremendous additional water pressure from the torrential rain, and the weak composition of the rock on one bank of the river.

Collapse of the Shimantan and Banqiao Dams (China, 1975) The largest industrial accident in world history was caused by the breach of the Banqiao and Shimantan irrigation dams in Central China in August 1975 due to Typhoon Nina. 26,000 people according to official estimates—but up to 83,000 according to unofficial data—were killed by the destruction of the dams and ensuing floods; 145,000 perished in the following months from disease and famine [328,329]. To put these disasters in context, by 2011, China had constructed up to 83,000 dams for irrigation, flood control and electricity production; according to information from the Department of Flood-Control and Disaster Reduction of the China Institute for Water Resources and Hydropower Research, 3481 dam collapse incidents occurred in the 50 years from 1954 to 2003 [330].

Collapse of Machhu Dam-II (India, 1979) The disaster occurred when the 4 km-long earthwall of the dam collapsed under water pressure after prolonged heavy rain in Western India. The cause of the disaster was a mistake in the calculation of the maximum potential water inflow to the reservoir of the dam, as a result of which the spillways to release excess water were too narrow. In the days before the disaster, the outflow of the spillways was 2–4 times less than the water inflow from the heavy rain. There had been warnings before the accident that the spillways would not allow enough water through in the event of serious rain, but they were ignored by the authorities [331]. The number of casualties was never accurately established—estimates range from 1800 to 25,000.

---

[326]Rinaldo Genevois, Monica Ghirotti, The 1963 Vaiont Landslide, Giornale di Geologia Applicata 1, 2005, pp. 41–52

[327]D. Sornette, A. Helmstetter, J.V. Andersen, S. Gluzman, J.-R. Grasso and V.F. Pisarenko, Towards Landslide Predictions: Two Case Studies. Physica A 338, 2004, pp.605–632

[328]Typhoon Nina–Banqiao dam failure, Encyclopædia Britannica

[329]David Longshore, Encyclopedia of Hurricanes, Typhoons, and Cyclones, Infobase Publishing, 2009, p.124

[330]Lu Zongshu, Shen Nianzu, Dams gone wrong: Is danger lurking in China's dams? Probe International, August 24, 2011, https://journal.probeinternational.org/2011/08/24/dams-gone-wrong-is-danger-lurking-in-china's-dams/

[331]Utpal Sandesara, Tom Wooten, Paul Farmer, No One Had a Tongue to Speak: The Untold Story of One of History's Deadliest Floods, Prometheus Books, 2011

Flooding of Blayais NPP (France, 1999) The extratropical storm Martin brought a combination of high tides and strong winds, flooding this coastal plant, cutting its off-site power supply and threatening some parts of the emergency core cooling systems (Fukushima Daiichi precursor).

Madras NPP and the Indian Ocean Tsunami (India, 2004) The great Sumatra–Andaman earthquake of December 26, 2004 had a moment magnitude of 9.2 and generated tsunamis which hit not only the coast of Thailand, Indonesia and Malaysia but also Sri Lanka, the Maldives, and parts of India. The waves damaged seawater pumps and the construction site of the nuclear power plant in Madras, forcing the emergency shutdown of the reactor [332]. Japanese nuclear engineers studied this case, which should have revealed some of the potential dangers of locating nuclear plants on the coast—but no significant action was taken in response. They did not appear to have learned from it until disaster struck at Fukushima-Daiichi [333,334].

Fukushima-Daiichi Nuclear Disaster (Japan, 2011) Early in the afternoon of March 11, 2011, a massive earthquake occurred 70 km from the east coast of the Tohoku region in Japan. With a moment magnitude estimated between 9.0 and 9.2, this was the largest earthquake ever recorded in Japan; according to the United States Geological Survey, it was the fifth largest recorded worldwide since 1900 [335]. The tsunami generated by the quake hit the coast of the prefectures of Iwate, Miyagi and Fukushima prefectures about 50 min after the main shock. Hundreds of kilometres of infrastructure were destroyed and more than 18,800 people died [336].

Critically, Japan had chosen to locate most of its nuclear power plants in the eastern part of the country, and five of these were located in the coastal disaster zone. Several were in the path of the tsunami but, at one plant, the damage caused was catastrophic, and would ultimately be classified as a major accident of level 7—the highest level—on the International Nuclear Event Scale. The Fukushima-Daiichi plant was owned by Japan's largest electric utility, the Tokyo Electric Power Co. (TEPCO). There were 6 reactor units at the plant, as well as large pools with spent nuclear fuel, but only Units 1–3 were in operation on the day of the earthquake—Unit 4 was being rebuilt, and 5 and 6 were shut down for routine inspection. The main shock of the quake triggered the emergency shutdown (SCRAM) system on all the operating reactors. The ground acceleration of the quake was beyond the

---

[332]Sobeom Jin, Sungjin Hong and Fumihiko Imamura, 2004 Indian Ocean Tsunami on the Madras Nuclear Power Plant, India, Transactions of the Korean Nuclear Society Spring Meeting, Chuncheon, Korea, May 25–26, 2006

[333]James M. Acton, Mark Hibbs, Why Fukushima Was Preventable, Carnegie Endowment for International Peace, March 2012, pp. 11, 22–23

[334]The official report of The Fukushima Nuclear Accident Independent Investigation Commission, The National Diet of Japan, Chap. 1. Was the accident preventable?, July 5, 2012, p.26

[335]Christina Nyquist, The March 11 Tohoku Earthquake, One Year Later. What Have We Learned? March 9, 2012, US Geological Survey

[336]Wolfgang Kröger, Fukushima: Need for Reappraisal of Nuclear Risks? ETH Zürich, Keynote SRA-Europe 21st Annual Conference, Zurich, June 18–20, 2012

design limits of the plant but, at the initial stage, the only damage was a leakage of coolant from Unit 1 [337]. The plant lost all offsite power because both the external transmission lines and the nearest transformer station were destroyed by the initial shock—but this in itself was not disastrous because onsite emergency diesel generators and batteries were available to provide power to cool the reactors [338].

But 51 min later, the tsunami reached the plant. The protective seawall was designed to withstand waves of up to 5.7 m [339]. Some vulnerable objects, like the seawater pumps, were located beyond the wall and at only 4 m above sea level; the diesel generators and batteries were 10 m above sea level, inside the reactor buildings [340]. But the Tohoku undersea quake generated tsunami waves of over 9 m in the open ocean, and as they approached the coast by the plant they had built up to a height of 14–15.5 m [341]. Thus, although the reactor buildings were strong enough to withstand both the tremors and the tsunami, the NPP lost all sources of electricity to cool the reactors of Units 1 and 2 and the spent fuel pool of Unit 4. Unit 3 had battery power for about 30 h, and the emergency diesel generators only provided emergency power to Units 5 and 6. Damage to the core of reactor 1, and the resulting meltdown of nuclear fuel, started only 3 h and 15 min after the tsunami struck. Reactor 3 started to sustain damage after 43 h, and reactor 2 after 76 h [342].

Units 1 and 2 were fuelled by low-enriched uranium (LEU) and Unit 3 was fuelled by mixed oxide (MOX) fuel that contained plutonium. At the time of the disaster, there were a total of 257 tons of nuclear fuel between the three functioning reactors, and 264 tons of spent fuel in the pool of Unit 4 [343,344]. Approximately 900–940 PBq of radioactive substances were ultimately released into the atmosphere [345,346]; this compares with 5200 PBq estimated to have been released in 1986 from

---

[337]The official report of The Fukushima Nuclear Accident Independent Investigation Commission, The National Diet of Japan, Executive summary, July 5, 2012, pp.17, 30

[338]The official report of The Fukushima Nuclear Accident Independent Investigation Commission, The National Diet of Japan, Executive summary, July 5, 2012, p.13

[339]Fact Finding Expert Mission of the Fukushima Dai-Ichi NPP Accident Following the Great East Japan Earthquake and Tsunami, IAEA mission report, 24 May – 2 June 2011, p.11

[340]The official report of The Fukushima Nuclear Accident Independent Investigation Commission, The National Diet of Japan, Executive summary, July 5, 2012, p.14

[341]Akira Izumo, Facts, Lessons Learned and Nuclear Power Policy of Japan after the Accident, Agency for Natural Resources and Energy, Ministry of Economy, Trade and Industry of Japan, January 24, 2012

[342]The official report of The Fukushima Nuclear Accident Independent Investigation Commission, The National Diet of Japan, Executive summary, July 5, 2012, p.13

[343]Overview of facility of Fukushima Daiichi Nuclear Power Station, TEPCO, http://www.tepco.co.jp/en/nu/fukushima-np/outline_f1/index-e.html

[344]The Status of Nuclear Fuel Stored at the Fukushima Daiichi and Fukushima Daini Nuclear Power Plants, Citizens' Nuclear Information Center (Japan), Jan. 31, 2013, http://www.cnic.jp/english/newsletter/nit154/nit154articles/03_nf.html

[345]The official report of The Fukushima Nuclear Accident Independent Investigation Commission, The National Diet of Japan, Executive summary, July 5, 2012, p.39

[346]Fukushima Accident, World Nuclear Association, updated 13 January 2014

Chernobyl. The Japanese government's report to the International Atomic Energy Agency (IAEA) stated that the total radiation released was 1/6 of the amount from the Chernobyl accident when converted to iodine. However, more than 150,000 residents had to leave their homes [347] because of radioactive contamination [348].

The Japan Center for Economic Research estimates that the removal and safe disposal of the melted fuel, and the cleanup of a huge area surrounding the plant which is heavily contaminated by radiation, will take 40 years and may cost as much as 20 trillion yen or around US$200 billion [349]—this would amount to 4.2% of the Japanese GDP. According to Juan Carlos Lentijo, leader of the International Atomic Energy Agency mission team, *"it will be nearly impossible to ensure the time for decommissioning such a complex facility in less than 30–40 years as it is currently established in the roadmap"* [350].

For over a year after the accident, it had a serious impact on Japan's energy supplies for industrial and domestic needs: by mid-May 2011, stringent safety inspections had taken all but 17 of the remaining 50 reactors in the country out of operation, and from May until July 2012, all Japan's reactors were suspended. The shutdown of the entire nuclear program had a double negative impact on the Japanese economy, reducing productivity and hugely increasing fossil fuel imports, and left the country with a US$134 billion trade deficit in 2013.

Japan's legislative body, the National Diet, established the Fukushima Nuclear Accident Independent Investigation Commission (NAIIC), which had legal authority to demand access to any documents or evidence required to establish exactly what happened. Investigators interviewed 1167 people and organized 900 h of hearings. The commission concluded that *"the accident at the Fukushima Daiichi Nuclear Power Plant cannot be regarded as a natural disaster. It was a profoundly manmade disaster – that could and should have been foreseen and prevented. The accident was clearly "manmade". We believe that the root causes were the organizational and regulatory systems that supported faulty rationales for decisions and actions, rather than issues relating to the competency of any specific individual. We found an organization-driven mindset that prioritized benefits to the organization at the expense of the public"* [351]. This conclusion rested above all on the fact that, prior to the disaster, several experts had warned that the Fukushima-Daiichi plant

---

[347] The evacuation map in the following official government site (http://www.meti.go.jp/english/ earthquake/nuclear/roadmap/pdf/141001MapOfAreas.pdf) shows the most dangerous area called Area3, which is defined as follows: *"Area3: Areas where it is expected that the residents have difficulties in returning for a long time"*.

[348] The official report of The Fukushima Nuclear Accident Independent Investigation Commission, The National Diet of Japan, Executive summary, July 5, 2012, p.38

[349] The Fukushima Nuclear Accident and Crisis Management — Lessons for Japan-US Alliance Cooperation, The Sasakawa Peace Foundation, Sep. 2012, p.38

[350] Mari Yamaguchi, IAEA: Japan nuke cleanup may take more than 40 yrs., Associated Press, April 22, 2013

[351] The official report of The Fukushima Nuclear Accident Independent Investigation Commission, The National Diet of Japan, Executive summary, July 5, 2012, p.21

was not adequately prepared for a high-wave tsunami, but the TEPCO hierarchy had failed to pass on these warnings [352].

North Anna NPP and Virginia Earthquake (USA, 2011)   Prior to the construction of the North Anna nuclear power plant (NPP) in Virginia, the utility company Dominion Energy Inc. had ordered a seismic study, which had confirmed the existence of a fault running through the site of the proposed plant. However, this fact was concealed from the public by both the utility and the US Nuclear Regulatory Commission in 1977: *"[V]irtually the entire Office of Regulation [of the Nuclear Regulatory Commission was] ...well aware of the fault and determined not to take any immediate action"* [353]. The utility was eventually fined for making false statements during the licensing process, but the plant was nevertheless erected on the proposed site; it was designed to withstand a 6.2 M earthquake. Decades later on August 23, 2011, there was a 5.8 M earthquake—the largest the East Coast had seen for 67 years—around 17 km from the plant. The quake triggered the emergency shutdown of the reactors, caused a loss of offsite power in addition to a failure of one emergency diesel generator which suffered a coolant leakage, cracked one of the spent nuclear fuel canisters and shifted spent nuclear fuel dry casks by 11 cm [354]. After less than 4 months, the reactors were restarted successfully; nevertheless, the accident prompted an emergency assessment of natural disaster risks at all American NPPs.

Hurricanes and American NNPs (USA, 2011, 2017)   Until 2017, Hurricane Sandy was the second most expensive weather event in US history after Hurricane Katrina, causing more than US$70 billion of damage. Sixteen reactors at 12 American NPPs were in the path of the storm through the East Coast of the country. Ultimately, three of them were forced to shut down and one was on alert due to high levels of water in the cooling water intake facility; this plant was on scheduled refueling and maintenance operations [355,356]. Concerning sewage management during the hurricane, up to 11 billion gallons of untreated and partially treated sewage were released into the environment [357]. During the hurricane season of 2017 and particularly during

---

[352]Dmitry Chernov and Didier Sornette, Man-made catastrophes and risk information concealment (25 case studies of major disasters and human fallibility), Springer, 2016

[353]North Anna Nuclear Plant Earthquake Risk: 1977 Memo Details Cover-Up Of Seismic Knowledge, The Huffington Post, January 06, 2012

[354]North Anna Independent Spent Fuel Storage Installation, Response to Earthquake, US Nuclear Regulator Commission, https://www.nrc.gov/docs/ML1200/ML12005A011.pdf

[355]Alex Kane, Nuclear Trouble: 16 Reactors in the Path of Hurricane Sandy, AlterNet, October 29, 2012

[356]Four Nuclear Power Plants Jolted by Hurricane Sandy, Environment News Service, October 30, 2012

[357]Alyson Kenward, Daniel Yawitz, Urooj Raja, Sewage Overflows from Hurricane Sandy, Climate Central, April 30, 2013

Hurricane Irma, three nuclear power plants in Florida were temporarily shut down and withstood the disaster without significant damage [358].

Failure of Grid Lines

Geomagnetic Storms and the Quebec Grid System (Canada, 1989) In 1859, the most intense solar superstorm in the history of scientific observation of the Sun, known as the Carrington Event, caused widespread failure of telegraph connections over North America and Europe. The second largest superstorm recorded occurred on March 1989 and resulted in a 12-h blackout for six million people in the Quebec region of Canada. These events are expected to occur again with a recurrence time of approximately 150 years and 50 years respectively [359].

Brazilian Blackouts (Brazil, 1999 and 2009) On March 11 1999, a lightning strike tripped a key transmission line in Southern Brazil, provoking a blackout for up to 97 million Brazilians. On November 10 2009, heavy rains and strong wind damaged three transformers on a crucial grid line, and the tripping initiated a cascade effect across the national energy system, leading to electricity disruption for up to 60 million people and forcing the shutdown of the hydropower plant at Itaipu—the second largest in the world in terms of installed capacity [360].

Italy and South Switzerland Blackout On September 28 2003, a storm initiated a flashover between a conductor cable and an inadequately cut tree, resulting in the tripping of a high-voltage transmission line between Switzerland and Italy. This caused a redistribution of the power flows and the subsequent 110% overload of another north-south transit line. Due to another flashover, this line also shut down 24 min later, followed by a series of cascading failures of other transmission lines in the border region. Corrective measures were taken including load shedding but, because of poor coordination among neighboring Transmission System Operators, the frequency and voltage drop could not be mastered and generation plants started to trip, giving rise to a total blackout through Italy affecting up to 56 million people. This blackout has been modelled as a cascade of failures in interdependent networks, where nodes in the power station network failed, causing a failure of the Internet

---

[358]Nicole Rodriguez, FPL nuclear facilities weathered Irma without sustaining damage, TCPalm, September 11, 2017

[359]Nicole Homeier and Lisa Wei, Solar storm Risk to the north American electric grid, Atmospheric and Environmental Research (AER), Lloyd's, 2013

[360]Jorge Miguel Ordacgi Filho, Brazilian Blackout 2009, Blackout Watch, ONS, Brazil, PAC, 36–37, March 2010, https://www.pacw.org/fileadmin/doc/MarchIssue2010/Brazilian_Blackout_march_2010.pdf

communication network, which in turn caused a further breakdown of power stations [361,362,363,364,365,366].

The 2006 European Blackout This major blackout occurred on a Saturday, November 4, 2006 and, for 2 h, more than 15 million users of the European Network of Transmission System Operators for Electricity did not have access to electricity. The cause of this major blackout was a miscommunication between the high voltage grid operator and the local transmission system operators concerning a planned routine disconnection of the Ems powerline crossing in Northwest Germany to allow a ship to pass beneath the overhead cables. The immediate reaction of the high voltage grid operator led to the line tripping out, with an electrical blackout cascading across Europe extending from Poland in the north-east, to the Benelux countries and France in the west, through to Portugal, Spain and Morocco in the south-west, and across to Greece and the Balkans in the south-east [367].

Hurricanes Katrina (23–31 Aug 2005), Irma (30 Aug–13 Sept 2017) and Maria (16 Sept–2 Oct 2017) (USA and Puerto Rico) In addition to the tremendous financial damage from Hurricane Katrina—more than US$100 billion—around 2.6 million people across the southern states experienced power outages. In 2017, more than a million people, 66% of all households connected to the electric grid on the island of Puerto Rico, lost power during Hurricane Irma [368]. There were forecasts that restoration after the disaster could last from 4 to 6 months because of the severe financial difficulties of the local electric power system [369]. Two weeks after Irma, its immediate successor Hurricane Maria led to a power outage of the whole of Puerto Rico, an island of 3.5 million inhabitants.

---

[361] https://en.wikipedia.org/wiki/2003_Italy_blackout

[362] Rainer Bacher and Urs Näf, Report on the blackout in Italy on 28 September 2003, Swiss Federal Office of Energy (SFOE), November 2003

[363] Final report of the Investigation Committee on the 28 September 2003 Blackout in Italy, UCTE Report April 2004 (https://www.entsoe.eu/fileadmin/user_upload/_library/publications/ce/otherreports/20040427_UCTE_IC_Final_report.pdf, accessed 21 Aug. 2017)

[364] Wolfgang Kröger, Securing the operation of socially critical systems from an engineering perspective: new challenges, enhanced tools and novel concepts, Eur. J. Secur. Res. 2, 39–55 (2017)

[365] Wolfgang Kröger and Enrico Zio, eds., Vulnerable systems, Springer (2011)

[366] Sergey V. Buldyrev, Roni Parshani, Gerald Paul, H. Eugene Stanley and Shlomo Havlin, Catastrophic cascade of failures in interdependent networks, Nature 464, 1025–1028 (2010)

[367] UCTE (union for the co-ordination of transmission of electricity), Final Report System Disturbance on 4 November 2006 (https://www.entsoe.eu/fileadmin/user_upload/_library/publications/ce/otherreports/Final-Report-20070130.pdf)

[368] Hurricane Irma & Hurricane Harvey Event Report (Update #22), US Department of Energy, September 7, 2017, https://energy.gov/sites/prod/files/2017/09/f36/Hurricane%20Harvey%20Event%20Summary%2022.pdf

[369] Leslie Josephs, Hurricane Irma could leave areas of Puerto Rico without power for up to six months, CNBC, September 6, 2017

In May 2018, Harvard university researchers published their calculations that the real death toll from Hurricane Maria was 70 times higher than official figures: 4645 American citizens died instead of 64. A third of the deaths occurred due to *"interruptions in medical care caused by power cuts and broken road links"* [370]. The report continued: *"The storm disrupted medical services across the island, and many households were left for weeks without water, electricity, or cell phone coverage. .... In addition to a significantly higher death toll, the study shows that the average household surveyed went approximately 41 days without cell phone service, 68 days without water, and 84 days without electricity following the storm. More than 30 percent of surveyed households reported interruptions to medical care, with trouble accessing medications and powering respiratory equipment being the most frequently cited challenges. In the most remote areas, 83% of households were without electricity for this entire time period"* [371].

Preventive measures against natural disasters include conducting a thorough survey of previous natural calamities at and around any proposed site retrospectively back to hundreds or even thousands years in the past, designing facilities to withstand the worst possible scenario, advance monitoring of weather, climate and land mass changes to inform a preventive response, and implementing rapid restoration measures after a disaster.

Blackouts Due to Political Instability, Wars, Clashes and Others

Political instability, intra- and inter-state wars can threaten the safe operation of utility infrastructure.

Blockade of Leningrad by German and Finnish Nazis (USSR, 1941–1943)
During World War II, German and Finnish forces surrounded Leningrad, now St. Petersburg, the second largest city of the Soviet Union, and swiftly destroyed the city's critical infrastructure—including power plants—in order to reduce military production output and make life insufferable for the civilian population. The Red Army was left with only one route to supply the city—through Lake Ladoga. This channel was used not only for supplying food and materials, but also to lay underwater electric cables for the military production sites of the city. The blockade lasted 872 days, but the Nazis could not conquer the city: more than 600,000 inhabitants died of starvation and more than 300,000 Soviet soldiers perished, but Leningrad did not surrender.

---

[370]Hurricane Maria 'killed 4600 in Puerto Rico, BBC, May 29, 2018

[371]Study Estimates Prolonged Increase in Puerto Rican Death Rate After Hurricane Maria, Harvard T.H. Chan School of Public Health, May 29, 2018, https://fxb.harvard.edu/2018/05/29/study-estimates-prolonged-increase-in-puerto-rican-death-rate-after-hurricane-maria/

Iraqi Attacks on Construction Site of Bushehr NPP (Iran, 1984–1987) On several occasions during the Iraq-Iran war, Iraqi air forces attacked the construction site of the Iranian reactor at Bushehr. They eventually destroyed the control building, and 11 civilians were killed [372].

Demonstration of Possible Threat to Krsko NPP (Slovenia, 1991) In July 1991, several days after Slovenia declared its independence from Yugoslavia, three Yugoslavian Air Force planes (controlled by Belgrade) flew over the now Slovenian Krsko NPP as an aggressive show of force to demonstrate their capacity to threaten the plant. The plant was temporarily shut down [373].

Israeli Attack on Jiyeh Power Station (Lebanon, 2006) During the Israeli-Lebanese war, the Israeli Air Force bombed the plant and damaged its oil tanks. The fire lasted up to 10 days because Israeli warplanes continued to threaten the plant, effectively preventing Lebanese firefighters from tackling the situation. The leak amounted to 20,000–30,000 tonnes, which is around 2/3 of the oil leaked during the infamous Exxon Valdez oil spill in Alaska. The oil slick polluted 170 km of coastline, reaching as far as Turkey, Syria and Cyprus.

Terrorist Action Against Baksan Hydroelectric Power Station (Russia, 2010) Four Islamist militants attacked Baksan hydroelectric power station in the North Caucasus. Two security guards were killed and two turbines were blown up.

Vasilikos Power Station and Explosion at the Evangelos Florakis Naval Base (Cyprus, 2011) The largest power plant in Cyprus—producing more than half the island's electricity—was severely damaged after an explosion at the naval base. It took more than a year to restart the plant.

Dams and Islamic State (2014, 2017) Islamic State militants had captured the two largest hydropower dams in Iraq—the Mosul Dam on the River Tigris and the Haditha Dam on the Euphrates. They threatened to open the spillways or even completely destroy the dams to flood the territories further downstream, which were under Iraqi government control. Later, the dams were liberated by the Iraqi Army and Kurdish Peshmerga forces. The Syrian Tabqa Dam—constructed on the River Euphrates in the 1970s to provide hydroelectric power to the city of Raqqa and irrigation to the surrounding farmland—became the site of a heavy battle in 2017 between ISIS and the Syrian Democratic Forces with the help of US special forces. The control room was destroyed in the fighting, leaving the dam unable to generate electricity [374]. There was also a struggle for control of the Syrian Tishrin and Baath dams between the government, pro-Western rebels and ISIS.

---

[372]Pierre Razoux, The Iran-Iraq War, Belknap Press: An Imprint of Harvard University Press, November 3, 2015

[373]Stritar, A., Mavko, B., Susnik, J. and Sarler, B., Some aspects of nuclear power plant safety under war conditions, Nuclear Technology, 1993, 101 (2), pp.193–201

[374]Emily Burchfield, Under Pressure: The Effect of Conflict on the Euphrates Dam, Atlantic Council, April 18, 2017

Pakistani Blackout (Pakistan, 2015) The Baloch Republican Army, a separatist group, attacked a 220 KV transmission line in the south of Pakistan, causing a cascade effect and disconnecting 140 million Pakistanis. The group also damaged gas lines and railways, and killed several state representatives [375].

Water Supply to Damascus (Syria, 2016–2017) At the end of December 2016, water was shut off to Damascus and its surrounding areas, following attacks on the two main sources—Wadi Barada and Ein El Fijeh springs. The cuts left an estimated 5.5 million people without access to safe water. Both the Syrian government and the rebels blamed each other for the attack, which inflicted extensive damage and led to severe water shortages for a month. The United Nations warned that targeting water sources constitutes a war crime [376].

Operators of utilities do not have the option of avoiding or escaping from political and military conflicts because of the immobility of their infrastructure. Instead, they must promote the idea among all conflicting parties that critical infrastructure should remain inviolable during clashes, and that it is in the interest of both sides to maintain the safe operation and integrity of the utilities for the sake of civilians.

Regulatory Changes (Environment Legislation, Technological Breakthroughs, Etc)

Changes in national energy strategies, new regulations on emissions and technological innovations can severely damage the financial situation of utilities because they can only recoup the immense cost of infrastructure over the long term, and there is only limited flexibility to respond to these challenges.

Sustainable Energy Promotion (Worldwide, Ongoing) In order to reduce emissions of carbon dioxide during electricity production, the governments of many countries are promoting hydro, solar, wind, wave, tide, geothermal and biofuel energies, accompanied by policies to improve energy consumption efficiency. This poses a direct threat to traditional thermal powered plants running on coal, fuel oil and natural gas and to producers of hydrocarbon mineral resources as well as nuclear power plants (although they are among the cleanest energy sources with zero carbon emissions). According to the agreement signed at the United Nations Climate Change Conference in Paris (also known as COP21) in December 2015, the main strategy to reduce CO_2 emissions is to work towards a complete halt in fossil hydrocarbon consumption worldwide over the long term. The medium-term goal is to bring coal consumption to an end. Within the framework of COP21, the

---

[375]Salman Masood, Rebels Tied to Blackout Across Most of Pakistan, The New York Times, January 25, 2015

[376]Patrick Wintour, UN warns of war crimes over disruption to water supply north of Damascus, The Guardian January 5, 2017

European Union set itself the goal of reducing carbon dioxide emissions by 40% by 2030, 60% by 2040 and 80% by 2050 (in order to reach the agreed reductions in carbon dioxide emissions, the EU should close all coal-fired plants by 2030 [377]). China has promised to reduce carbon dioxide emissions by 60–65% per unit of GDP by 2030 in comparison with 2005. In the Oil and Gas subchapter, we have mentioned the severe damage that the implementation of this policy has already inflicted on the traditional energy sector—for example the bankruptcy of the coal industry and coal-firing plants in the United States. On the other hand, from the point of view of sustainable energy producers, such changes in energy structure could be beneficial: usually governments will subsidize development that helps them make the transition away from fossil fuels (which has indeed happened on a significant scale in the EU [378]). Nevertheless, the reliability of sustainable energy production is still in question: for instance, during the Australian summer of 2017, there were several blackouts in South Australia caused by heavy consumption of energy during the heat, at a time of the year when the output from wind generation was low and traditional thermal power could not compensate quickly enough for the shortfall [379]. Misjudgments on how to achieve a nationwide transition from coal to gas in Australia, during a period when much of the natural gas from the country's offshore fields was being exported through LNG delivery, provoked a shortage of natural gas for domestic customers and led to price rises. Thus the Australian Prime Minister Malcolm Turnbull remarked that *"On a continent filled with gas, we are paying around three times for gas as another firm would in the United States"*, and called for deregulation to allow for the extensive exploration and development of onshore gas resources [380].

Nuclear Phase-out After Fukushima Disaster (Ongoing) After the Fukushima disaster, the Japanese government dramatically changed its position on the role of nuclear energy in the country's energy balance, aiming to reduce its contribution

---

[377]Gero Rueter, EU needs to shut all coal plants by 2030 to meet climate goals, DW, February 22, 2017, https://www.dw.com/en/eu-needs-to-shut-all-coal-plants-by-2030-to-meet-climate-goals/a-37665345

[378]A recent report by the Council of European Energy Regulators shows that the weighted average subsidy paid to renewable generators in EU 26 in 2015 was €110 / MWh. The maximum was €184 / MWh in the Czech Republic and the minimum €16.2 / MWh in Norway. This should be compared with the wholesale price of electricity in Europe, which lies in the range from €40 to €60 / MWh, Thus, renewables are costing on average about 3 times as much as conventional power (wholesale≈50, subsidy≈110, total≈160). In fact, this estimation is a lower bound as it ignores the "system costs" to expand the grid itself and to provide a portfolio of balancing and backup services for these very intermittent energy sources. In particular, there is need for an almost 100% backup from conventional energy sources, which face a dwindling market share and thus rising costs. Status review of renewable support schemes in Europe, Council of European Energy Regulators, Ref: C16-SDE TF-56-03, 11-04-2017.

[379]Katharine Murphy, Christopher Knaus, South Australian blackout blamed on thermal and wind generator failures, plus high demand, The Guardian, February 15, 2017

[380]Rebekah Ison, PM blames states for gas shortage 'crisis', Australian Associated Press, March 9, 2017

from 35% in the 2010s to zero by 2035. Before the accident, the country had been planning to get more than 50% of its energy from nuclear energy by 2030. As we have outlined above, in 2011–2012, Japan faced a serious shortage of electricity for industrial and domestic needs: shortly after the accident, only 17 out of the remaining 50 nuclear reactors in the country were still operating while the rest were shut down for safety checks, and from May until July 2012, all Japanese nuclear reactors were suspended. This led to an additional spending of US$40 billion on hydrocarbon fuel imports. Higher electricity prices and increased CO_2 emissions are also of concern. After a change of government, the political will began to grow again for reinstating the nuclear industry as a major source of energy in Japan. In June 2014, the three major business lobbies urged the Industry Minister to expedite the restart of the country's nuclear reactors. *"The top priority in energy policy is a quick return to inexpensive and stable supplies of electricity"*, they said [381].

After the nuclear disaster in Japan, the German government, with the support of environmentalist parties, decided to shut down 8 of its 17 reactors with the target of a complete nuclear energy phase-out by 2022. However, this strategy has already made Germany the largest polluter in Europe in terms of particle and CO_2 emissions, since it has forced the country to resume energy generation from coal to compensate for the phased-out nuclear plants [382]. It could also lead to increased imports of cheap Russian natural gas to the EU: the prospects for the alternative option of importing expensive US shale energy through LNG carriers remain unclear while global energy prices are kept down by Saudi Arabia. Other countries in Europe also considered a possible nuclear phase-out, but decided not to follow the radical German shift in energy policy. After the presidential elections in South Korea in 2017, the new government announced a gradual phase-out of coal and nuclear power plants, and the state promotion of natural gas and renewables as the principal energy sources for this Asian nation in the coming decades [383].

Utilities have to make a serious investment in lobbying for their interests with politicians and key decision-makers. In some cases, utilities could ask governments for compensation for their help in implementing planned energy policy changes. Two examples are the claim of Swedish electric power company Vattenfall against the German government for almost €4.7 billion in damages, and the ruling on 5 December 2016 by the German Federal Constitutional Court that nuclear plant operators affected by the accelerated phase-out of nuclear power following the Fukushima disaster are eligible for *"adequate"* compensation.

---

[381]Fukushima Four Years Later—A Tale of Three Countries, Nuclear Energy Institute, https://www.nei.org/News-Media/News/News-Archives/Fukushima-Four-Years-Later—A-Tale-of-Three-Countri

[382]Europe's dark cloud (how coal-buring countries are making their neighbours sick), WWF European Policy Office, Sandbag, CAN Europe and HEAL in Brussels, Belgium, June 2016

[383]South Korea's President Moon says plans to exit nuclear power, Reuters, June 19, 2017

Other Crises

Spent Nuclear Fuel (Worldwide)   As of 2017, there are 446 nuclear power reactors in operation worldwide with a total net installed capacity of roughly 390 GWe [384]. There are 60 more reactors currently under construction. Nuclear power stations generate 11% of the world's total electricity [385]; but in some countries, they account for a much larger share, for example in France where they supply nearly 80% of the country's energy needs. While they have some advantages over other sources, nuclear power generation facilities produce about 200,000 m³ of low and intermediate level radioactive waste and 10,000 m³ of high level waste—including spent fuel designated as wasteevery year worldwide [386]. Low and intermediate level waste is usually disposed of in near-surface disposal facilities, which require engineering and management control for up to 300 years (!). The majority of high level radioactive waste produced in the last 40–50 years is still kept at nuclear power stations and waiting to be moved to geological repositories, which are envisioned to be hundreds of meters underground in a stable geological formation, to be stored for a million years or even longer. This represents a challenge for the world: in the last 100 years, humanity has created a huge number of dangerous materials and facilities, with such a long lifetime that many of them will outlive the societies that invented these materials and operated those facilities. There is the danger that materials dangerous enough to threaten life on the Earth could be left out of supervision and control [387,388]. A recent book by one of the present authors outlines a way out of this conundrum by launching an ambitious R&D nuclear program to make nuclear energy much safer and cleaner [389].

Massive Resettlement During Development of Hydropower Plants (Worldwide)   The construction of hydropower plants and their supply reservoirs requires the requisition of land and the wholesale migration of its former inhabitants to new locations. In some cases, this relocation provokes social upheaval, conflicts with local residents and legal battles lasting for years. For instance, 1.3 million people were relocated in China during the construction of the Three Gorges Dam [390]—the largest hydropower plant in the world in terms of installed capacity.

---

[384]The Database on Nuclear Power Reactors, International Atomic Energy Agency, https://www.iaea.org/pris/

[385]Nuclear Power in the World Today, World Nuclear Association, January 2017

[386]International Atomic Energy Agency, 2006, http://www.iaea.org/Publications/Factsheets/English/manradwa.html

[387]Ibid

[388]D. Sornette, A civil super-Apollo project in nuclear R&D for a safer and prosperous world, Energy Research & Social Science, 2015, 8, pp. 60–65

[389]D. Sornette, W. Kröger and S. Wheatley, New Ways & Needs for Exploiting Nuclear Energy, Springer, 2018

[390]Thousands being moved from China's Three Gorges – again, Reuters, August 22, 2012

Customer Payment Discipline (Worldwide) Because utilities provide electricity, water, sewage, and other services, in advance, in some cases they encounter difficulties in demanding payments for consumed goods from poor customers, bankrupted enterprises, public service or community organizations financed by governments with a budget deficit, and so on.

Thermal Pollution of Power Plants and Water Shortage (Worldwide) Nuclear and thermal plants use water as a coolant in their electricity production process. Usually, outflow water from the plants is warmer, and contains less oxygen, than the inflow water—resulting in potential threats to the wildlife in the lake or river supplying the plant. It is ironic that the suffering in regions stricken by drought is often compounded because power plants have to suspend production due to lack of water. Numerous suspensions have occurred in India: recent instances include shutdowns at Chandrapur thermal power station in 2011, Parli power station in 2015, and at the stations at Farakka and Raichur in 2016. In 2017, a water shortage was recorded at several hydropower plants in Russian Siberia and at the South Ukraine nuclear power plant.

Impact of Utility Infrastructure on Nearby Communities Pacific Gas and Electric (PG&E), the largest supplier of electricity and natural gas in California, filed for bankruptcy in January 2019 following lawsuits from residents of Butte County. In November 2018, the Camp Fire—the deadliest wildfire in history of the state—had swept the county, claiming 85 lives and burning out more than 18 thousand buildings. There are suspicions that the fire was triggered by sparks caused by the failure of a poorly-maintained high-voltage line owned by PG&E: after the fire, the company admitted to the regulator that *"it had found a hook designed to hold up power lines on the tower was broken before the fire, and that the pieces showed wear"* [391]. The rapid spread of the fire was exacerbated by strong winds, and grass which was tinder-dry after months of drought. Several days before the disaster, the utility had warned clients about potential power shutdown after forecasts of very strong wind and low humidity. Many Californian utilities shut down to prevent wildfires in the drought-prone state. But PG&E did not stop transmission in Butte County until disaster had already struck. The compensation bill to victims and their families was expected to exceed US$7 billion, and the utility had only $1.5 billion available.

Unexpected Surge of Electricity Consumption Caused by Mining of Cryptocurrencies According to PricewaterhouseCoopers, servers used for mining cryptocurrencies require 22 TWh/year—equivalent to the electricity consumption of Ireland. By comparison, Google's servers worldwide consumed just a quarter of this (5.7 TWh) in 2015 [392]. Cryptocurrency mining has led to unexpected levels of electricity consumption in some regions of China, where 3/4 of global mining

---

[391]Thomas Peele, Power interrupted: State regulators tackle rules to guide when PG&E and other utilities should cut electricity to avoid sparking fires, Bay Area News Group, December 13, 2018

[392]Why bitcoin uses so much energy, The Economist, July 9, 2018

servers are located, and caused much higher carbon emissions because mining farms use cheap electricity from nearby coal power plants. The process has been recognized by the Chinese National Development and Reform Commission as an activity which must *"be eliminated immediately"* because it is *"wasting energy"* and *"polluting the environment"* [393].

Key Risk Mitigation Measures in Utilities

- Ensuring non-stop 24-7-365 procurement of utility products for customers, or fast recovery in the event of disruption.
- Maintaining good and very cooperative relations with authorities: governments tightly regulate utilities at every stage, from the design, construction and licensing of infrastructure through to its operation; they also determine energy strategy, set environmental legislation, decide for or against deregulation and even finance/subsidize the construction of national critical infrastructure.
- Developing a reliable and safe production process in order to reduce workplace incidents and industrial disasters, especially in civil nuclear and hydroelectric power.
- Continuous recruitment and retention of technical staff able to operate utilities precisely according to instructions, with a deep understanding of the equipment and the production process, and/or to repair the equipment promptly and skillfully.
- Close cooperation with suppliers of the sophisticated equipment for utility units in order to operate it correctly according to its design: there needs to be continuous exchange of risk information concerning the current behavior of the equipment in order to detect "beyond design" deviations and prevent accidents.
- Ongoing search for reliable and cost-effective supplies to maintain the lowest possible utility production costs, along with the development of fuel and spare parts reserves in the event of supply chain disruption.
- Ensuring access to long-term and low-cost investment resources to finance the construction and modernization of capital-intensive utility infrastructure.

---

393 Zheping Huang, China, home to the world's biggest cryptocurrency mining farms, now wants to ban them completely, South China Morning Post, April 9, 2019

2.4   Construction

This subsector includes the following industries according to ISIC Rev. 4:

- 41—Construction of buildings
- 42—Civil engineering
- 43—Specialized construction activities

General Description, Key Features and Industries Included in this Subsector

The work within this subsector comprises project–oriented activities to develop new built space by integration of certain land, capital and labor, or to renovate or demolish existing structures. A given project is usually characterized by long-term design and construction phases (several years) and the constructed building has often a long useful lifetime (20–100+ years). Construction is highly cyclical and volatile because of a high dependence on the national economic situation, on the central bank discount rate (since the size of mortgage payments is connected to banking interest rates) and on future economic growth—all of which are uncertain. The business is critically dependent on low-cost and long-term credit. Municipal and regional authorities play an important role since questions about land rights, construction of surrounding infrastructure and relations with neighbors of a planned building all fall under their jurisdiction. High competition characterises this subsector as a result of the relative ease of entering the market: all construction equipment for a given project can be rented, there is a wide range of suppliers, there is high mobility of labor, and so on. The unique and specific design of each project makes it impossible to automate most of the tasks involved. Therefore, the influence of employees in the construction sector is higher than in other industrial sectors: many other industrial processes lend themselves more easily to automation because of the similarity and repeatability of operations. It is possible that the development of 3D (additive) printing construction technologies could reduce the influence of builders in the medium term.

Critical Success Factors for an Organization Within the Subsector

The main critical factors in this subsector are the quality and appropriateness of design of the buildings to be constructed, the quality of construction and workmanship and the reliability of the buildings constructed over the long term. The budget and schedule performance of construction projects are also important in ensuring that they remain economically viable and in maintaining good relationships with clients. Good relations with municipal authorities and regulators allow a building organization to continue acquiring land for future construction, to develop public infrastructure near the construction site, to get prompt approval of architectural proposals and structural drawings, and to quickly resolve conflicts with neighbors over new buildings. There is also the crucial need for long-term access to low-cost capital. Maintaining a low occupational accident rate during construction work avoids excessive external control over the site from regulators, and maintains the morale of the builders and the reputation of the company. Companies in this subsector have to constantly search for and implement innovative and cost-effective construction solutions.

Stakeholders in this Subsector [394]

- Customers—30%
- Municipal authorities, regulators and neighbors of the building under construction/renovation—20%
- Investors—20%;
- Employees—15%
- Suppliers—10%
- Other—5%

Typology of Common Risks, Main Features of Major Accidents and Risk Mitigation Measures Within the Subsector

Budget Overrun and Schedule Delays on Construction Projects

This subsector has a common problem—delays with construction project leading to overrunning budgets and liquidity shortage. The inability to finish construction on time provokes surges of customer dissatisfaction and damages reputation. Robert Strange McNamara is credited for the eponymous law according to which, in frontier areas, the cost and the time estimated initially for the project should both be multiplied by about 3 to represent reality [395]. This is also known more generally as the planning fallacy [396], according to which predictions about how much time

---

[394]Our informed appraisal of the influence of each audience on a typical organization within the subsector (100% = combined influence of all audiences)

[395]Bourdaire J.M., R.J. Byramjee, R.Pattinson, Reserve assessment under uncertainty -a new approach", Oil & Gas Journal, June 10, 1985, pp. 135–140

[396]Kahneman, Daniel; Tversky, Amos, Intuitive prediction: biases and corrective procedures, TIMS Studies in Management Science, 1979, 12, pp. 313–327

will be needed to complete a future task suffer from an optimism bias and underestimate the real time needed [397]. In many projects, there are large uncertainties, and to get a project approved, the minimum cost and minimum duration are usually presented. In reality, it is typically the average cost and duration that are observed, which are several times larger.

There are a number of prominent examples of underestimation of cost and completion time. Thus the Sydney Opera House cost 14 times its projected budget and took 10 years longer than planned. The Central Artery/Tunnel Project in Boston became the most expensive highway project in the US, suffering both delay and budget overrun: it cost not $6 billion but $14.6 billion in 2006 prices ($21.93 billion including interest) and instead of being commissioned in 1998 it was completed in 2007. The Scottish Parliament cost £414 million instead of the projected £40 million to build, and took 3 years longer than agreed. Completion of the Zenit-Arena—the main football stadium in St. Petersburg, Russia—was planned for 2009 with a budget of 6.7 billion rubles, but due to delays and budget overruns a new deadline was imposed to be 2018 with expenses of more than 43 billion rubles. Construction started on reactor #3 at Olkiluoto NNP in Finland in 2000 with an estimated budget of €3 billion, and commission was planned in 2010, but after a few years, construction costs had shot up to €8.5 billion and the launch was postponed until 2018–2020; the contractors were sued over the overrun, which will make the project unprofitable for both the builder and the operators of the plant. A similar situation characterizes the construction of the Flamanville NPP in France, with an initial cost of €3.3 billion that has inflated to €12.4 billion accompanied by large commissioning delays. Reconstruction of Berlin's Brandenburg Airport should have cost no more than €3 billion and taken 5 years, enabling the airport to be reopened in 2011. However, expenses are now estimated at €6.9 billion and it will not be operational until 2020–2021.

---

[397]The following simple quantitative model illustrates the planning fallacy. Consider a project that is made of N tasks, requiring respectively a time t_1, t_2, .., t_N. Thus, the total time to complete the project is $T = t_1 + t_2 + \ldots + t_N$. For simplicity, let us assume that the random time t_i are i.i.d. (independent and identically distributed) according to a generic probability distribution function (pdf) taking the form of a power law $f(t) = m\, t_0^m / t^{1+m}$ for $t_0 < t$. In other words, t_0 is the minimum time to complete any of the N tasks, but there is the possibility for large fluctuations in the task completion time. Since each to the N tasks has a random duration, the total duration T to complete the project is also random and its pdf is easily shown to be given by $g(T) = m\, N^m\, t_0^m / T^{1+m}$ for $Nt_0 < T$. From this, the probability G(T) that the time to complete the project be larger than or equal to T is simply given by $G(T) = (Nt_0/T)^m$. Now, most planners will optimistically allocate a time $k\, t_0$, with $k > 1$, for each task, where $k-1$ is a safety margin assumed to be sufficient to address unforeseen issues or difficulties. Then, the estimated time for completion is $T_p = N(kt_0)$. The probability that the project takes more time than the prediction T_p is thus $G(T_p) = (Nt_0/T_p)^m = 1/k^m$. Ambitious planners would put a 20% over time flexibility (k = 1.2), leading to the following estimates> For m = 1.2, 80% of the projects take longer than planned; for m = 1.5, 76% of the projects take longer than planned; for m = 2, 69% of the projects take longer than planned; for m = 3, 58% of the projects take longer than planned. Typical values for the exponent m are usually around 1 to 2, which rationalise the ubiquitous planning fallacy.

Modern project management methodology and instruments, combined with well-designed blueprints, a thorough geological survey of the potential construction site, and adequate supply chain management should allow for more accurate time and budget forecasting even on complex projects. It is also necessary to develop a frank assessment of the uncertainties within the projects and report not only estimations of cost and durations but also their uncertainties in the decision-making process.

Real Estate Bubbles and Economic Shocks

The construction sector is heavily dependent on domestic economic growth and financial policy. For example, a major real estate and construction boom occurred in the United States from 2002 to 2007, when the financial sector was deregulated and the subprime mortgage "*securitization pipeline*" was created. This was a series of financial sleights of hand that allowed banks and financial organizations to provide financing to "subprime" borrowers—who would traditionally not have been seen as credit-worthy—without apparently putting themselves at too great a financial risk. Moreover, the Bush administration reduced the down payment required to apply for a mortgage from 3% of the price of the house being purchased to just US$500 (the "*Zero Down Payment Initiative*"). This prompted a massive demand for mortgages from millions of subprime borrowers, leading to years of rising house prices and a nationwide construction boom [398]. The eventual failure of the securitization pipeline and resulting burst of the American real estate bubble in 2007–2008 initiated the global financial and economic crisis of 2008–2009 [399], which caused the most severe recession in over 50 years and severely harmed the US construction industry. There was also a construction boom in Japan in the 1980s, followed by the so-called "*two lost decades*" of poor performance of the national economy and construction sector in the 1990s and 2000s. The current overheated Chinese real estate market is also a headache for the national government.

In order to forecast the movement of the real estate market and national economic development, construction companies need to continuously monitor and reassess macroeconomic data, the national central bank's position on interest rates, the strength and stability of the currency and the real estate market, the credit conditions the financial sector are setting to borrowers, and general trends in earnings.

---

[398]D. Sornette and P. Cauwels, 1980–2008: The Illusion of the Perpetual Money Machine and what it bodes for the future, 2014, Risks 2, pp. 103–131, http://ssrn.com/abstract=2191509

[399]Markus K. Brunnermeier, Deciphering the Liquidity and Credit Crunch 2007–2008, Journal of Economic Perspectives, 2009, 23(1), pp. 77–100

Crises Induced by Mistakes in Building Design and Poor Quality of Materials or Construction Work

The reputation and long-term business of a construction company will be damaged most severely if a building collapses because of design errors, poor-quality construction materials or bad workmanship. An accident like this leads to a criminal trail against the architects, terminates their careers and can push the architectural bureau and even the construction company to bankruptcy, due to the damaged reputation and huge compensation claims from relatives of victims and the owners of the collapsed building. There can even be claims from the owners of other buildings designed and constructed by the same companies, because the investigation of such collapses usually includes scrupulous analysis of the design and construction of all buildings by the suspect organizations.

Joelma Building Fire (Brazil, 1974) The fire occurred in a 25-story building in Sao Paulo after a short-circuit in the air-conditioner on the eleventh floor. At least 179 people perished in the fire, which was exacerbated by the highly combustible materials used in the construction of the building: plastics, wooden walls, etc. Finding incombustible materials to build skyscrapers is a challenge for the whole construction industry. Numerous conflagrations have been caused by builders using inappropriate materials, some of the worst being the Transport Tower, Astana, Kazakhstan, 2006; the Olympus Tower, Grozny City, Russia, 2013; the Torch skyscraper, Dubai, UAE, 2016; the Address Downtown Dubai tower, Dubai, UAE, 2017 and Grenfell Tower, London, UK, 2017.

Citigroup Center in Manhattan (USA, 1977) This distinctive skyscraper complex was already complete when it became known to the chief architect that the subcontractor had changed the way the internal bracing framework was connected—and to his horror, he realized this made the whole building vulnerable to collapse if hurricane force winds hit New York, which was likely at least every 16 years. Large scale emergency plans were made in complete secrecy and, in order to mitigate the risk, a team of construction workers immediately set about welding special plates onto all the vulnerable connections.

Hyatt Regency Walkway Collapse (USA, 1981) On July 17, 1981, more than a thousand people gathered for a tea-dance party in the lobby of the Hyatt Regency Hotel in Kansas City, Missouri. Tens of them observed the dance from the walkways of the hotel. Unexpectedly, two walkways collapsed onto the main floor, killing 114 visitors and injuring more than 200. An inquiry after the accident concluded that there had been a mistake in the design of the rods and locking tabs which suspended the collapsed walkways from the ceiling. The design of the rods had been changed by the manufacturer without the formal approval of the designers of the project—though during the investigation, it became clear that even the original design would not have been strong enough to support the number of people the walkways might have to carry in exceptionally busy circumstances. After the disaster, many of the key players in the construction of the hotel lost their licenses and faced legal action from victims.

Collapse of the Hotel New World (Singapore, 1986) This six-storey hotel collapsed on March 15 1986, taking the lives of 33 people. The ensuing inquiry found that the design of the building had been developed by an unqualified draughtsman: when designing the supporting structure, he had estimated the weight of the building's contents and inhabitants (the live load) but not the weight of the building itself (the dead load). In addition, the managing director of the building had intervened in its design and construction, stipulating the use of cheaper materials and installing additional heavy equipment after the building was completed (he was one of the people who perished during the collapse). There were early warnings of potential collapse when cracks were discovered throughout the building, but people were not evacuated in time.

L'Ambiance Plaza Collapse (USA, 1987) 28 construction workers perished in the collapse of a nearly completed 16-story residential building in Bridgeport, Connecticut. The Plaza was built using the lift slab construction method, where each new concrete floor or roof slab was cast on top of the previous slab, then raised up to the next level with hydraulic jacks. Using this method put extra stress on the floor slabs, causing cracking of the concrete which ultimately led to the collapse. After the accident, this construction method was banned while a nationwide investigation of the safety of the method took place.

Massive Collapse of High-Rise Buildings During the Armenian Earthquake (USSR, 1988) On December 7, 1988, a devastating earthquake, measuring 6.8–7.0 on the Richter Scale, occurred in the then Soviet republic of Armenia. More than 25,000 people died and more than half a million lost their homes. The highest mortality rate was registered not in older, lower-rise buildings—many of which survived the disaster—but in collapsed nine-storey buildings, which had been erected relatively recently. Despite the fact that Armenia is located in the Caucasus Mountains, formed by the collision of the Arabian and Eurasian tectonic plates, Soviet engineers had underestimated the seismic risk of the area when they were planning new housing. They reckoned that an earthquake of greater than level 7 on the Medvedev-Sponheuer-Karnik scale (MSK-64) was not possible in Armenia. This misjudgment, combined with the pressure of demand for accommodation in the republic, influenced the decision to compromise on quality of construction in favour of speed and economy: in some cases, the percentage of sand in the concrete was above the legal maximum. The results were catastrophic. For example, in Leninakan (now Gyumri), the second largest city of the republic—where the earthquake hit 10 on the MSK-64 scale—94% of the nine-storey buildings collapsed [400,401].

---

[400]Lessons and questions emerge from Armenian quake, Science News, January 21, 1989
[401]"Buildings-killers" of the Spitak earthquake, Sputnik, April 12, 2016

Savar Building Collapse (Bangladesh, 2013) 1127 workers at garment factories died when the Rana Plaza building collapsed on them due to the illegal construction of additional floors and deployment of heavy machinery, which generated excessive vibration.

Haiti Earthquake (2010) This disaster claimed at least 100 thousand lives. The earthquake, with a moment magnitude of 7.0, killed one in 15 of the people affected; in the same year, an earthquake in Chile with a magnitude of 8.8, 500 times more powerful (in energy released) than that in Haiti, killed only one in every 595 people affected. According to the United Nations Office for Disaster Risk Reduction, the cause of the tremendously high mortality figures in Haiti was "*poorly built buildings with limited regulations*" [402], illustrating the adage well-known to geologists and seismologists: "*Earthquakes don't kill people, buildings do*".

The main accident prevention measures at the design and construction stage include the involvement of independent expertise in the design of a building, the choice of materials and the quality of construction work, comprehensive testing of materials, strong government oversight of design and construction, continuing professional development of architects, engineers and state regulators, careful selection of construction staff, and the worldwide exchange of information about innovative solutions and good practice.

Accidents Caused by Unstable Foundations

Sometimes geologists, architects and builders misjudge the stability of the ground they are planning to build on; this can lead to collapse during erection of the buildings, or later when other factors compound the existing instability.

Highland Towers Collapse (Malaysia, 1993) On December 11, 1993, 48 residents died after the collapse of one of three 12-storey apartment blocks in the vicinity of Kuala Lumpur. The buildings were constructed at the base of a steep terraced hill, and their foundations were reinforced by poorly designed retaining walls, with a drainage system to divert a stream that had flowed through the site. Later, the hillside above the blocks was stripped of vegetation for a new building project, overloading the existing drainage system; the ground around the blocks became increasingly waterlogged. After a prolonged period of heavy rain, a mudslide overwhelmed the retaining walls and flooded the car park at the base of the blocks. The weight of the mud pressed against the already weakened foundation of one of the blocks, and the pilings of the building were no longer solidly supported by the ground around them. Finally, the foundations gave way and the block collapsed. Later, the other two blocks were evacuated permanently, amid concerns that the disaster could be repeated.

---

402 2010–2011 World Disaster Reduction Campaign, United Nations Office for Disaster Risk Reduction, November 8, 2010, https://www.youtube.com/watch?v=PqrUJXgwrnU

Lotus Riverside Complex Collapse (China, 2009) One block of a nearly completed complex in Shanghai fell after builders had been excavating soil to create an underground car park on one side of the block and storing the soil on the opposite side. During heavy rains, the stored soil was saturated with water, putting huge pressure on the foundation and concrete pilings of the block. The block collapsed, killing one of the site workers.

To respond to this challenge requires comprehensive analysis of the soil under any planned new structure, serious protective measures if the ground is found to be unstable, and constant monitoring of soil movement in any environment where the stability of foundations could be affected by changes nearby.

Accidents During the Construction Process

There are likely to be minor incidents on any construction site during the implementation of a project. However, in some cases accidents during construction can lead to severe injuries, structural damage and even the deaths of builders or neighbours.

Panama Canal Construction (1880–1914) During the most ambitious building project of the late nineteenth century, more than 30,000 workers died—mainly from yellow fever and malaria, a minority from construction incidents.

Willow Island Disaster (USA, 1978) and Fengcheng Power Station Scaffold Collapse (China, 2016) During the construction of a cooling tower for the coal-fired Pleasants power plant, a load of unpoured concrete fell and pulled over the crane which was hoisting it. Some of the most recently set concrete started to collapse, pulling down with it scaffolding which was bolted to the failing concrete and killing 51 builders. In 2016, a quite similar event cost the lives of 74 Chinese workers at Fengcheng Power Station, when a scaffolding platform collapsed.

Garley Building Fire (Hong Kong, 1996) and Shanghai Fire (China, 2010) These fires claimed the lives of 41 and 58 people respectively, and had a similar cause: a blaze of sparks during welding work, which set fire to scaffolding erected around the buildings.

Mecca Crane Collapse (Saudi Arabia, 2015) There have been hundreds of cases of crane collapses worldwide, but this disaster in the Muslim holy city of Mecca is exceptional because it happened when the city was packed for the hajj—the annual pilgrimage of believers. 111 people died and around 400 were injured when a poorly supported crane collapsed onto the floor of the Grand Mosque during strong winds.

Las Vegas Construction Workers' Strike (USA, 2008) During the development of the City Centre Las Vegas, which cost US$9.2 billion, six builders died. After the sixth death in 2008, workers walked out in protest, demanding improved safety conditions on construction sites.

Gotthard Base Tunnel (Switzerland, 2000–2012) Despite the good reputation of Swiss engineers and builders, fatal accidents have occurred even on Swiss construction sites. Thus, during the 12-year construction of the world's longest railway tunnel beneath the Swiss Alps, nine builders were killed. Previously, 19 had lost their lives during the digging of the St. Gotthard Road Tunnel—also at the time the longest ever built.

Improving construction site safety requires the careful selection of construction staff, and ongoing employee preparation—this must include safety training to ensure the right balance between construction productivity and occupational safety, and to explain how safety measures are changing with new construction technology. Construction companies also need a sophisticated compensation program, which motivates staff to comply with safety rules and to respond to critical feedback on their working practices.

Crises Induced by Conflict with Local Residents

During any construction process, residents living nearby will suffer from the noise of construction equipment and vehicles, from dust and dirt, from vibration levels which can sometimes damage nearby buildings, from the polluting exhaust of heavy machinery, from parking shortages, and possibly from the bad behavior of some of the site workers: littering, pilfering, debauchery, and even loud music at night if workers are staying on site.

In order to prevent disputes with local residents, a construction company should be in good communication with local authorities, key figures in the local community, and the residents of nearby buildings themselves. The company should make a commitment to comply strictly with agreed working hours on the site, to compensate for any damage to residential buildings in the event of the construction affecting structures nearby, to keep good control of its builders, to reconstruct damaged roads or pavements, and to respect the wishes and complaints of local residents.

Key Risk Mitigation Measures in Construction

- Maintaining the highest quality of work over the whole life-cycle of a construction project (from the design of a building to the beautification of the surrounding area after its completion).
- Realistically assessing the required budget, resources, and schedule at the initiation stage of a construction project, and keeping good control over these while the project is under way.
- Maintaining strong relations with municipal and regional authorities to ensure access to land rights, construction permits, infrastructure connections and the goodwill of those living near a planned building.

(continued)

- Delivering a low occupational incident rate through the careful selection and training of staff and the use of safe and well-functioning equipment.
- Deeply understanding the economic situation, monetary policy and trends in the banking sector to inform an accurate assessment of the prospect of a construction business in the medium term.
- Securing and maintaining access to low-cost and long-term financial resources.

2.5 Agriculture

This sector includes the following industries according to ISIC Rev. 4:

- 01—Crop and animal production, hunting and related service activities
- 02—Forestry and logging
- 03—Fishing and aquaculture

General Description and Key Features of the Sector

Modern agriculture, in developed and in some more advanced developing countries, does not resemble the old traditions of farming: growing crops and breeding livestock organically, catching fish and cutting wood through the manual work of several peasants united in a family farm. Nowadays, large-scale highly automated agriculture holdings do not "cultivate" food, but "produce" it in a manner more akin to advanced industrial production than traditional farming: products of a uniform quality, size and characteristics are obtained by genetic interference and the influence of external factors is minimized by advanced fertilization, use of pesticide and herbicide treatment and artificial feeding of animals, poultry and fish. Even the nomenclature of the ISIC classification system reflects this: the key industry in the agriculture sector is defined in ISIC Rev. 4 as "crop and animal production". The implementation of industrial principles and automation has multiplied productivity in the sector during the last 100 years: for instance, in the United States, corn yield has increased from 20 bushels/acre in 1920 to 200 bushels/acre nowadays. A 100 years ago, a US farmer produced enough food to feed 8–15 people, compared to 140 people nowadays [403]. Vertical integration, popular in the industrial sector—especially the oil, gas and mining industries—is also a popular trend in modern agriculture. Currently in the United States, four agriculture holdings produce 85% of all the beef, 65% of all pork and almost half of all chicken; the holdings not only breed livestock, but manufacture meat products for retail and corporate consumers

---

[403]Michael Pollan, Farmer in Chief, The New York Times, Oct. 9, 2008

[404]. And on a global scale, three companies control 55% of seeds, 51% of agrochemicals, 49% of farm equipment and 31% of fertilizers [405]. The pricing of some positions within the sector is established on the commodity exchange. Because agriculture is a matter of national security, the sector depends on national food security policy, legislation on genetically modified organisms, price interventions, and quotas for food production and food export. The business activity is by nature localised, because performance depends heavily on local climate conditions. Municipal and regional authorities are also important because land rights, the construction of surrounding infrastructure, local employment, etc. are under their jurisdiction.

Analysis of the key features of this sector confirms that there are many similarities between corporate agriculture and the production of raw materials, especially in the oil, gas and mining subsectors. They share a similar profile of stakeholders; both involve highly localized production; both are highly regulated because of their importance to national security, the demands they make for access to land and water resources, their need for government subsidy and the fact that government may wish to restrict their exports; both cause localized damage to the environment from production—in the case of agriculture through pollution by fertilizers, pesticides, and herbicides and through soil erosion, overfishing, and deforestation; prices for some products in both sectors can be fixed in similar ways; both involve high automation to increase productivity; and both are under (entirely legitimate) pressure from environmental organisations. Because of these parallels, we have included this sector in the chapter on industrial production.

Critical Success Factors for an Organization Within the Subsector

Product safety and quality. Understanding customer needs and growing products required by the market. Low production cost. Explaining to authorities why agriculture needs government support, which includes restrictions on foreign imports, direct subsidies, access to cheap land and surrounding infrastructure, special tax regimes, etc. Finding convenient location(s) with a suitable climate for agricultural activity, or overcoming the influence of adverse weather conditions on yield. Application of advanced technologies for cultivation and fighting disease to ensure a good harvest. Low losses of during cultivation, transportation and storage of agriculture products. Access to long-term and low-cost capital to mitigate seasonal factors and survive lean years. Motivating field staff to follow exact instructions for safe cultivation.

---

[404]Christopher Leonard, How the Meat Industry Keeps Chicken Prices High, Slate, March 3, 2014

[405]Mega-Mergers in the Global Agricultural Inputs Sector: Threats to Food Security & Climate Resilience, ETC Group, September 2015, http://www.etcgroup.org/sites/www.etcgroup.org/files/files/etcgroup_agmergers_22oct2015.pptx_.pdf

Stakeholders in this Subsector [406]

- Customers: food processing companies, distributors (wholesale traders), exporters, retail networks and direct retail customers—30%
- Authorities (local, regional and national levels)—20%
- Suppliers of seeds and young animals, hardware vendors and service providers—15%
- Employees—15%
- Investors—10%
- Local communities—5%
- Other—5%

Typology of Common Risks, Main Features of Major Accidents and Risk Mitigation Measures Within the Sector

Biological Risks and Product-Related Crises

The most damaging crises in the sector occur when agricultural products are found to have a negative impact on the health of consumers because of the presence of aggressive biological agents, chemical contamination (from antibiotics to pesticides), inadequate labeling or the usage of genetically modified organisms whose safety is still in dispute. The main accusations directed at agriculture companies and even governments following agricultural accidents are usually connected with low quality of production, violations of regulations on safe cultivation or negligence towards disease prevention measures. Animal disease outbreaks can destroy livestock and even pose a direct threat to humans either through the mutation of viruses, as in the case of bird flu, or through the consumption of contaminated food.

Fusarium banana wilt (worldwide) According to the UN Food and Agriculture Organization (FAO), bananas are the most traded fruit in the world, the eighth most important food crop in the world and the fourth most important food crop in less-developed countries. For many decades, the soil-borne Fusarium fungus, commonly known as Panama disease or banana wilt, has posed *"the world's greatest threat to banana production"* because it is resistant to fungicides and can be active in soil for more than 30 years [407]. The fungus destroys the banana plant by attacking the xylem vessels which are responsible for the transportation of water and nutrients from the roots to the leaves. Between 1903 and the 1950s, the Tropical Race 1 (TR1) strain of the fungus wiped out the Gros Michel cultivar which had dominated the

---

[406]Our informed appraisal of the influence of each audience on a typical organization within the subsector (100% = combined influence of all audiences)

[407]FAO plants new efforts to protect bananas under disease threat, FAO, October 3, 2019, https://news.un.org/en/story/2019/10/1048532

world banana trade. Eventually, plantations of Gros Michel bananas in Latin America, the Caribbean and South Asia suffered heavy damage. The United Fruit Company—which had a near monopoly in banana production, owned many banana plantations in Central America and was the largest importer of fresh fruit to North America and Europe at that time—was almost bankrupted due to the Panama disease. The company was forced to close several divisions in Panama (1926), Honduras (1939), Costa Rica (1940, 1956), Nicaragua (1942), and Guatemala (1955); they also faced major losses in Jamaica [408]. Cavendish, a cultivar which is resistant to TR1, replaced the Gros Michel and became the most traded subgroup of bananas worldwide, accounting for around 90% of global trade. However, despite their resistance to TR1, Cavendish bananas are vulnerable to a new strain of the Fusarium fungus, TR4. Over recent decades TR4 has destroyed banana plantations in South East Asia, Northern Australia, the Middle East and Mozambique. In August 2019, a first case of TR4 contamination was registered in Columbia. There is a risk of fast spread of TR4 across Latin America because of the constant movement of teams of plantation workers across different countries and weak enforcement of quarantine measures in some less-developed countries. The spread of this strain of the fungus in Latin America—where most bananas for the global market are produced—could hurt many agricultural companies and small farmers, reduce the revenue of whole economies dependent on banana exports, and drive up retail prices on bananas globally. Goldfinger is a hybrid banana cultivar which is resistant to TR4 and to black sigatoka, another prevalent destructive fungus; it could replace Cavendish, but Goldfinger bananas do not have the same commercial qualities as Cavendish [409].

"Mad cow disease" Outbreak (UK, 1980s) In the mid 1980s, in a context where the UK domestic meat industry was highly profitable, the British government withheld information from the public at all stages and even tampered with evidence of the existence of bovine spongiform encephalopathy (BSE, nicknamed "mad cow disease") in British beef [410]. The disease led to more than 200 deaths in the UK and other counties over the following decades [411]. Similar concealment occurred during the Irish pork crisis in 2008, when a supplier delivered fodder contaminated by dioxins to pork and beef farms, causing some of the meat produced there to be unfit for consumption. When the contamination was discovered, the Irish authorities tried to downplay the seriousness of the situation—especially in relation to the export of Irish meat.

---

[408]Steve Marquardt, "Green Havoc": Panama Disease, Environmental Change, and Labor Process in the Central American Banana Industry, The American Historical Review, Feb., 2001, Volume 106, No. 1, pp. 49–50

[409]Marty McCarthy, Banana fungus TR4 fought by Australian scientists with 'mutant' plants, ABC, Oct 14, 2016, https://www.abc.net.au/news/2016-10-15/scientists-battle-to-beat-banana-fungus-before-it-spreads/7929772

[410]Richard Lacey, Mad Cow Disease: The History of BSE in Britain, Cypsela, 1994, p. xx

[411]Variant Creutzfeldt-Jakob disease, The World Health Organization, Febrary 2012

E. coli O157:H7 Bacteria in Dole Vegetables (United States, 2005–2006) Three people died and 205 fell ill in an outbreak of E. coli infection in 2006. The bacteria were traced to a batch of spinach from a farm in California, packed under the brand of Dole Baby Spinach. Previously in 2005, the same bacteria had been found in Dole lettuce.

Salmonella in Eggs from Iowa Farms (United States, 2010) More than 80 billion eggs are produced annually in the United States. In 2010, salmonella bacteria in eggs from two Iowa producers led to 2000 cases of illness. The producers in question— Wright County Egg and Hillandale Farms, both connected with DeCoster, the third largest egg producers in the country—were forced to recall 550 million eggs from shelves, restaurants and other service points. The main cause of the outbreak was the flagrant violation of sanitary rules: *"FDA investigators found salmonella all over the farms, along with filthy conditions including dead chickens, insects, rodents and towers of manure"* [412].

Salmonella in Cargill Turkey (United States, 2011) 136 people were infected by Salmonella Heidelberg and one died due to consumption of ground turkey produced by Cargill Meat Solutions Corporation. The infections continued even after the company had recalled more than 16,000 tons of ground turkey products, and a second recall was needed 2 months later [413].

E. coli O104:H4 Outbreak (Mainly Europe, 2011) More than 4000 people fell ill and 50 died in this outbreak of E. coli infection. The infection started and was mainly focused in northern Germany, but affected 15 other countries. Initially, the German food safety authorities thought that Spanish cucumbers were the initial cause of the outbreak, but later it was determined that fenugreek seeds imported from Egypt for sprouting at an organic farm in Saxony were responsible [414].

From Strategic Food Management and Subsidies to Global Pandemics and World Health Risks (United States) After World War II, realising the strategic nature of food supplies, the US government decided to subsidise the production of corn. As a consequence, production skyrocketed, far exceeding the demand for human consumption. Corn producers quickly found a market for the excess, selling to cattle producers. The concentrated energy in corn (as with soy and other types of so-called *"by-product feedstuff"*) led to increased cattle production in concentrated animal feeding operations. As corn is a high-starch, high-energy food, it cut the time needed to fatten cattle, increased production efficiency and yield from dairy cattle, and reduced the area needed to support the energy requirements of large

---

412Disgraced egg industry titan charged over 2010 salmonella outbreak, The Associated Press, May 21, 2014

413Multistate Outbreak of Human Salmonella Heidelberg Infections Linked to Ground Turkey, US Centers for Disease Control and Prevention, November 10, 2011

414Outbreak of Escherichia coli O104:H4 Infections Associated with Sprout Consumption — Europe and North America, May–July 2011, US Centers for Disease Control and Prevention, December 20, 2013

herds. The problem is that the large ruminant animals, endowed with four-compartment stomachs to digest the cellulose of grass and plants, do not naturally use grains in their diet. So cattle fed mostly with grain developed lower (more acidic) colonic pH—and because cattle are a natural reservoir for the pathogenic Escherichia coli (E. coli) bacterium, they soon developed an acid-resistant strain. Grain-fed cattle have a million times more acid-resistant E. coli than cattle fed on hay [415]. The more acidic human gut is a natural barrier to food-borne pathogens, but E. coli can survive at pH 2.0 if they are grown under mildly acidic conditions—such as the more acidic intestines of corn-fed cattle. There have been outbreaks of diarrhea in the United States caused by highly virulent strains of enterohemorrhagic E. coli O157:H7. These strains produce toxins similar those made by the Shigella bacterium which causes dysentery, and infection can result in death, particularly in young children and the elderly. Various foods have been implicated in outbreaks, including ground beef, raw milk, apple cider, and most recently fermented hard salami.

To combat this risk, antibiotics were widely introduced to control E-Coli in cattle. It was quickly found that antibiotics have the apparent side-benefit of being growth promoters for food animals. At low doses, they are now added to almost all animal feeds and are considered to improve the quality of the product, with a lower percentage of fat and a higher protein content in the meat. Other benefits of the use of antibiotic growth-promoters include the control of other zoonotic pathogens such as Salmonella, Campylobacter and enterococci. But this generalised use of antibiotics in the production of cattle is now recognised to have a dangerous unintended consequence. It has created a selection pressure for pathogenic bacteria that have become resistant to antibiotics—including some of those used in clinical or veterinary practice—thus compromising the continued efficacy of antimicrobial chemotherapy [416].

Moreover, the overproduction of cheap corn used to feed cattle has effectively subsidised meat by making it relatively cheap: from 1995 to 2005, 73.8% of all US federal food subsidies went to meat and dairy [417]. As a consequence, fruit and vegetables are relatively more expensive, making them inaccessible to the poorest in US society—who are therefore much more likely to be unhealthy [418]. Fast food is the cheapest food one can buy in the US, yet the cost of that food to the environment

---

[415]Francisco Diez-Gonzalez, Todd R. Callaway, Menas G. Kizoulis, James B. Russell, Grain Feeding and the Dissemination of Acid-Resistant Escherichia coli from Cattle, Science, 1998, 281 (5383), pp. 1666–1668

[416]EFSA (European Food Safety Authority) and ECDC (European Centre for Disease Prevention and Control), 2014. The European Union Summary Report on antimicrobial resistance in zoonotic and indicator bacteria from humans, animals and food in 2012. EFSA Journal, 2014, 12(3)

[417]Catherine Rampell, Why a Big Mac Costs Less Than a Salad, The New York Times, March 9, 2010, https://economix.blogs.nytimes.com/2010/03/09/why-a-big-mac-costs-less-than-a-salad/

[418]Barbara A. Laraia, Tashara M. Leak, June M. Tester and Cindy W. Leung, Biobehavioral Factors That Shape Nutrition in Low-Income Populations - A Narrative Review, American Journal of Preventive Medicine, 2017, 52 (2), Suppl. 2, pp. S118-S126

and public health is probably the highest [419]. Ultimately, a heavily meat-based diet requires more energy, land, and water resources than a lacto-ovo-vegetarian one—which, in this sense, is more sustainable [420]. To convert grain protein into meat protein, it takes roughly 7 kg of grain to produce a 1-kg gain in live weight for cattle in feedlots [421]. But this figure needs to be treated with caution, as the efficiency with which different animals convert grain into protein varies widely [422].

In summary, a strategic decision to ensure food supply security has played a significant role in bringing on a general obesity *"epidemic"* [423] with dramatic and increasing health impacts [424], in the frightening rise of bacterial antibiotic resistance, and in a wasteful overproduction of food that may add to the stress on future human sustainability.

Biological Risks of Beef (United States, 2000s–2010s) In 2015, the non-profit organization Consumer Reports ran tests for bacteria in American ground beef. The researchers procured more than 200 kg of beef from 103 groceries, supermarkets and natural food stores in 26 cities. The results were grim: all of the beef was contaminated by enterococcus and/or non-toxin-producing E. coli, which implies contamination of the meat by cow faeces; almost 20% was contaminated by Clostridium perfringens, which leads to around a million cases of food poisoning every year in the United States alone; 10% contained potentially dangerous Staphylococcus aureus bacteria; and 1% contained salmonella, which causes 1.2 million infections and 450 deaths annually. In three samples, there was a strain of Staphylococcus aureus that is resistant to methicillin and other antibiotics—these bacteria kill 11,000 Americans annually. Extrapolating these results to all beef on sale in American stores and restaurants, they clearly represent a huge danger to tens of millions of consumers.

The main cause of the contamination of such a high proportion of meat is the very fast slaughter process—up to 400 cows per hour—in modern meat processing plants. Quite apart from its disgusting brutality, this mechanized slaughter does not allow for the efficient sanitization of the resulting meat, which picks up bacteria from the hides or digestive tracts of the cows and during the mixing of meat from different cows. Ground beef contaminated by E. coli O157 bacteria was the main cause of

---

[419]Bonnie Ghosh-Dastidar, Deborah Cohen, Gerald Hunter, Shannon N. Zenk, Christina Huang, Robin Beckman and Tamara Dubowitz, Distance to Store, Food Prices, and Obesity in Urban Food Deserts, American Journal of Preventive Medicine, 2014, 47 (5), pp. 587–595

[420]David Pimentel and Marcia Pimentel, Sustainability of meat-based and plant-based diets and the environment, The American Journal of Clinical Nutrition, 2003, 78 (suppl), pp. 660S–663S

[421]conversion ratio of grain to beef based on Allen Baker, Feed Situation and Outlook staff, ERS, USDA (see http://www.earth-policy.org/books/pb2/pb2ch9_ss4)

[422]Tim Worstall, It Does Not Take 7 kg Of Grain To Make 1 kg Of Beef: Be Very Careful With Your Statistics, Bloomberg, September 3, 2012

[423]Obesity Update 2017 report, Organisation for Economic Co-operation and Development, http://www.oecd.org/health/health-systems/Obesity-Update-2017.pdf

[424]The GBD 2015 Obesity Collaborators, Health Effects of Overweight and Obesity in 195 Countries over 25 Years, The New England Journal of Medicine, 2017, 377 (1), pp. 13–27

nearly 80 outbreaks between 2003 and 2012, in which 1144 people were infected and five died. A representative of Food & Water Watch was pessimistic about the prospects for improvement: *"USDA [the US Department of Agriculture] has a presence in these plants to do inspections—though it's against the companies' wishes. The economic power of the Big Four [Tyson Foods, Cargill Meat, JBS USA, National Beef, which produce 80% of beef in the US] gives them a lot of political weight to push back against USDA inspectors' efforts to enforce existing rules and to fight against any tighter safety standards being enacted. The sheer volume of beef that big-company plants crank out means that a quality control mistake at a single plant can lead to packages of contaminated beef ending up in stores and restaurants across 20 or 30 states"* [425].

African Swine Fever (China, Since 2018)   In August 2018, the first case of African swine fever infection in China's pig population was recorded in the northern province of Liaoning. Within just a year, the disease had spread throughout China. The fever became a national food disaster: according to the Ministry of Agriculture, over the following year 140 million (!) animals were culled, bringing the pig population down by 32% [426]. The outbreak caused direct losses of $140 billion to the national agriculture sector [427]. Because pork accounts for 60% of all meat consumed in the country, the retail price shot up by 70% within a year of the outbreak. To increase pork supply and control retail prices, the Chinese authorities were forced to open national food reserves and start importing the meat from uncontaminated countries. The CEO of Tyson Foods, the largest meat-packing company in the United States, observed: *"In my 40 years in the business, I've never seen an event that has the potential to change global protein production and consumption patterns as African swine fever does"* [428]. Previous experience of dealing with the fever in Spain shows that eradicating the virus on a national scale could take 35 years. In the case of highly populated China, where people, domestic animals and wild boars live closely together, this task could be very difficult. From China, the virus has spread to other countries in Southeast Asia (Vietnam, Laos, Thailand and Cambodia).

Questionable Development of Genetically Modified Organisms   The world population exploded in the twentieth century, from around 1.6 billion in 1900 to 6.7 billion by 2000. It passed the 7 billion mark in 2011, and continues to increase by about *"one Germany"* (84 million people) per year. However, Thomas Malthus' famous prediction—that global famine would inevitably result because agricultural

---

[425] Andrea Rock, How Safe Is Your Ground Beef? Consumer Reports, December 21, 2015, https://www.consumerreports.org/cro/food/how-safe-is-your-ground-beef

[426] Jim Long, China pig herd continues massive decline, August 21, 2019, https://thepigsite.com/news/2019/08/jim-long-pork-commentary-china-pig-herd-continues-massive-decline

[427] Sun Liangzi, Tang Ziyi, Deadly Pig Disease Has Cost China More Than $140 Billion: Professor, Caixin, September 25, 2019, https://www.caixinglobal.com/2019-09-25/deadly-pig-disease-has-cost-china-more-than-140-billion-professor-101466143.html

[428] Aporkalypse now, The Economist, May 25, 2019

output could not possibly grow at the same exponential pace—did not in fact come to pass. What prevented this was the Green Revolution: a spectacular surge of scientific and technological applications in agriculture including a scientific approach to crop breeding through the careful selection of seeds, the increasing use of chemical fertilizers and pesticides, and large-scale development of the infrastructure for irrigation. Together, these changes maximized yields and freed millions of people from back-breaking agricultural labor. The selective breeding element of the Green Revolution led in due course to more sophisticated interventions, a *"gene revolution"* that further increased harvests and reduced production costs.

Over recent decades, a new field of biotechnology has developed, creating plants whose genetic material (DNA) has been manipulated far beyond what would occur through natural mutation [429]. In developing genetically modified organisms (GMOs), biotechnologists are typically aiming to enhance: (I) the crop's capacity to withstand a specific herbicide—so that when crops are dusted with that herbicide, all plants except the GMOs will die; (II) its resistance to insects—the genetic manipulations will turn the plants into living insecticides, which are toxic for plant pests; or (III) other special characteristics such as drought tolerance, shelf life, appearance, size, etc. Obviously, such fine-tuning can confer huge economic advantages. But the long-term impact for human health of consuming GMOs is hard to predict, and even scientific opinion—let alone public opinion—is sharply divided [430].

Firstly, while the long-term sustainability of introducing GMOs remains in question, their profitability in the short-term is all too compelling. For example, the Director of Corporate Communications for Monsanto, the leading multinational in agrochemicals and agricultural biotechnology, explained to a New York Times columnist: *"Monsanto should not have to vouch for the safety of biotech food. Our interest is in selling as much of it as possible. Assuring its safety is the FDA's job"* [431]. Claire Robinson, a researcher who worked with two leading genetic engineers to compile an evidence-based examination of the field entitled "GMO Myths and Truths", concluded that *"Claims for the safety and efficacy of GM crops are often based on dubious evidence or no evidence at all. The GMO industry is built on myths. What is the motivation behind the deception? Money. GM crops and foods are easy to patent and are an important tool in the global consolidation of the seed and food industry into the hands of a few big companies. We all have to eat, so selling patented GM seed and the chemicals they are grown with is a lucrative business model"* [432].

---

[429] 20 questions on genetically modified foods, World Health Organization, 2002

[430] John Fagan, Michael Antoniou, Claire Robinson, GMO Myths and Truths, An evidence-based examination of the claims made for the safety and efficacy of genetically modified crops and foods, Earth Open Source, 2014, second ed., pp. 22–23

[431] Phil Angell (Monsanto) cited in Michael Pollan, Playing God in the Garden, The New York Times Magazine, October 25, 1998

[432] GMO Myths and Truths, Earth Open Source, May 19, 2014, http://earthopensource.org/index. php/reports/gmo-myths-and-truths

Secondly, government regulators have an altogether too cozy relationship with the industry. People still working in the industry have been given leading roles in regulatory bodies, and thus been directly involved in the regulation of GM products. The regulation of the GM food-supply chain in the US was in the hands of senior government representatives during the Clinton, Bush and Obama administrations; many of those senior regulators came straight from the boards of Monsanto and other biotech companies [433,434]. And in Europe, the same conflicts of interests have been rife, with the industry's top management filling many executive roles in regulatory bodies [435,436,437].

Thirdly, the regulatory authorities seem more interested in deregulation, in protecting the industry from unnecessary restriction rather than protecting consumers from possible danger. Neither the FDA nor the EFSA (European Food Safety Authority) conduct their own independent tests on GM products: instead they are simply accepting research either carried out internally by biotech companies or funded by them. They appear to take the view that, since it is in the industry's interest not to have health scares, its major players will protect their business by ensuring the safety of their products. The FDA states in a policy document that *"Ultimately, it is the food producer who is responsible for assuring safety"* and the EFSA writes *"It is not foreseen that EFSA carry out such [safety] studies as the onus is on the [GM industry] applicant to demonstrate the safety of the GM product in question"* [438]. Professor David Schubert of the Salk Institute for Biological Studies, an independent research body, commented: *"One thing that surprised us is that US regulators rely almost exclusively on information provided by the biotech crop developer, and those data are not published in journals or subjected to peer review... The picture that emerges from our study of US regulation of GM foods is a rubber-stamp 'approval process' designed to increase public confidence in, but not ensure the safety of, genetically engineered foods"* [439]. Thus, the FDA's

---

[433]Documentary "Food Inc.", Director: Robert Kenner, 2008 (1:16:10–1:17:40) and Documentary "Seeds of Death: Unveiling The Lies of GMO's", Directors: Gary Null, Richard Polonetsky, 2012 (0:05:15–0:08:30)

[434]John Fagan, Michael Antoniou, Claire Robinson, GMO Myths and Truths, An evidence-based examination of the claims made for the safety and efficacy of genetically modified crops and foods, Earth Open Source, 2014, second ed., pp. 58–59

[435]EU Commission shortlists ex-Monsanto employee for EFSA Management Board, Munich Corporate Europe Observatory and Testbiotech, March 8, 2012

[436]Frederick William Engdahl, The Toxic Impacts of GMO Maize: Scientific Journal Bows to Monsanto, Retracts anti-Monsanto Study, Global Research, December 6, 2013

[437]Christophe Noisette, Roumanie – OGM: un ex de Monsanto, ministre de l'Agriculture, Inf'OGM, Febrary 2012

[438]John Fagan, Michael Antoniou, Claire Robinson, GMO Myths and Truths, An evidence-based examination of the claims made for the safety and efficacy of genetically modified crops and foods, Earth Open Source, 2014, second ed., p. 56

[439]Ibid, p. 59

assertion that GMOs are *"generally recognized as safe"* actually rests on biased tests carried out within the industry [440]. In fact, even conducting independent research is far from straightforward: a group of leading American agricultural entomologists complained to the EPA that *"No truly independent research can be legally conducted on many critical questions"*. A New York Times report explained their concern: *"The problem, the scientists say, is that farmers and other buyers of genetically engineered seeds have to sign an agreement meant to ensure that growers honor company patent rights and environmental regulations. But the agreements also prohibit growing the crops for research purposes. So, while university scientists can freely buy pesticides or conventional seeds for their research, they cannot do that with genetically engineered seeds. Instead, they must seek permission from the seed companies. And sometimes that permission is denied or the company insists on reviewing any findings before they can be published, they say"* [441].

This protectionism is clearly illustrated by the fate of a recent research project led by Gilles-Eric Seralini, professor of molecular biology at the University of Caen in France, on the long-term effects of GMOs on animals. In 2009, the EFSA approved Monsanto NK603—a GM maize developed to be resistant to Monsanto's glyphosate-based herbicide Roundup. The regulators did not demand independent testing of the product, maintaining that *"data provided [by Monsanto] are sufficient and do not raise a safety concern"*. They went further, stating that *"The EFSA GMO Panel is of the opinion that maize NK603 is as safe as conventional maize. Maize NK603 and derived products are unlikely to have any adverse effect on human and animal health in the context of the intended uses"* [442]. Monsanto's tests prior to the release of the maize had been over a relatively short period but, on the basis of these, they had declared that *"NK603 is as safe and nutritious as conventional corn currently being marketed... Roundup Ready corn plants containing corn event NK603 were shown to be as safe and nutritious as conventional corn varieties and to pose no greater environmental impact than conventional corn varieties"* [443]. Professor Seralini was skeptical of the EFSA approval, and embarked on an independent—and secret—test on the new Monsanto maize. It had already been grown extensively in Canada using Roundup, as recommended, to dust the crop. For the test, 200 rats were fed with NK603 maize exposed to varying quantities of Roundup; the process took 2 years and cost US$3 million. Seralini's conclusion was that the rats which had been fed Monsanto GM maize for the entire test period showed higher than normal rates of cancer; however, consuming the GM maize for shorter periods of up to 90 days did not increase cancer rates. Seralini published the

---

440 Ibid, p. 61

441 Andrew Pollack, Crop Scientists Say Biotechnology Seed Companies Are Thwarting Research, The New York Times, Feb. 19, 2009

442 Frederick William Engdahl, The Toxic Impacts of GMO Maize: Scientific Journal Bows to Monsanto, Retracts anti-Monsanto Study, Global Research, Dec. 6, 2013

443 Safety Assessment of Roundup Ready Corn Event NK603, Monsanto, September 2002, pp.3–17

results in November 2012 in the Journal of Food and Chemical Toxicology, under the title "*Long term toxicity of a Roundup herbicide and a Roundup-tolerant genetically modified maize*" [444]. There was worldwide support for this article from journalists and consumers. But the regulators, the industry and some of the scientific community put tremendous pressure on Seralini, accusing the professor of failing to meet scientific standards in the design, reporting and analysis of the study; the article was retracted by the journal in November 2013. In 2017, it surfaced that the editor-in-chief of the Journal of Food and Chemical Toxicology had consulting relationships with Monsanto. However, in June the following year, another journal, "*Environmental Sciences Europe*", republished the findings [445]. Clearly, Seralini's observations and claims about the potential long-term danger of GM maize and Roundup—for both animals and humans—have huge implications for public health: there is a strong case for initiating new tests in order to prove or disprove them. But remarkably, neither regulators nor other research teams in the field have so far taken the trouble to do so. Meanwhile, many governments worldwide are supporting the development of GMOs—apparently, they have decided it is worth the risk of implementing innovative but unproved technologies, in order to ensure a steady supply of cheap food for their ever-growing populations.

Managing these risks should include: continuous internal monitoring of the quality of agricultural production and of livestock disease statistics; independent, including governmental, monitoring of agricultural production; quarantine measures for preventing spreading of diseases; advanced claim management to ensure the timely identification of any risks that may threaten the safety of products over their whole life-cycle; and ongoing research for new safe technological solutions in agriculture and disease prevention.

Weather-Related Crises

Flood, strong winds, drought, hail, sudden frosts, grasshopper and caterpillar plagues, natural fires in forests and ocean storms could all pose a serious threat to crops, domestic animals, timber logging and fishing activity. A severe natural event can destroy crops and livestock, damage food reserves, disrupt the food supply chain, raise food prices, and in extreme cases even cause famine in the affected area. FAO, which analyzed the impact of 78 medium and large-scale disasters in 48 developing countries between 2003 and 2013, found that 22% of all losses inflicted by

---

[444]Gilles-Eric Séralini, Emilie Clair, Robin Mesnage, Steeve Gress, Nicolas Defarge, Manuela Malatesta, Didier Hennequin, Joël Spiroux de Vendômois, RETRACTED: Long term toxicity of a Roundup herbicide and a Roundup-tolerant genetically modified maize. Food and Chemical Toxicology, 2012, 50(11), pp. 4221–4231

[445]Gilles-Eric Séralini, Emilie Clair, Robin Mesnage, Steeve Gress, Nicolas Defarge, Manuela Malatesta, Didier Hennequin, Joël Spiroux de Vendômois, Republished study: long-term toxicity of a Roundup herbicide and a Roundup-tolerant genetically modified maize, Environmental Sciences Europe, 2014, 26:14

natural disasters are in the agriculture sector. The worst damage from the natural disasters studied was to crops (42% of all damage and losses) and livestock (36%). Of crop losses, 60% were due to floods, 23% to storms and only 15% to drought. For livestock, 85% of all losses were caused by droughts, 8% by floods and only 3% by storms. 70% of losses of fishery were connected with tsunamis, 16% with storms and only 9% with floods. And for forestry, 89% of all losses were caused by storms, 5% by floods and only 3% by tsunamis [446].

The worst flood of the last century in terms of casualties occurred in central China in 1931, when up to four million people died from drowning, disease and subsequent famine. Torrential snow and rain storms caused two of China's major rivers, the Yangtze and the Huai, to overflow, wreaking wholesale destruction on agriculture infrastructure. The severest storm in terms of casualties took place in East Pakistan (now Bangladesh) in 1970: the Great Bhola Cyclone claimed at least 300,000 lives in areas of low land that were completely vulnerable to the huge tidal waves that swept in from the ocean. Nearly 90% of the country's marine fishing industry was severely affected, 65% of its annual industrial capacity was destroyed, 46,000 fishermen perished and 9000 boats were lost [447]. In 1974, Bangladesh faced severe famine when rice plantations were destroyed by heavy rainfall during the monsoon season and subsequent floods along the Brahmaputra River. In the 1930s in the American Great Plains, severe droughts and poor cultivation techniques led to the massive erosion of topsoil over 400,000 km^2, which was blown away by winter winds. The disaster was known as the Dust Bowl and provoked the displacement of one million farmers in Kansas, Colorado, New Mexico, Oklahoma and Texas in the early 1930s, and 2.5 million people after 1935 [448]. In the more recent past in the United States, severe droughts occurred again in 1988 causing losses of US$41.2 billion (adjusted to 2017 values), in 2012 with losses of $31.8 billion, and in 1980 ($29.7 billion) [449]. There was a particularly severe winter in Mongolia in 2009–2010, when the temperature dropped below −40 °C; between 7 and 12 million cattle perished. During the following summer, a record-breaking heat wave in the western part of Russia provoked gigantic wildfires and a drought. According to Munich Re, around 56,000 people died from the effects of the smog and heat wave cause by the fires [450]. More than nine million ha of crops were lost. Total damage from the wildfires and drought was estimated at between US$15 billion and $50 billion [451].

---

[446]The impact of natural hazards and disasters on agriculture and food security and nutrition, FAO, May 2015

[447]Neil Frank, S. A. Husain, The deadliest tropical cyclone in history? Bulletin of the American Meteorological Society, June 1971, pp.438–439

[448]Donald Worster, Dust Bowl: The Southern Plains in the 1930s, Oxford University Press, p. 49

[449]US Billion-Dollar Weather and Climate Disasters, NOAA National Centers for Environmental Information, 2017, https://www.ncdc.noaa.gov/billions/

[450]Natural catastrophes 2010. Analyses, assessments, positions, Munich Re, February 2011, p.27

[451]A. Shapovalov, D. Butrin, $15 billion lost in Russian fires, Kommersant, August 10, 2010

Responding to the kind of weather risks we have discussed here requires a wide range of strategies. Advance forecasting of weather or of plagues of locusts and caterpillars, and good communication when an extreme natural event looks highly probable, enables those in the affected areas to be prepared for the event if it occurs. Adequate insurance must be available for weather-related risks. The cultivation of drought-resistant crops, advanced pesticide treatment, and preparation of livestock and land can mitigate the impact of droughts; in flood-prone regions, contingency arrangements should be in place for quick transportation to a safe zone. Due to the geographical distribution of agricultural production, the impact of hazardous natural events varies widely over different locations and from 1 year to another—so large agricultural holdings can recover from a natural event more easily than a single farm, because their assets are spread over different locations and diversified across a range of agriculture business. Special fire prevention measures are vital to the protection of forests: development of mineralized strips, carefully planned forest clearance, and investment in equipment (including air support) and firefighting staff. Fishing fleets should be equipped with advanced weather radar and tsunami warning systems.

Depletion of Natural Reserves

The degradation of agricultural land, deforestation, and the depletion of fishery reserves pose a serious threat to conventional agriculture. Thus, according to one UN body, around 40% of agricultural land in the world has been seriously degraded by intensive crop farming; in Central America, this measure is 75%, in Africa 20% and in Asia 11% [452]. Nevertheless, according to the FAO nearly 30% of the world's agricultural land is currently taken up with producing food that is ultimately never eaten because the agriculture and retail industries are unable to collect, store, deliver or sell it to the customer; at least 45% of fruit and vegetables will never be eaten [453].

Over the 25 years from 1990 to 2015, global forest coverage has been reduced from 4128 million ha to 3999 million ha—a loss greater than the area of South Africa. Deforestation has been unequal: the largest decline of forests has been in the tropics of South America (Brazil), equatorial Africa (Nigeria, Tanzania), Indonesia and Myanmar, while the area of forested land in China has actually increased [454]. Forest areas are usually cleared for food crops, for planting oil palms—a huge amount of palm oil is used in food manufacturing—for coffee plantations or for residential use.

In the 1990s, tens of thousands of fishermen from Eastern Canada, Greenland, Portugal, Spain and other countries lost their jobs and their businesses through the overfishing of cod on the Newfoundland Grand Banks, which were previously

---

[452] Ian Sample, Global food crisis looms as climate change and population growth strip fertile land, The Guardian, August 31, 2007

[453] Make #Not Wasting, FAO, http://www.fao.org/3/a-c0088e.pdf

[454] Global Forest Resources Assessment 2015, FAO, Second edition, 2016, pp.3,17

known for their extremely rich cod stocks. Similar crises occurred after the overfishing of sardines in California and Peruvian anchovies in the Pacific. The FAO warns that 90% of the world's marine fish stocks are now fully exploited [455].

Biomass is also a favored source of 'renewable' energy in Europe [456], with plantations for biofuels displacing ecologically diverse ecosystems that could have absorbed carbon more efficiently, or compete with land that could have been used for food production. With current biofuels targets set by the US and Europe, the amount of land used to create fuel rather than food could increase significantly. If so, food prices could rise—as happened in 2010–2011, leading to the so-called 'Arab spring' [457]. Hundreds of millions of people would be pushed towards starvation [458], with the potential to trigger widespread social instability. Where the cultivation of biofuels is not permitted to displace food production, biofuel production moves into previously uncultivated land such as forests or grasslands [459], leading to loss of biodiversity [460], deforestation and the actual net increase of emissions [461] in Europe.

Nevertheless, because flora and fauna recover so much faster than mineral resources, the restoration of a world depleted by intensive land exploitation, deforestation [462] and overfishing could be an achievable task for the agriculture sector. It would, however, require urgent and coordinated action from governments, scientists, industry and local communities.

[455]Mukhisa Kituyi, Peter Thomson, 90% of fish stocks are used up – fisheries subsidies must stop emptying the ocean, WEF, July 13, 2018, https://www.weforum.org/agenda/2018/07/fish-stocks-are-used-up-fisheries-subsidies-must-stop/
United Nations Secretary-General's Special Envoy for the Ocean, United Nations

[456]Biomass, European Commission, https://ec.europa.eu/energy/en/topics/renewable-energy/biomass

[457]Lagi, M., K.Z. Bertrand and Y. Bar-Yam, The food crises and political instability in North Africa and the Middle East, 2011, https://arxiv.org/abs/1108.2455

[458]Biofuels: Fueling Hunger?, ActionAid International USA, https://www.actionaid.org.uk/sites/default/files/publications/biofuels_fuelling_hunger.pdf

[459]Biomass, European Commission, https://ec.europa.eu/energy/en/topics/renewable-energy/biomass

[460]F. Stuart Chapin III, Erika S. Zavaleta, Valerie T. Eviner, Rosamond L. Naylor, Peter M. Vitousek, Heather L. Reynolds, David U. Hoope, Sandra Lavorel, Osvaldo E. Sala, Sarah E. Hobbie, Michelle C. Mack and Sandra Díaz, review article: Consequences of changing biodiversity, Nature, 2000, 405, pp. 234–242

[461]Walsh, B., Even Advanced Biofuels May Not Be So Green, Time, 2014, http://time.com/70110/biofuels-advanced-environment-energy; Steer, A. and C. Hanson, Biofuels are not a green alternative to fossil fuels, The Guardian, July 29, 2015

[462]Balanced "primordial" forests take however several centuries to recover, see e.g. Peter Wohlleben, The Hidden Life of Trees: What They Feel, How They Communicate (Discoveries from a Secret World), Greystone Books, 2016

Other Risks

Volatility of food prices can severely affect the stability of the agriculture business. Because of competition within the sector between different agricultural producers, wholesalers and retail networks are in a dominant position in the procurement of food, and can dictate purchasing prices; therefore negotiation on prices, volumes and merchandising can be fierce.

Because agriculture is a matter of national security, national governments **may limit the export activity of agricultural producers or set state procurement prices**, neither of which are good news for food producers. There is also tough **competition for access to leased land for crops or livestock, permission to log forests, and more generous government quotas for industrial fishery**. There is strict state regulation over the development of more productive crops, animals or fish—for example, the use of genetically modified organisms is restricted in many countries.

Supply-chain crises can affect the sector in many ways: energy price rises, shortages of fuel for machinery fleets, conflicts over licensing with seed suppliers, the failure of spare parts suppliers or machinery producers to enable the prompt repair of fleet, or the inability of contractors to provide additional outsourced forces during the peak demand of harvest. Staff strikes or labor disputes could disrupt the functioning of an agriculture company. There are also local logistics-related risks that can disrupt the storage of crops, vegetables, meat or fish: dilapidated warehouses, overproduction, power outages, and so on.

Farmers can be ruined by the **impossibility of accessing long-term capital** due to the position of financial institutions unwilling to provide loans to risky businesses, and the absence of government support for national agriculture.

Political instability and intrastate or interstate wars can greatly disrupt the normal business processes of agriculture production. Sanction wars can also threaten agricultural producers by restricting exports to other countries, as in the case of Russian tit-for-tat sanctions against the EU over the crisis in Ukraine in 2014, when European food producers were banned from exporting to Russia.

Finally, the sector can be forced to adapt in response to **pressure from environmental organizations opposed to unsustainable agriculture**—for instance, the clearing of forests for oil palm or coffee plantations, excessive use of pesticides, or the discharge of untreated waste products into reservoirs for drinking water.

Key Risk Mitigation Measures in Agriculture

- Ensuring the stable quality of raw materials and safety of products for consumers.
- Establishing fast and reliable supply chains to distribute products to customers, along with the operation of depositories for the careful storage of harvest.

(continued)

- Developing close relations with the authorities at all levels to ensure access to affordable land and surrounding infrastructure, lobby for subsidies, foster government support in the struggle with foreign competitors, etc.
- Finding convenient locations with a suitable climate for agricultural activity.
- Supporting ongoing research and development of innovative and safe agriculture technologies, which could increase productivity and mitigate adverse weather conditions.
- Selecting carefully the suppliers of seeds, young animals/plants, equipment, fuel, etc.
- Keeping the ability to find long-term and low-cost financial resources to mitigate seasonal factors.

Chapter 3
Specific Features of Risk Management in the Service Sector

3.1 Trading

The subsector includes the following industries according to ISIC Rev.4:

- 45—Wholesale and retail trade and repair of motor vehicles and motorcycles
- 46—Wholesale trade, except of motor vehicles and motorcycles
- 47—Retail trade, except of motor vehicles and motorcycles

General Description and Key Features of this Subsector

This subsector is an important intermediary between manufacturers of goods and millions of consumers of those goods, providing a direct sale service through extensive networks of warehouses/stores or remote sales through doorstep delivery. The main criteria for choosing a trading service point include its location, prices on goods, range and quality assurance of products, opening hours for purchasing, and the affability, efficiency and quality of service provided by staff. Quality of infrastructure also affects customer satisfaction: convenience of access to the store, parking, good maintenance of the facility (from cleaning the toilets and floor to air conditioning), uninterrupted operation of equipment, and so on. Trading is one of the labor-intensive subsectors, since every stage of the process—from transportation, handling, and storage to merchandizing, ordering, billing and refunds—depends on the manual work of service staff. Nevertheless, the subsector is starting to implement automated solutions: on-line ordering and sales, ERP [1] systems to connect traders

---

[1]Enterprise resource planning (ERP), which is the integrated management of core business processes, often in real-time and mediated by software and technology.

© Springer Nature Switzerland AG 2020
D. Chernov, D. Sornette, *Critical Risks of Different Economic Sectors*,
https://doi.org/10.1007/978-3-030-25034-8_3

with suppliers, RFID tags [2], and so on. The relative ease of entering the market with different business models leads to ruthless competition between players—price wars, promotions, and competition through excellence of service—all of which benefit customers.

Critical Success Factors for an Organization Within the Subsector

Success for a trading company boils down to reliably providing customers with the variety of products they need at reasonable prices. This could include: providing *"one stop"* shopping at locations convenient to customers, either by smart placement of stores/warehouses or by offering doorstep delivery; ensuring customer satisfaction during each purchase, and thus repeat orders in the long term and via word-of-mouth recommendations; motivating countless staff to serve customers with a heartfelt desire to satisfy their expectations; convincing suppliers to provide the lowest possible wholesale prices; and having the flexibility to adapt quickly to changes in economic situation and consumer spending power.

Stakeholders in this Subsector [3]

- Customers—40%
- Employees—30%
- Suppliers—20%
- Other—10%

Typology of Common Risks, Main Features of Major Accidents and Risk Mitigation Measures Within the Subsector

In our discussion of the risks involved in this subsector, we will not explore the risk of misjudgments at the level of business strategy: choosing an inappropriate location for a service point, misjudging target audiences and customer trends, preference and taste, inadequate pricing in the service network, mistakes in marketing and promotion of services, or cash flow problems due to low marginality in the business and

---

[2]Radio-frequency identification (RFID), which uses electromagnetic fields to automatically identify and track tags attached to objects.

[3]Our informed appraisal of the influence of each audience on a typical organization within the subsector (100% = combined influence of all audiences)

spiraling costs. There are also factors beyond the control of any individual company, such as the influence of an economic recession (or, even worse, depression) or the consequences of tough competition in the subsector.

Customer-Related Crises Induced by Poor Quality of Service

This subsector has a small number of large-scale crises, along with countless instances of individual customer dissatisfaction for unacceptable service: these "*micro-crises*" occur daily. Poor customer service from a trader's staff causes the most common crises within the subsector: unresponsiveness to personal needs, inflexibility in options, making promises to customers that cannot be delivered, slow servicing and check-out, mistakes in processing orders, poor workmanship during vehicle repair, delays in delivery or refusal to take back a product if the customer is not satisfied. Sometimes bad customer service amounts to fraud, as in the sale of goods known to be substandard—overdue, re-frozen, repaired, or with defects which are concealed during transportation—or the conscious over-billing of customers. Even a single instance of such outrageous disrespect to a customer could lead her to look for a new service provider, in spite of positive experience in the past.

Responding to these challenges requires a deep analysis of consumer preference and behavior, the constant recruitment of new workers in a subsector with high staff turnover, and ongoing staff preparation. Service staff need training in customer relations, and in the specifics of retail servicing—merchandising, cash register, in-depth knowledge of stock and so on. There should also be advanced motivation programs for employees for excellence in customer service, and continuous monitoring of the quality of service, of customer complaints and of requests through an advanced customer relationship management (CRM) system.

Supply Chain Crises

Problems with the supply chain could arise through poor control over the quality of goods supplied to the network, insufficient price pressure on suppliers to provide the lowest prices for customers, delays in ordering, poorly functioning outsourced IT systems, an imbalance between stock levels and customer demand, or inflexibility from suppliers in responding to the specific needs of individual customers.

IT and the Supply Chain Collapse of FoxMeyer (USA, 1993–1996) This company was one of the leading drug wholesalers in the country with annual sales of US $5 billion. Between retail customers at their own Ben Franklin Retail Stores and other pharmacies, and corporate supply to hospitals and clinics, sales exceeded half a million items daily. Aiming to increase the efficiency of its storage at distribution centers and customer supply, FoxMeyer developed a cutting-edge enterprise resource planning (ERP) system. They chose the R/3 system from the German software company SAP, which had previously been used only in manufacturing, and brought in Anderson Consulting (now Accenture) to supervise the integration of the new system. In total, the contract cost $65 million, but FoxMeyer expected to save up to $40 million annually by cutting costs and increasing supply chain efficiency. However, during the roll-out of the project, it became clear that R/3, far from improving operations as expected, seemed to be creating new problems. For instance, the system could handle only 10,000 transactions per day in each warehouse, while the previous IT system could process 420,000 items daily. Then there were data errors, computing mistakes in sales histories, insufficient flexibility in the system to respond to changes, a shortage of in-house staff qualified to operate the new system, high turnover among consultants... and ultimately, complaints from customers. FoxMeyer was forced to spend up to $100 million on improving the system but eventually, in 1996, the company filed for bankruptcy: they had to cover $34 million in losses from uncollectable shipping and inventory costs, also caused by errors in the IT system. The fact that such minimal losses were enough to push FoxMeyer over the edge stemmed from the low marginality of the wholesalers' business, which had dropped from 5.5% in 1980 to just 0.35% in 1996 [4].

Tohoku Earthquake and Supplies of Japanese Cars and Spare Parts (Worldwide, 2011) The earthquake not only damaged Japanese coastal infrastructure and provoked the nuclear disaster, but also suspended production at several car manufacturing plants because of structural damage, destruction of transport infrastructure and shortage of power across the whole country. Because of the prevalence of the "*just in time*" strategy, many car producers around the globe did not hold significant inventory and were forced to suspend or reduce car production when supplies of critical parts from Japan ran out. IHS Global Insight estimated that 4.2 million cars were never produced because of the earthquake—90% of them from Japanese producers [5]. This had a knock-on effect on car sales and repair businesses in many countries, with some models and original Japanese spare parts being unobtainable during the summer of 2011 [6,7,8].

Meat Adulteration Scandal (Ireland and UK, 2012–2013) Several meat producers were found to have supplied beef adulterated with horsemeat and pork, with neither retailers nor final customers being informed. The crisis affected major retail chains including Tesco, Co-op and Aldi, as well as several wholesalers.

---

[4]FoxMeyer case: a failure of large ERP implementation, The Pharmaceutical Distribution Industry, http://higheredbcs.wiley.com/legacy/college/turban/0471229679/add_text/ch08/fowmeyer.pdf

[5]Japanese Automakers and Suppliers Shift Production to Weekends, Idle Plants on Weekdays, IHS Global Insight, May 19, 2011

[6]Toyota warns US dealers of spare parts shortage, Just-Auto, March 29, 2011, http://www.just-auto.com/news/toyota-warns-us-dealers-of-spare-parts-shortage_id109980.aspx

[7]Graeme Roberts, Earthquake stock shortage still biting, Just-Auto, July 6, 2011, http://www.just-auto.com/news/earthquake-stock-shortage-still-biting_id112466.aspx

[8]Shortage adds to slowing Australia sales, Just-Auto, June 3, 2011, http://www.just-auto.com/news/shortage-adds-to-slowing-australia-sales_id111656.aspx

Preventing such shortages of goods and spare parts, and enabling prompt delivery and rapid repairs, requires inter-organizational coordination between a trader and its suppliers. This could include investment in IT systems to keep track of business processes, supply chains and inventories, cooperation on marketing activity, and good relationships with backup suppliers.

Occupational Safety and Infrastructure-Related Crises

Poor occupational safety and inadequate maintenance of service infrastructure leads to accidents: transport accidents during delivery of goods or supplies, crashes between loaders and delivery trucks, vehicles falling during repair, and so on. Sometimes, it can even cause larger scale disasters such as store fires or the collapse of supermarkets.

Transport Accidents According to the consultants Advanced Safety & Health, 30% of the 400 annual fatalities among retail workers in the United States, and 51% of the 200 annual deaths in the wholesale business, are caused by transport incidents [9].

Other Occupational Accidents The figures mentioned above also show that 9% of deaths both in retail and wholesale trade occur through falls. Contact with objects or equipment causes 7% of deaths in retail and 23% in wholesale trade; 3% (retail) and 7% (wholesale) are from exposure to harmful substances or environments; and fires/explosions cause 1% of deaths in retail and 4% in wholesale trade.

Fires Less frequent—but the most dangerous infrastructure crisis in terms of one-time casualties—is fire at a retail service point occupied by thousands of customers, or in a busy warehouse. The severest fire in the subsector occurred in the Ycua Bolanos supermarket in Asuncion, Paraguay in 2004, where 464 people perished when fire broke out in a fast-food section. The main reason for the tremendous number of casualties was the astonishing decision by store management to close doors in order to prevent looting by the escaping customers [10]. There have been all too many similar fires with multiple casualties. The L'Innovation department store fire in Brussels, Belgium in 1967 killed more than 320; the cause remains unclear. The Linxi department store fire in Tangshan, China in 1993 claimed 79 lives, and was caused by the negligence of workers during welding [11]. A fire at the Kebon Kembang shopping mall in Bogor, Indonesia in 1996 left 77 people dead, most of whom were employees of the mall; this time the cause was an electrical

---

[9]Warehouse and Retail Injuries and Deaths by the Numbers, Advanced Safety & Health, LLC, April 24, 2012

[10]Six Charged in Paraguay Supermarket Fire, The Washington Post, August 4, 2004

[11]Fire Kills 79 People In Department Store In Northeast China, The New York Times, February 16, 1993

short-circuit [12]. A firework display set off a chain reaction and spread to neighboring firework stalls in the Mesa Redonda shopping center in Lima, Peru, killing 290 customers and vendors [13]. The fire at Zhongbai Commercial Plaza in Jilin, China in the winter of 2004, which led to the deaths of 53 people, originated from an illegally constructed boiler room [14]. 46 people died when a backup generator caught fire at the Nakumatt supermarket in Nairobi, Kenya in 2009 [15]. And an explosion at a retail petrol station in Accra, Ghana in 2015 led to an inferno that took around 250 lives.

Sampoong Department Store Collapse (South Korea, 1995) 502 people died and more than 1000 were injured when this luxury department store collapsed in Seoul. The disaster was caused by misjudgments by the management team of the building, whose decisions were not adequately overseen by regulators. The building was originally constructed as a four-storey apartment complex, but it was adapted into a department store. In the process, several columns were taken out to install new elevators and the diameters of some of the remaining columns were reduced. The management team of the store also decided to construct several restaurants, creating a fifth floor on the originally four-storey building; at this point, the original construction company protested against the changes, but it was fired and replaced by the owners. Seating in a traditional Korean restaurant requires a warm floor, because guests are seated on cushions. The construction of heated pipes under the fifth floor increased the stress within the building. In addition, the team installed a heavy air-conditioning system at the top of the building by dragging three 15-tonne air-conditioning units across the roof instead of using cranes [16]. The movement caused cracks in the ceiling of the fifth floor and damaged one of the columns. Once installed, the AC system caused constant heavy vibration, which gradually exacerbated the already damaged column and ceiling over the following 2 years. On the morning of the collapse, one of the fifth-floor columns had completely separated from its floor-plate. Even then, the management team would not evacuate the whole building, contenting themselves with restricting access to the fifth floor and having the air-conditioning system switched off; this in a building where 1000 staff worked, and which 40,000 customers visited daily [17]. When the building finally collapsed, it was early evening, when crowds of shoppers flooded the department store. The owner of the building was sentenced to 7.5 years in prison. Twelve Seoul

---

[12]Blaze Kills 77 in Indonesian Shopping Mall, Associated Press, March 30, 1996

[13]Death foretold, The Economist, January 3, 2002

[14]Fire investigations reveal China's lax safety standards, Radio Australia, March 29, 2012

[15]Lester Young, Rowena Orosco, Stephen Milner, Double Fire Tragedy of Kenya, Eplasty, 2010, #10

[16]Colin Marshall, Learning from Seoul's Sampoong Department Store disaster – a history of cities in 50 buildings, day 44, Guardian, May 27, 2015

[17]Chris McLean, Brooks Anderson, Carl Peterlin, Kyle Del Vecchio, Sampoong Department Store, 2015, https://failures.wikispaces.com/Sampoong+Department+Store

bureaucrats also faced imprisonment for corruption and negligence towards construction safety.

Poor Working Conditions at Amazon Warehouses (USA, 2011, 2016–2018)
Amazon is the world's largest online store. There have been several controversies about the working conditions of packers at some of the company's warehouses. In the summer of 2011, the employees of a warehouse located in Breinigsville, PA, complained about temperatures of up to +45 °C in the warehouse, where hundreds of packers were making up parcels for Amazon customers. The inevitable summer heat was made unendurable by two decisions on the part of the warehouse management. Firstly, they were reluctant to install a proper cooling system; secondly, for fear of theft by staff or intruders, they kept the warehouse doors closed at all times, which prevented fresh air coming into the premises. Rather than avoiding an overheating crisis by installing air-conditioning, they installed 13 additional fans in the warehouse and opted for hiring paramedics to treat heat-stricken staff during June and July 2011 (15 of the 1600 workers at the warehouse experienced heat-related symptoms) [18]. That Amazon thought they could literally get away with running a sweatshop is probably explained by the scarcity of work in this area: employees had little choice but to tolerate such unsafe conditions if they wanted to keep their jobs [19]. In 2016, an undercover journalist for the Business Insider news website was hired as a packer at one of Amazon's UK warehouses. He later revealed some outrageous facts about working conditions there: packers were forced to pee into bottles and trash cans in order not to leave their post, for fear of being punished for low productivity; the typical working day was 10.5 h, with a lunch break of only 15–20 min; workers were penalized for taking days off sick. The journalist described the culture prevailing in Amazon warehouses as being like *"prison"*. After the article was published, Business Insider began to receive similar complaints from packers in the United States and elsewhere in Europe about appalling working conditions within the Amazon logistics network. The company replied: *"We don't recognise these allegations as an accurate portrayal of activities in our buildings ... Amazon ensures all of its associates have easy access to toilet facilities, which are just a short walk from where they are working. Associates are allowed to use the toilet whenever needed. We do not monitor toilet breaks"* [20]. At the end of 2018, workers in some European divisions of Amazon organized strikes in order to improve working conditions. A statement described the current situation as *"frankly inhuman. [Workers] are breaking bones, being knocked*

---

[18]Spencer Soper, Inside Amazon's Warehouse, The Morning Call, Sep 18, 2011, https://www.mcall.com/business/mc-xpm-2011-09-18-mc-allentown-amazon-complaints-20110917-story.html

[19]Alexander Lyon, Case Studies in Courageous Organizational Communication: Research and Practice for Effective Workplaces, Peter Lang, 2016, pp. 264–267

[20]Shona Ghosh, The undercover author who discovered Amazon warehouse workers were peeing in bottles tells us the culture was like a 'prison', Business Insider, April 18, 2018

unconscious and being taken away in ambulances. Enough is enough. Time [Amazon] workers were treated like humans, not robots" [21].

Measures to minimize the risk of full-scale workplace disasters include: the careful selection of service staff and drivers; regular training on occupational safety; regular safety checks; proper and timely investment in service infrastructure; fire prevention programs, ranging from well designed and located fire-emergency doors and alarms to the installation of sprinkler systems; full coordination with state emergency services to run mutual fire evacuation drills, and ensure independent assessment and communication of risks; and the provision of clear safety instructions for clients. As for exploitative working conditions like those in the last case, these are probably only avoidable through more stringent regulation and independent inspection.

Other Crises

Assaults and Thefts According to information from Advanced Safety & Health, more than half of annual employee deaths in retail trade are due to assaults by robbers; in wholesale trade, the figure is only 6% [22]. Shoplifting costs around 1.4% of US retail sales—in 2014, losses from shoplifting were $44 billion [23]. Tackling retail theft involves enhanced coordination between the trader's in-house security service and state law enforcement agencies for crime prevention and punishment, ongoing training of security officers, cashiers and other service staff, installation of monitoring and anti-theft systems, and industry-funded publicity of the penalties for shoplifting. Nevertheless, a prejudiced and officious security service could generate other problems by causing annoyance to legitimate customers, or even making a disaster far worse—as in the case we have described at the Ycua Bolanos supermarket in Paraguay, when over-zealous security guards closed doors to prevent shoplifting and prevented the escape of tens or hundreds of people, leading to their deaths.

Staff-Related Conflicts (Strikes, Lawsuits) If a company creates an environment where staff feel exploited or even discriminated against, they will not be motivated to serve customers warmly, fairly, and with pleasure. Such ill-treatment can lead to strikes, disputes over pay and even lawsuits. One such conflict has been running for decades between Wal-Mart and its 1.4 million employees. The staff complain that Walmart is a sweatshop, with inadequate pay, poor working conditions, enforced overtime, and no health insurance for some groups of workers such as part-time staff. They maintain that the company discourages employees from union membership,

---

[21]'We're not robots': Amazon employees protest across Europe on Black Friday, RT, November 23, 2018

[22]Warehouse and Retail Injuries and Deaths by the Numbers, Advanced Safety & Health, LLC, April 24, 2012

[23]Shoplifting, other fraud cost US retailers $44 billion in 2014: Survey, Reuters, June 24, 2015

though Walmart claim they are open to hearing staff grievances directly and that union intervention is therefore unnecessary [24]. Popular clothing manufacturers and retailers Abercrombie & Fitch were accused of discrimination against their Hispanic, Asian and African personnel. Employees of these races, according to some staff members, were not allowed to work in sale rooms and were offered only stockroom work, while their white colleagues got the lucrative sale room jobs [25]. In 2016, Macy's Inc. faced a strike of 4500 workers; the retailer was forced to raise wages, provide a more affordable health care plan and allow a more employee-friendly timetable [26].

In order to avoid such violent displays of staff displeasure, management should monitor the mood and analyze the claims of workers, and take preventive action by improving working conditions and perhaps pay, promoting competition between different teams of workers with benefits to the most productive groups, and so on.

Cyber Risks (Worldwide Problem) Ensuring the cyber security of e-commerce, keeping transactions safe for customers and protecting personal information became more important challenges for retailers with the growth of cyber-related crimes [27]. Thus in 2007, the discount chain T.K. Maxx admitted that at least 45 million of its customers could be exposed to fraud, after their debit or credit card details were stolen in an attack on its parent company's computer system. In 2014, hackers stole the personal data of 70 million customers of the discount retailer Target—including not only names, addresses and phone numbers, but credit and debit card numbers [28]. That same year, the furniture retailer Home Depot confirmed that information about 56 million credit and debit cards had been stolen, and EBay was unable to prevent the leakage of information about 145 million accounts—although fortunately the hackers this time were unable to access credit card information [29]. Because of the sensitivity of such data breaches for customers and the huge possible damage, management at trading companies need to prioritize investment into the enhancement of cyber security, the selection and training of their own highly-qualified IT staff, and close cooperation with IT security providers to create solutions against hacking.

---

[24]Neil Irwin, Walmart's lesson: Treat employees well, and the company does well, The Record. com, Oct 26, 2016, https://www.therecord.com/news-story/6932197-walmart-s-lesson-treat-employees-well-and-the-company-does-well

[25]Steven Greenhouse, Clothing Chain Accused of Discrimination, The New York Times, June 17, 2003

[26]Rachel Abrams, Macy's and Union Have a Deal, Averting a Strike, The New York Times, June 16, 2016

[27]Spencer Wheatley, Thomas Maillart and Didier Sornette, The Extreme Risk of Personal Data Breaches and The Erosion of Privacy, European Physical Journal B, 2016, 89 (7), pp.1–12

[28]Jia Lynn Yang, Amrita Jayakumar, Target says up to 70 million more customers were hit by December data breach, The Washington Post, January 10, 2014

[29]Jim Finkle, Hackers raid eBay in historic breach, access 145 million records, Reuters, May 22, 2014

Restrictive Government Regulation of Trading In some countries, governments intervene in the business of retail companies with protectionist or anti-monopoly regulation. They may require traders to work only or preferentially with domestic suppliers, limit the growth of dominant players in order to support smaller local retailers and generally increase competition, issue special licenses for alcohol sale, stipulate limits on the number, size and even location of stores, or strictly control fire prevention and occupational safety. The development of a constructive relationship with local, regional and national authorities, and strict adherence to the laws and regulations they impose, are important for any successful retailer.

Key Risk Mitigation Measures in Trading

- Finding and keeping reliable product manufacturers and persuading them to sell/provide their products to the trader at the lowest possible prices on the market.
- Continuous recruitment and retention of staff who can serve clients with heartfelt care, and with vigilance in the matter of occupational safety.
- Maintaining service infrastructure in safe and clean condition.
- Sourcing of plots and buildings in locations convenient for customers and retention of the lucrative rights on this infrastructure: negotiation with lessors on affordable rates, low expenses and taxes, good relations with local authorities and communities, etc.
- Close cooperation with law enforcement to fight shoplifting, prevent theft by staff and avoid possible assaults on trading facilities.

3.2 Transportation

This subsector includes the following industries according to ISIC Rev. 4.

- 49—Land transport (not including transport via pipelines)
- 50—Water transport (not including transportation of energy-related substances by tankers)
- 51—Air transport
- 52—Warehousing and support activities for transportation
- 53—Postal and courier activities

General Description and Key Features of the Subsector and Incorporated Industries

This subsector focuses on the conveyance of people and goods by all the different modes of transport. Personnel are still critically important for the safe and smooth running of transportation and postal services: it may still be decades before it is possible to substantially eliminate the human factor, because current automotive control solutions still require final approval from the pilot/driver. It is still personnel error that causes most crisis situations within the subsector. For example, in global commercial aviation between the 1950s and 2000s, pilot errors led to 53% of fatal accidents, mechanical failures caused 20% of accidents and weather only 12% [15% remaining from other causes] [30]. According to Boeing data from 2007, approximately 80% of airplane accidents occur through human error—usually from pilots, air traffic controllers, or mechanics—and only 20% are due to equipment failure [31]. Transportation accidents on a carrier's network usually have limited direct impact on the carrier: compensation to affected customers, write-off of a damaged vehicle, and so on. However, an accident brings indirect damage to the brand of a carrier, influences customer decision-making, leads to reduction of competitive advantage and reduces the investment attractiveness of the carrier. The subsector is highly dependent on the wider economic situation because demand for transportation services depends so much on the intensity of economic activity. There is high competition and low marginality because of the interconnectivity of current transport systems; the recent global trend to liberalize transport legislation also allows providers to offer services more easily over a wide geographical area including different countries.

Critical Success Factors for an Organization Within the Subsector

The most critical factor for the subsector is the human factor: *"Your employees come first... Start with employees and the rest follows from that"* [32]. This definitive statement of the vital importance of people in transportation business comes from Herb Kelleher, co-founder of Southwest Airlines, the largest low-cost airline in the world and a pioneer of many innovations in the industry. This includes the recruitment and retention of highly qualified staff (from pilots/drivers to mechanics); the ability to provide a fast and punctual transportation service on its carrier's route network; and the ability to motivate employees to transport customers safely and comfortably with satisfaction, and to carry freight carefully. Other factors include: timely investment in transportation equipment for proper and safe operation;

---

[30]Causes of Fatal Accidents by Decade (percentage), PlaneCrashInfo.com, http://www.planecrashinfo.com/cause.htm

[31]William Rankin, The MEDA process is the worldwide standard for maintenance error investigation, Aero quarterly (Boeing), 2Q 2007, http://www.boeing.com/commercial/aeromagazine/articles/qtr_2_07/AERO_Q207_article3.pdf

[32]Herb Kelleher Quotes, https://www.brainyquote.com/quotes/quotes/h/herbkelleh575183.html

ongoing work to optimize costs without impacting safety by improving the fleet and transportation model, so as to increase the carrier's competitive advantage; finding long-term and low-cost financial resources (including government backing) to smooth out seasonal fluctuations in demand and invest in transport infrastructure; and maintaining good relations with the authorities to influence state policy on liberalization, environmental legislation, distribution of international air routes, and so on.

Stakeholders in this Subsector [33]

- Customers—40%
- Employees—30%
- Regulators—10%
- Suppliers—5%
- Partners—5%
- Investors—5%
- Other—5%

Typology of Common Risks, Main Features of Major Accidents and Risk Mitigation Measures Within the Subsector

Crises Induced by the Actions of Pilots, Attendants, Controllers and Other Staff

Isolated and non-systemic incidents of bad customer service (impolite behavior of flight/ship/train attendants, low quality of food, delays, dirty or unhygienic facilities, loss of luggage, pilfering by staff, etc.) have a limited impact on a transportation company's business—but they generate unsatisfied customers who may post their experience on social media, decide not to travel with that provider in the future, and ultimately deter other potential customers.

Because of the penetration of social media, cases of bad customer service can become known nationwide in a matter of hours: thus in 2008, United Airlines staff broke a guitar belonging to Canadian singer Dave Carroll, who later released a protest song called "United Breaks Guitars". Over the next 7 years, more than 17 million people watched the clip on Dave's YouTube channel [34].

---

[33] Our informed appraisal of the influence of each audience on a typical organization within the subsector (100% = combined influence of all audiences)

[34] United Breaks Guitars, https://www.youtube.com/watch?v=5YGc4zOqozo

Another instance of horrific customer service that "*went viral*" occurred in April 2017 when United Airlines was forced to seat four staff members for another flight onto a fully booked plane headed from Chicago to Louisville. The company offered tickets for the next available flight, hotel accommodation for the night and a compensation voucher to any passenger willing to give up their seat. Finally, three passengers agreed to accept the offer, but no fourth volunteer came forward. Instead, David Dao—a doctor who had patients waiting at his final destination the day after the flight—was forcedly removed from the plane by airport security officers, with blood all over his face, a broken nose and the loss of two teeth. This alarming episode was filmed on smart phones by other passengers and immediately uploaded onto American social media: at least seven million people watched one or another of the clips in less than 24 h and the incident was broadcast nationally on traditional media. The adverse publicity caused an outcry, and United's capitalization fell by $770 million [35].

A third notorious example was the negligence of a FedEx employee in 2011, who "*delivered*" a computer monitor by simply throwing it over the gate, despite the fact that the customer was waiting at home with the front door wide open. The monitor screen was shattered—but unfortunately for the employee, his neglectful behavior was recorded on the customer's CCTV camera [36]. Unsurprisingly, the furious customer uploaded the film to social media, where it was viewed by more than nine million people. Of course, these cases are the exception—most bad customer service is not widely publicized—but such minor incidents can still gradually wear down a company's reputation and the loyalty of its customers, ultimately threatening its financial stability. Strikes can also have a very damaging impact on customer loyalty: if management seem unable to solve wage disputes with their staff without strikes and the resulting upheaval to thousands of passengers, those passengers are naturally less inclined to choose that carrier over its competitors next time they travel.

Crises like these are at best annoying and at worst a serious disruption to people's lives; but undoubtedly the severest crisis any shipping company can face is a crash where passengers are killed or injured.

Tenerife Aircraft Collision (Spain, 1977) The worst disaster in aviation history in terms of casualties occurred on March 27, 1977. 583 people were killed when a KLM Boeing 747 jumbo jet collided with a Pan-Am jumbo on the runway of Tenerife's Los Rodeos airport. The main airport for the Canaries, Las Palmas Airport on Gran Canaria, had been closed due to terrorist action, and the majority of flights were redirected to the smaller Tenerife airport, where ultimately there was an aircraft traffic jam. When operation was resumed at Las Palmas airport, the KLM and Pan-Am jumbo jets began to prepare for forthcoming take-off. Neither plane was able to use the main taxiway, which was blocked by other parked aircraft, so both

[35]Rob Wile, Here's How Much United Airlines Stock Tanked This Week, TIME, April 14, 2017

[36]FedEx Guy Throwing My Computer Monitor, https://www.youtube.com/watch?v=PKUDTPbDhnA

had to taxi on the airport's only runway. During taxiing, the pilots of the Pan-Am plane missed the subsidiary taxiway into which they had been ordered to turn off the runway by the airport controller. Consequently, they had not left the runway when the KLM jumbo began to accelerate for take-off. In their turn, the pilots of the KLM flight initiated take-off without clear confirmation from the airport controller, believing that the runway was clear and confirmation for the take-off had been given. The causes of the collision mainly relate to human factors: failure of communication and the language barrier between the American and Dutch pilots and the airport's Spanish controller. The situation was exacerbated by poor visibility in the airport due to a sudden fog—and by the impatience of the hundreds of passengers, eager to complete the flight as quickly as possible, which implicitly put pressure on the pilots. In addition, there was time pressure from the work schedule of the KLM pilots, which also influenced the decision-making of the KLM crew. After the disaster, KLM took responsibility and paid US$110 million in compensation to the families of the victims.

Alexander Suvorov River Cruise Ship Collision (USSR, 1983)   On June 5, 1983, the Alexander Suvurov collided with a rail bridge girder on the Volga River near Ulyanovsk, resulting in the death of 176 people. An investigation found that the ship's mate had not given the right course to the steersman, and the steersman in his turn had mistaken a section of the bridge obstructed by the girder as a navigable gap, because there was a trackman's hut on the girder which resembled the alarm panel indicating a navigable passageway. Moreover, the bridge was not illuminated adequately by signal lights.

SS Admiral Nakhimov Collision (USSR, 1986)   On August 31 1986, during a voyage from Novorossiysk to Sochi, the ocean liner Admiral Nakhimov was hit by the bulk carrier Peter Vasev. The accident, which claimed the lives of 423 people on the liner, mainly occurred because the pilots and captains of the two ships could not adequately coordinate their mutual actions to prevent a collision, despite favorable weather conditions and good visibility.

Herald of Free Enterprise Ferry Capsize (Belgium-UK, 1987)   193 people died after the Herald of Free Enterprise car ferry capsized just after leaving Zeebrugge harbor on its way to Dover on March 6, 1987. The main cause of the disaster was that the bow doors were left open, and incoming water flooded the car deck and destabilized the ship. An investigation concluded that the crew member responsible for closing the bow doors had been asleep during departure. The officers in charge did not check the closure of the bow doors, assumed that they were sealed and took the decision to leave Zeebrugge.

MV Doña Paz Ferry Collision and Fire (Philippines, 1987)   With up to 4400 casualties, this was the largest peacetime maritime accident in history—though it is still almost unknown by comparison with the sinking of the Titanic in 1912, in which 1517 people lost their lives. It occurred on the night of December 20, 1987, when the Doña Paz collided en route to Manila with the oil tanker MT Vector. After the collision, oil leaking from the damaged tanker burst into flames. The fire spread to the ferry and leaked gasoline and diesel onto the water. Terrified passengers tried to find shelter inside the ship but ultimately died in the fire. Others, who jumped into the water, were badly burned by the flaming petrol and diesel and attacked by scores of sharks. Only 24 people survived the disaster. The number of casualties is still in question, because the ferry was seriously overloaded and many of the passengers were not listed on the official manifest, having bought cut-price tickets. Officially a maximum of 1493 passengers and 59 crew were allowed on the ship, but post-disaster estimates are that more than 4400 people were on board. Survivors testified that, just before the collision, the majority of the crew were having a pre-Christmas party, drinking beer and watching TV, with only one person remaining on the bridge. Meanwhile the tanker, due to technical problems with its rudder, was approaching in zigzag fashion—confusing the watch on the ferry by alternately showing its port (red) and starboard (green) lights and thus giving the impression first that it was passing on the left, then heading straight for the ferry, then passing on the right, and so on [37]. Moreover, neither ship communicated with the other during the approach, nor did they send an SOS after the collision. Because almost all of the crews of both ships were killed—only two sailors from the Vector survived and, at the moment of the accident, they were both asleep—and both ships sank, the investigators could not reconstruct the full picture of the tragedy. However, it did come to light that the license of the tanker had expired, and that the vessel was unseaworthy. Ultimately, the court resolved that the crew of the tanker were responsible, and the tanker's owners liable, for the accident.

Paddington Train Collision (UK, 1999) On October 5, 1999, the collision of two trains in London led to the deaths of 31 people and injured more than 200. The main cause of the accident was that the driver of one of the trains misread a signal just before the track section where the collision took place. He had only 2 weeks of driving experience and, passing the signal in low morning sunlight, assumed that it was yellow when in reality it was red.

Überlingen Mid-Air Collision (Germany-Switzerland, 2002) On July 1 2002, a Bashkirian Airlines plane with eight crew and 60 passengers on board, mostly children, collided with a DHL freighter in the sky above Southern Germany. A controller at the Zurich air traffic control center, operated by the Swiss company Skyguide, had mistakenly ordered the Bashkirian Airlines flight to descend in order to avoid a collision, while the automatic traffic collision avoidance systems on the planes dictated that the Bashkirian flight should climb and the DHL should descend. At the critical moment of decision, the controller was distracted by a repair team, which had arrived at the control center to make modifications to the radar and telephone systems; some crucial instruments were temporarily unavailable. Every-one aboard both planes died. Incensed by what he saw as the disregard of Skyguide

---

[37]Asia's Titanic, The National Geographic Channel, August 25, 2009

towards the victims, a father who had lost his wife and two children in the crash killed the controller 19 months later.

Costa Concordia Sinking (Italy, 2012) On January 13 2012, roughly 100 years after the sinking of the Titanic, the Italian cruise ship Costa Concordia collided with rocks off the island of Giglio with 4229 people on board—3206 passengers and 1023 crew. The ship sank killing 32 people—27 passengers and 5 crew—more than half of whom were 60 years old or older. That evening, the Concordia had left the port of Civitavecchia and was heading for Savona. The captain, Francesco Schettino, gave orders to deviate from the scheduled route and skirt Giglio island in order *"'to kill three birds with one stone': to please the passengers, salute a retired captain on Giglio and do a favour to the vessel's head waiter, who was from the island"*. Later, the cruise company Costa Crociere assessed this maneuver as *"unauthorised, unapproved and unknown to Costa"* [38]. It was at least the third such maneuver under Schettino's leadership; the previous one was in August 2011 and followed a route significantly further from the coast.

On the night of the tragedy, the liner was dangerously close to the rocky shore of the island. This was due partly to navigational error—the misjudgment that the intended sailing route was at a safe depth—and partly to the difficulties of communication in English between an Italian captain and an Indonesian helmsman, who did not understand the captain's order to turn until a critical few seconds too late. During the investigation of the accident, the captain insisted that the *"rock under water. . . was not on charts"*, while a Costa Crociere executive refuted this claim: *"I can confirm to you that this rock does appear on maps. The captain made an independent decision to change the course specified by Costa. The ship was obviously too close to the shore, and the captain's assessment of the emergency did not correspond with Costa standards"* [39].

The collision led to a 53 m-long breach of the ship's floor, flooding of the engine rooms and failure of the ship's rudder and electrical power generation. The captain hesitated to declare a general emergency in the early stages of the disaster. Even the crisis center of Costa Crociere heard only reassuring reports from the crew, whose apparently calm manner belied the gravity of the accident [40]. As for the Italian police and coastguard, they were alerted about the mishap not by the crew, but through mobile calls to emergency lines from the cruise passengers. In the radio discussion that followed with the coastguard, the captain referred only to a blackout, though he was perfectly aware that the blackout was caused by the flooding of three

---

[38]Captain on trial: Costa Concordia's Francesco Schettino, BBC, February 11, 2015

[39]Fall of an Italian cruise captain, Euronews, January 17, 2012

[40]Report on the safety technical investigation "Cruise Ship Costa Concordia. Marine casualty on January 13, 2012", Marine Casualties Investigative Body of Italian Ministry Of Infrastructures And Transports, pp.30–31

adjacent compartments of the hull [41]. He convinced state representatives that all was under control: *"We have a blackout and we are checking the conditions on board"* [42]. Because critical information about the collision was not immediately communicated to the emergency services, it took much longer for patrol boats, helicopters, and rescue to arrive at the disaster site. Thousands of passengers, who had gathered in panic near the lifeboats to await immediate evacuation, were also deliberately misled about the real cause of the accident: *"The situation is under control. Go back to your cabins"* [43]. Eventually, 45 min after the collision, a general emergency alarm was broadcast—but some cruise guests did not understand the message due to the language barrier and stayed in their cabins.

The drifting ship was turned by wind and current and finally grounded near the island with serious heeling [44,45]. The captain and some senior officers of the cruise crew left the bridge 45–55 min after the order to abandon ship, although hundreds of passengers were still on board [46]. This callous disregard for the guests in their charge was not confined to the senior officers: some crew members pushed guests aside to take free seats in the lifeboats. In spite of urgent requests from the coastguard captain to go back on board and complete the evacuation procedure, Schettino and his subordinates did not comply. The captain stayed at a hotel on the island, where he was finally arrested the following morning. Even when a full evacuation was finally declared, not all the lifeboats could be deployed because of the heeling of the ship, the blackout and an inbuilt structural problem with the balancing pumps. Only around 70% of those on board—2930 people out of the 4197 evacuated—used the boats and life-rafts during the evacuation [47]. More than 1200 people were rescued with the help of emergency and municipal services. Were it not for favorable weather conditions that washed the liner ashore, and the delay in starting the evacuation because of the captain's indecision, the number of casualties could have far exceeded the 32 killed and 157 injured. The Italian media dubbed Schettino *"Captain*

---

[41]Report on the safety technical investigation "Cruise Ship Costa Concordia. Marine casualty on January 13, 2012", Marine Casualties Investigative Body of Italian Ministry Of Infrastructures And Transports, p.17

[42]Report on the safety technical investigation "Cruise Ship Costa Concordia. Marine casualty on January 13, 2012", Marine Casualties Investigative Body of Italian Ministry Of Infrastructures And Transports, p.31

[43]Terror at Sea: The Sinking of the Concordia, Channel 4, Directors: Paul O'Connor, Marc Tiley, January 31, 2012

[44]Report on the safety technical investigation "Cruise Ship Costa Concordia. Marine casualty on January 13, 2012", Marine Casualties Investigative Body of Italian Ministry Of Infrastructures And Transports, p.4

[45]Costa Concordia: What happened, BBC, February 10, 2015

[46]Report on the safety technical investigation "Cruise Ship Costa Concordia. Marine casualty on January 13, 2012", Marine Casualties Investigative Body of Italian Ministry Of Infrastructures And Transports, p.34

[47]Report on the safety technical investigation "Cruise Ship Costa Concordia. Marine casualty on January 13, 2012", Marine Casualties Investigative Body of Italian Ministry Of Infrastructures And Transports, p.25

Coward" and the prosecutor characterized him as a *"careless idiot"* [48]. He was found guilty of multiple manslaughter and sentenced to 16 years in prison. Five officers from the vessel were also convicted of lesser charges. The cruise company Costa Crociere lost up to €1.5 billion in legal fees, compensation payments and salvage. The accident shook up the entire global cruise industry, prompting improvements in safe navigation, and better training of crew members in the safe evacuation of thousands of passengers from today's gigantic cruise ships.

Santiago de Compostela Rail Disaster (Spain, 2013) 79 passengers died and more than 140 were injured during the derailment of a high-speed train, which took a bend too fast near the railway station at Santiago de Compostela. The driver of the train was having a work-related phone conversation just before the disaster, had a lapse of concentration and failed to brake in time to bring the speed down to an appropriate level. He was accused of professional recklessness. Quite a similar case had occurred before, at Amagasaki near Osaka in Japan, on April 25 2005. An inexperienced driver recklessly took a commuter train up to a speed of 116 km/h, on a section of track with a limit of 70 km/h. During the subsequent derailment, 107 people died and more than 500 were injured. The driver had mistakenly passed a red light earlier on the route, tripping an automatic stop system that brought the train to a halt; then just before the derailment, he had overshot the platform at the previous station and had to reverse the train, causing a 90-s delay. With a tough time schedule and harsh penalties on drivers for any delay, he was in a rush to make up the time.

German Wings Suicide Pilot Crash (Germany/France, 2015) The co-pilot of a flight from Barcelona to Dusseldorf had been suffering from severe depression and suicidal tendencies for some time. While the captain was out of the cockpit, the co-pilot made a fast descent and crashed the plane into a mountain in the French Alps, killing 150 people. It is likely that suicide pilots also caused the crash of SilkAir #185 in 1997 (104 deaths), EgyptAir #990 in 1999 (217 deaths), LAM Mozambique Airlines #470 in 2013 (34 deaths), and Malaysia Airlines #370 in 2014 (239 deaths).

Singapore Self-Driving Taxi, Tesla Autopilot Challenge and Uber Self-Driving Taxi Crash (Singapore and USA, since 2016) There is a global scientific race to construct a genuinely self-driving vehicle in order to reduce human factor mistakes during driving. This challenge will probably be solved over the next few decades, but at the moment these technologies still have limitations. Thus in August 2016, Singapore launched the first self-driving taxi, but a human driver is still necessary to oversee the decisions of the car computer and quickly intervene in the event of a computer error during driving.

Tesla, which installs autopilots in its cars, faced a law suit from several customers in April 2017. The plaintiffs maintained that *"Tesla has delivered software that*

---

[48]Captain on trial: Costa Concordia's Francesco Schettino, BBC, February 11, 2015

causes vehicles to behave erratically. Contrary to what Tesla represented to them, buyers of affected vehicles have become beta testers of half-baked software that renders Tesla vehicles dangerous if engaged... Autopilot is essentially unusable and demonstrably dangerous" [49]. Tesla rejected the accusations: *"[W]e have never claimed our vehicles already have functional 'full self-driving capability', as our website has stated in plain English for all potential customers that 'it is not possible to know exactly when each element of the functionality described above will be available, as this is highly dependent on local regulatory approval'. The inaccurate and sensationalistic view of our technology put forth by this group is exactly the kind of misinformation that threatens to harm consumer safety"* [50].

But by 2020, fully self-driving Google cars, under the brand name Waymo, will be on sale to customers according to the executive of the project [51]. From 2009 until February 2016, Google's cars drove more than two million kilometres, and 18 crashes occurred; 17 of these were in fact caused by human error and only one by an error in the automatic self-driving system [52].

On the night of March 18, 2018, a pedestrian crossing a road with her bicycle in Tempe, Arizona, was killed by an Uber self-driving taxi during its testing stage. The car did not recognize the pedestrian as being in danger because its software assessed that she was not in the way of the car. All self-driving cars are required to have a driver on board to deal with emergency cases and software mistakes, but the driver did not react in time to prevent the collision.

Until fully fledged automotive solutions are implemented, current measures to mitigate human error in driving/flying/navigation include rigorous selection of personnel, ongoing staff training to improve professional skills to meet world-class standards and deal with worst-case scenarios, monitoring of the quality of service and customer claims, management development of advanced customer care skills and crisis management knowledge among service staff.

Crises Induced by Malfunction of Equipment

Equipment malfunctions are a problem common to all transportation companies, but in some cases these defects can have disastrous consequences if they are not detected and eliminated in time.

---

[49]Dean Sheikh, John Kelner, and Tom Milone v. TESLA MOTORS, INC. The United States District Court Northern District of California, San Jose division, Case 5:17-cv-02193 Document 1 Filed 04/19/17, pp.1–2

[50]Matthew DeBord, Tesla owners criticizing Autopilot have unrealistic expectations, Business Insider, April 21, 2017

[51]Ina Fried, Google Self-Driving Car Chief Wants Tech on the Market Within Five Years, Recode, March 17, 2015

[52]Alex Davies, Google's Self-Driving Car Caused Its First Crash, Wired, February 29, 2016,

Japan Airlines Boeing 747 Crash (Japan, 1985) On August 12 1985, Japan Airlines Flight 123 collided with a ridge of Mount Takamagahara, near Osaka in Japan. 32 min earlier, a sudden decompression of the salon had caused the tail of the aircraft to break off. 520 passengers and crew perished in the accident; only four passengers survived. The investigation found that in 1978 the plane, a Boeing 747, had made a difficult landing with its nose too high, damaging the tail. Boeing staff had been called in to repair the defective aft pressure bulkhead, but—as became clear after the crash—the repair was substandard and did not comply with the design specifications for this part of the plane. It was only a matter of time before metal fatigue would cause the faulty repair to fail, leading to the explosive decompression of the salon, destruction of the vertical stabilizer and rudder, and critical failure of the hydraulic systems to maneuver the plane. Finally, after 7 years and more than 12,000 take-offs and landings, the failure happened. In terms of casualties, it was the largest single-plane accident in aviation history.

MS Estonia Sinking (Baltic Sea, 1994) This roll-on/roll-off ferry was crossing the Baltic Sea from Tallinn in Estonia to Stockholm in Sweden during a stormy autumn night between September 27 and 28, 1994. Stormy weather was common at this time of year, and all other scheduled ferries were at sea that night. For more than 14 years, the ship had been operating in the Baltic without incident, but during this crossing the visor—the moveable section of the bow which was raised to allow vehicles onto or off the car deck—was damaged by waves, and the locks and hinges holding it in place failed. These had been subjected to more stress than usual because of the crew's decision to cross the stormy water at high speed in order to avoid delays. However, the visor was not designed to withstand such stress [53]. The bridge did not respond quickly enough to the watchman's reports of loud noises coming from the bow—caused by the gradual destruction of locks and hinges as the waves struck. Finally, the visor separated from the ferry, the ramp fell open and water flooded into the car deck; the ship lost stability, capsized and sank. Of the 989 passengers and crew who were on board during the crossing, 852 people died. Remarkably, the European ferry industry had had many near-miss incidents with the visors becoming detached from Ro-Ro vessels over more than two decades leading up to the disaster. For instance, in January 1993, the Estonia's sister ship DIANA II "*nearly lost her bow visor between Trelleborg and Rostock on the same stormy night in January 1993 when the Polish ferry JAN HEWELIUSZ sank*" [54]; but regular and rigorous inspections of ferries of this type, and the distribution of risk-related information among crews, were not implemented by most operators. It was only after this disaster that the safety of ships with visors was reassessed, and new recommendations on the management of passenger evacuation came into force.

---

[53]Ulrich Jaeger, Scientists Unveil Cause of Estonia Ferry Disaster, Spiegel, January 10, 2008

[54]The German 'group of experts' (Dr. Peter Holtappels and Captain Werner Hummel), Investigation report on the capsizing on 28 September 1994 in the Baltic Sea of Ro-Ro Passenger Vessel, Chapter 33 "Bow door failures and other incidents of ro-ro vessels", http://www.estoniaferrydisaster.net/estonia%20final%20report/chapter33.htm

Baku Subway Fire (Azerbaijan, 1995) This subway conflagration in the capital of Azerbaijan occurred when one of the railway motors of a train caught fire. The disaster was exacerbated by the driver's decision to stop the train in the narrow tunnel of the metro and cut off its electricity supply. 289 people died, either from smoke inhalation or crushed in the panic by other passengers.

The Mont Blanc Tunnel Fire (France-Italy, 1999) On March 24, 1999, there was a conflagration in the road tunnel under Mont Blanc in the Alps which claimed the lives of 39 people. The disaster occurred when a Belgian freight lorry, carrying 9 tons of margarine and 12 tons of flour, caught fire in the middle of the tunnel. If the driver had kept going to the end of the tunnel, the disaster could perhaps have been averted, but he decided to stop to try and deal with the fire, allowing more oxygen to reach the smoldering engine. Before long, the fire had spread to the tons of highly flammable oil-based margarine in the back of the truck, and become uncontrollable: it would burn for more than 2 days. There was also a slow reaction to the smoke from tunnel operators, who did not immediately stop new cars coming into the burning tunnel.

Al Ayyat Train and al-Salam Boccaccio 98 Ferry Disasters (Egypt, 2002, 2006) On February 20 2002, a night train was heading from Cairo to Luxor. The explosion of a cooking cylinder in wagon #5 set off a massive fire that spread to seven carriages and killed at least 383 people according to official data. The driver of the train did not detect the fire until too late, greatly increasing the number of casualties. Because the train was overcrowded, some passengers travelling without tickets, the exact number of passengers in the burned wagons was unknown; and since most of the casualties were burned to ash, the true death toll cannot be established. A similar case occurred in Pakistan in October 2019, when passengers cooked breakfast on gas stoves inside one of the wagons of the running train (at least 75 died, most of the deaths were caused when people jumped from the train to escape the flames). Fire also caused the sinking of the Egyptian ferry al-Salam Boccaccio 98, which was cruising from Saudi Arabia to Egypt on Febrary 3, 2006; more than 1000 people died in this disaster.

Igandu Train Disaster (Tanzania, 2002) An overcrowded passenger train with 22 carriages and more than 1200 passengers was bound from Dar es Salaam to Dodoma on June 24, 2002. Near the top of a hill called Igandu, the brakes failed completely due to a technical malfunction and the train rolled backwards—rapidly picking up speed—to the bottom of the hill, where it slammed into a freight train waiting to make its ascent. In the collision, 281 people were killed and more than 400 injured.

Grounding of Airplane Fleets (Worldwide) For any airline, the grounding of its entire fleet of one type of craft when a defect has been discovered can cause a serious interruption to business, because the sudden fleet shortage makes it almost impossible to keep to the schedule. For example, all McDonnell Douglas DC-10 flights were suspended after the crash of American Airlines Flight 191 in 1979, when the plane lost its left engine during take-off. Aviation authorities assumed there must be a design fault in the DC-10 model, and grounded the whole fleet for more than a month; it eventually emerged that the crash had in fact been caused by improper engine maintenance procedures. All of the new Boeing 787 Dreamliners were grounded in 2013, after several fire incidents with lithium-ion batteries during the plane's first year of service. Correcting the design problem and getting approval from the authorities to resume flights took Boeing more than 3 months. In 2016, the Russian Sukhoi Superjet 100 was grounded after the discovery of metal fatigue in "*one element of the tail portion*". The Russian Aeroflot and Mexican Interjet airlines, the main operators of the model, suspended all flights on this type of plane 2 weeks after the grounding. In 2019, all Boeing 737 MAX 8 and 9 were grounded on suspicion of failures within the Maneuvering Characteristics Augmentation System, which were the probable cause of two air crashes in Indonesia and Ethiopia. Three hundred and eighty four planes were recalled to update the software within the system installed on these models. According to Boeing, it prepared up US $5.6 billion compensations to address potential claims from airlines affected by the grounding.

Delta IT Outage (Worldwide, 2016) Delta was forced to cancel more than 2000 flights during August 8-10, 2016 due to a 5-h computer outage caused by a power surge at the airline's command center in Atlanta. The IT system went down, preventing Delta staff from boarding passengers and luggage from all around the world because they could not access the passenger information system. The cost of the outage is estimated at up to US$150 million [55]. Major IT system failure also struck at Los Angeles International Airport in 2007 due to trouble with software at the US Customs and Border Protection, and at Heathrow Terminal 5 during its opening in 2008; IT issues affected the Australian airline Virgin Blue in 2011, United Airlines and American Airlines in 2015, Southwest Airline in 2016 and British Airways in both 2016 and 2017.

Response measures include timely investment in transportation fleet, close inter-action with vehicle manufacturers to exchange information about common defects, and regular servicing and replacement of equipment. The human factor, including the critical importance of skilled and vigilant mechanics, makes it essential to select the right personnel and to provide ongoing training of engineering and service staff according to the best practice recommended by the fleet manufacturers.

---

[55]Chris Isidore, Delta: 5-hour computer outage cost us $150 million, CNN, September 7, 2016

Crises Induced by Unfavourable Weather Conditions

Bad weather or natural disasters can make transportation impossible. In some cases, these hazards can even destroy vehicles, causing great numbers of fatalities.

Liziyida Bridge Rail Disaster (China, 1981)   This railway bridge over the Dadu He river in Sichuan province was destroyed by a landslide just before a passenger train arrived at the bridge from the Nainaibao railway tunnel. The driver had no time to brake, and several carriages including the locomotive fell into the river, with others derailed. The disaster claimed 360 lives.

Delta 191 and Downburst (USA, 1985)   A Lockheed L-1011 TriStar plane was finishing its flight from Florida to Texas, when during landing at Dallas/Fort Worth International Airport the crew unexpectedly met with a downburst or microburst: a temporary very intense downward wind [56]. Because the downburst occurred near the ground, the crew did not have the space to respond to this natural hazard, and during the subsequent hard landing 135 people died. Reflecting on the dramatic statistics—from 1964 until 1985, 26 air crashes took place due to downbursts, costing more than 500 lives—the US authorities initiated the installation of Doppler weather radar at airports and on board commercial airplanes, to detect downbursts in time to take evasive action.

Sumatra–Andaman Undersea Quake Tsunamis and the Queen of the Sea Train Disaster (Sri Lanka, 2004)   A packed express train, travelling from Colombo to Galle on the south-west coast of Sri Lanka on December 24 2004, was hit by two powerful tsunami waves generated by the devastating Sumatra–Andaman undersea quake. The calamity occurred roughly 2 h after the undersea quake, and neither crew nor passengers knew anything about the danger, or the damage that had already been wreaked on the coastline of Thailand, Indonesia, Malaysia and the eastern coast of Sri Lanka. When the waves struck, the train was on a particularly vulnerable section of the line, located only a few hundred meters from the coast. The overcrowded carriages were derailed including the locomotive; most of the carriages were washed away far inland, some dragged into the Indian Ocean. The exact number of casualties is still unresolved because of unrecorded fare dodgers, but at least 1700 people died in the disaster.

Sinking of Ferries Due to Severe Weather (Indonesia and the Philippines)
Between them, these countries comprise around 20,000 islands: 260 million Indonesians live on more than 6000 inhabited islands and around 100 million Filipinos live on 2000 islands. The near-equatorial location makes for a tropical climate with frequent typhoons and severe storms (especially in the Philippines). Because maritime transportation is the cheapest and most convenient way of conveying people and freight between the islands, this is the dominant mode of transport in both

---

[56]Downburst Wind Awareness, US National Oceanic and Atmospheric Administration, http://www.weather.gov/cae/downburst.html

countries. Not surprisingly, both have a long history of serious maritime accidents, many caused by severe weather. Thus in October 1988, the Filipino MV Doña Marilynin sank during Typhoon Unsang with 389 deaths; in December 2006, the Indonesian MV Senopati Nusantara sank during a storm with the loss of 500 people; in June 2008, the Filipino MV Princess of the Stars capsized during Typhoon Fengshen, killing 846 people; and in January 2009, the sinking of the Indonesian MV Teratai Prima cost the lives of more than 250 people. There have been numerous other cases with fewer casualties, the majority of which never become known outside the countries concerned.

Interruption of Caribbean Cruise Business Due to Hurricanes Harvey and Irma (USA, 2017) 15,000 cruise ship passengers had problems disembarking at Port Galveston (Texas) because of Hurricane Harvey. At least 40,000 passengers were affected by Hurricane Irma, which destroyed many Western Caribbean islands and damaged the coastline of Florida [57].

An effective response to extreme weather challenges could include the installation of advanced radar and weather monitoring equipment to allow decisions to be informed by more accurate forecasting; a better system of weather/natural hazard warning signals, with faster transmission; and wider discretion for pilots and drivers to make their own decisions on taking their vehicles out of unfavorable weather conditions.

Overloading and Improper Storage of Dangerous Goods

The pursuit of profit at the expense of safety often leads to violations of safety regulations on the proper storage of dangerous goods, and these can affect not only transportation vehicles but nearby residents, other vehicles and transportation, and storage infrastructure.

Texas City Explosion (USA, 1947) On April 16 1947, 581 people perished when the SS Grandcamp exploded with 2100 metric tons of ammonium nitrate on board. Many shortcomings were identified in the production, labeling, transportation, storage, and loading of ammonium nitrate, which while widely used as a fertilizer is also highly explosive.

Los Alfaques Campsite Disaster (Spain, 1978) At least 217 people died when a truck carrying highly flammable propylene exploded near the campsite of Los Alfaques in Catalonia on July 11, 1978. The truck was overloaded—with 23 tons of propylene in its tank, almost four tons over the design limit—and it was not equipped with emergency pressure release valves. The death toll was greatly increased by the decision to route the ill-fated truck through highly populated

---

[57]Melanie Lieberman, Some cruise lines give refunds — others, free drinks and Wi-Fi, Travel + Leisure, September 6, 2017

areas near the coast rather than taking the main road further inland, in order to avoid paying a toll.

Arzamas Freight Train Explosion (USSR, 1988) On June 4 1988, 91 people died, 1500 were injured and more than 150 buildings were destroyed in an explosion in the city of Arzamas. Three goods wagons, containing 121 tons of high explosives for mining, blew up on a freight train bound for Kazakhstan. The cause of the explosion was never established.

Overloading of Ships On September 26 2002, 1863 people died when a ferry capsized in the Atlantic, 20 km off the coast of Gambia. The ship was designed as a river ferry for 550 passengers, but it was in the open ocean with 2000 people on board. The reason for this overcrowding was simple: surface transport between Ziguinchor and Dakar was several times more expensive, much more dangerous and longer than the direct ferry to the capital. Also, the shortage of vessels on this popular route meant that everyone tried to hop on the ferry, which only sailed twice a week. The ship had only one of its two diesel engines working properly. The competence of the crew was also in doubt because most of them had started in the Senegalese Navy, where they had sailed much smaller boats than Le Joola. They were probably far less concerned than they should have been about the overcrowding because the ferry had made a previous trip with more than 3000 soldiers on board. However, on the day of the disaster, there was insufficient water and fuel ballast, a poorly balanced cargo, large numbers of people sleeping on the upper decks and most of the portholes open; all of these factors made the ferry unstable in the rough seas, and liable to sink quickly if it did capsize [58].

The other two cases from Tanzania and South Korea also involved capsizing caused by overloading. Thus on September 10 2011, the ferry MV Spice Islander I capsized and sank between the islands of Pemba and Zanzibar, leaving 1573 dead or missing. The boat was designed to take 645 passengers, but on the day of the disaster 2470 were on board. On 16 April 2014, the MV Sewol sank on its way from Incheon to Jeju, drowning 302 people. The ferry was carrying 3608 tons of cargo—nearly four times its licensed weight of 987 tons—but had too little water in its ballast tanks [59]. On 20 September 2018, MV Nyerere, a passenger ferry on Lake Victoria in Tanzania, capsized due to overloading and the reckless actions of the captain. The ship was designed to carry 100 passengers; on that tragic day it was carrying more than 270 people, 228 of whom perished.

The main approaches to mitigate the risk posed by unsafe storage or overloading are stronger government control over loading and storage in ports and depots, along with the implementation of stricter internal corporate policy: staff should be clearly prohibited from setting out with overloaded vehicles or carrying improperly stored dangerous goods, with rewards for complying with the policy and severe punishment for violations.

---

[58]Pat Wiley, The Sinking of the MV Le JOOLA: Africa's Titanic, 2013

[59]Doomed Sewol carried three times its cargo limit, Korea Joong Ang Daily, May 22,014

Crises Caused by External Factors or Third Parties

Terrorist Acts by Means of Vehicles and on Transport-Related Infrastructure (Worldwide Problem) The largest terrorist event in terms of casualties, and the most sophisticated in its use of airplanes, was undoubtedly 9/11 (2001), when up to 3000 people died. However there have been many other terrorist attacks on transportation or transport infrastructure: in 1980, the attack on Central Station at Bologna (Italy) killed 85 people; in 1985, a Sikh separatist group exploded a bomb on board an Air India Boeing 747 (329 people died); in 1988, Libyan agents planted a bomb in a Pan Am Boeing 747, which crashed on the town of Lockerbie in Scotland (270 people died); in 1995, the Aum Shinrikyo movement in Japan released sarin at five locations on the Tokyo subway (13 died and from 4000 to 6000 were injured); in 2014, a bomb exploded on board SuperFerry 14 near Manila in the Philippines, setting fire to the ferry and partially sinking it (116 died); in 2004, four suburban trains were blown up in the Spanish capital (192 perished and up to 2000 were injured); in 2005, Islamist extremists detonated suicide bombs on the London underground and in a double-decker bus (52 died); in 2006, Muslim radicals planted bombs on commuter trains in Mumbai, India (209 died and more than 700 were injured); in 2006, Madrid–Barajas Airport was bombed by the Basque separatists ETA (2 people were killed); in 2011, an Islamist suicide bomber attacked Domodedovo airport near Moscow, Russia (37 died and 170 were injured); in 2013, months before the Winter Olympic Games in Sochi, suicide terrorists attacked a train station and trolleybus in Volgograd, Southern Russia (34 were killed); in 2015, an Islamic State bomb destroyed a Russian A321 Airbus, with 224 people on board, in the sky above Northern Sinai in Egypt; in 2016, Islamist radicals attacked the airport and a metro station in Brussels (32 perished); and in 2017, an Islamist suicide bomber blew himself up on the Saint Petersburg subway (14 victims), in an attack resembling those on the Moscow metro in 2004 and 2010 (a total of 81 perished in the two attacks).

Sabotage Transportation companies can also face tragic consequences from acts of sabotage or suicide motivated simply by individual madness or despair, with no veneer of supposed political justification. For instance, in 1990, a fire on the ferry "Scandinavian Star", possibly initiated by a Danish passenger with a previous conviction for arson, resulted in the deaths of 158 people. And in 2004, 192 people died in the subway in Daegu, South Korea, when an arsonist suffering from depression and suicidal tendencies set fire to a train. Several factors conspired to increase the number of casualties: another train arrived from the opposite direction, neither driver dealt with the accident effectively, the plastic upholstery produced large amounts of toxic smoke and evacuation measures were chaotic.

Mistaken Attacks on Civil Aircraft by Armed Forces Since we have previously mentioned the Iraq-Iran war of the 1980s, we will keep our description of this airliner accident short. The American cruiser USS Vincennes, armed with a sophisticated anti-aircraft system, was part of the US Navy's presence in the Persian Gulf during the oil tanker war. On July 3 1988, the crew mistakenly shot down an Iran Air commercial flight from Bandar Abbas in Iran to Dubai in the United Arab Emirates. The American Navy explained that they had mistaken the plane for an F-14 fighter jet. All 274 people on board the plane died. Afterwards, the US government paid millions of dollars in compensation to Iran, and to the relatives of non-Iranian passengers.

On September 1, 1983, Korean Air Lines Flight 007 was shot down near Sakhalin island killing 269 people, when a Soviet anti-aircraft jet mistakenly attacked the Korean passenger liner assuming that it was an American spy aircraft; the Korean plane had not responded to requests from the Soviet Air Force for several hours while flying without permission up to 500 km into Soviet air space.

During military exercises in 2001, Ukrainian anti-aircraft forces mistakenly attacked a Russian Siberia Airlines plane flying over the Black Sea from Tel-Aviv in Israel to Novosibirsk in Russia, killing 78 people.

And on July 17, 2014, Malaysia Airlines Flight MH17 was shot down in unclear circumstances while en route from Amsterdam to Kuala Lumpur through the civil war zone of Eastern Ukraine, killing 298 people. Over the following years, both sides of the Donbass conflict blamed each other for the disaster. But whoever was at fault, for civil aviation worldwide this calamity was a pivotal event: since then airlines have tried to route their flights away from military conflict zones with anti-aircraft systems, to avoid being involved in potential provocation and geopolitical struggle as an accidental victim.

Firozabad Rail Disaster (India, 1995) and Level Crossing Collisions (Worldwide) On August 20 1995, the brakes of the Kalindi Express passenger train were damaged in a collision with a cow, and the train was forced to stop near the city of Firozabad. Later, the Kalindi Express was hit from behind by the incoming Purushottam Express, killing 358 people. Worldwide, there are thousands of collisions every year between trains and cars, buses, lorries or pedestrians at level crossings. For instance, in the United States in 2016, according to preliminary statistics, there were 2025 collisions with 265 deaths and 798 injured [60].

Big Bayou Canot Bridge Collision and Train Disaster (USA, 1995) On September 22, 1993, a towboat with six heavy barges got lost in the fog and accidentally collided with the Big Bayou Canot railway bridge in Alabama. The collision shifted a girder by around a meter and bent the rails but without breaking them—arguably the most dangerous situation, since in the event of a broken rail the driver of any incoming train will receive an automatic warning signal. The high-speed Amtrak train Sunset Limited reached the damaged bridge soon after the towboat collision and was derailed. The accident claimed 47 lives and more than a hundred were injured. Emergency services took several hours to reach the remote crash site.

---

[60]Crossing Collisions & Casualties by Year, Federal Railroad Administration, March 5, 2017, https://oli.org/about-us/news/collisions-casulties

Concorde Crash (France, 2000) From January 1976 until July 2000—in more than 24 years of operation—the Concorde supersonic passenger airplane never crashed. But on July 25 2000, disaster struck during take-off from Paris Charles De Gaulle airport. Five minutes before the Concorde took off, a Continental Airlines McDonnell Douglas DC-10 had departed to Newark from the same runway and lost a metal wear strip from one of its engines. When the Concorde began its take-off, a sharp edge of this strip ripped one of its tires and a 4 kg piece of fractured tire damaged the left wing of the jet. This part of the wing was designed to withstand the impact of a 1 kg piece of broken tire; there had been 57 cases of ruptured tires over the 24 years, including some near-miss incidents similar to the take-off from Paris [61]. Because of the sudden pressure change in the left wing, one of the fuel tanks depressurized and fuel began to leak rapidly. Another part of the destroyed tire cut an electric cable in the undercarriage compartment, causing a short circuit that ignited the leaking fuel. Shortly after leaving the runway, the burning plane crashed on a hotel near the airport. A total of 113 people died. In 2010, a Parisian court found Continental Airlines and its mechanics team criminally responsible for the crash; then in 2012, a French appeals court overturned the first ruling, but upheld civil responsibility and the obligation to pay compensation to relatives of the victims.

Piracy In terms of the threat of pirates, the most dangerous shipping areas are the Horn of Africa and the Gulf of Aden (especially off the Somali coast), and the coastal waters off West Africa and near the Strait of Malacca. In 2005, Somalian pirates attacked the cruise ship Seabourn Spirit heading towards Mombasa, Kenya. In 2008, the MS Astor, MS Nautica and MS Athena all came under siege. In 2009, the cruise ship MSC Melody was under threat of capture with 1500 people on board. In 2010, the SS Oceanic faced the same situation. All of these attempts failed because of the active resistance of the crew and the ships' security services. According to calculations by the One Earth Future Foundation, Somali piracy cost the global shipping community between US$5.7 and $6.1 billion in 2012 [62].

Occupy Movement and Suspension of American Ports (USA, 2011) During nationwide protests against Wall Street bankers, activists tried to disrupt the operation of several ports on the west coast of the United States such as Oakland, Long Beach, Portland, Longview and Tacoma.

Cyber Attacks on Transport Companies and Infrastructure Between 2011 and 2013, computer systems to control container shipping at the Belgian port of Antwerp came under cyber attack from drug traffickers, who were sending narcotics in container shipments of bananas from South America [63]. Personal information about 750,000 customers on Japan Airlines' frequent-flier program was leaked

---

[61] Jon Henley, Concorde crash 'a disaster waiting to happen', The Guardian, August 17, 2000

[62] The economic cost of Somali piracy 2012, One Earth Future Foundation, 2013

[63] Tom Bateman, Police warning after drug traffickers' cyber-attack, BBC, October 16, 2013

after a virus attack on 23 computer terminals in September 2014 [64]. In July 2016, hackers got hold of secure data about frequent fliers with Vietnam Airlines. In addition, two Vietnamese airports at Ho Chi Minh City and Hanoi came under cyber attack resulting in the delay of more than 100 flights; a similar attack took place in Warsaw with LOT Airlines in June 2015. In June 2017, A.P. Moller-Maersk, the largest container shipping company in the world with 16% of the global market, faced a severe cyber attack. The Petya virus affected container shipping, port and tug boat operations, oil and gas production, drilling services, and oil tankers and caused up to US$300 million in lost revenue [65,66]. In November 2017, it became known that Uber had paid hackers US$100,000 for their silence about stealing the personal data of 50 million clients and seven million taxi drivers in October 2016. Uber had already been fined by the American authorities for concealing a previous data breach after a hacker attack in 2014 [67].

Transportation companies need to make a more serious assessment of this threat, and coordinate their response with other carriers, local authorities where transport infrastructure is located, domestic and international regulatory bodies, in order to avoid conflict areas.

Other Crises

Struggle for Bilateral Air Services Agreements (Worldwide) The assignment of international air passenger and cargo routes is still the responsibility of national authorities in many countries. Governments try to support their domestic air carriers by limiting the number of airlines that can conduct flights into national airports from abroad. Generally, a government will approve international carriers on a given route if their own national carriers are allowed similar rights and an equivalent share of flights on that route. Competition between rival national carriers for the available flights can be fierce. Some countries operate a more liberalizing system called the Open Sky policy, allowing carriers from both countries to conduct unlimited flights in their skies and to any airport, but the flip side of this kind of policy is that it reduces the marginal value of flights because of the tough price competition between different carriers. Even within the Open Sky policy, air carriers can be in conflict: for example, there has been an ongoing struggle between American, Delta, and United Airlines, Emirates Airlines and Etihad and Qatar Airways for flights between Europe and the United States. The Middle Eastern companies carry out their operations not

---

[64]Megumi Fujikawa, Japan Airlines Reports Hacker Attack, The Wall Street Journal, September 30, 2014

[65]Jacob Gronholt-Pedersen, Maersk says global IT breakdown caused by cyber attack, Reuters, June 27, 2017

[66]Danny Palmer, Petya Ransomware: Cyberattack Costs Could Hit $300 m For Shipping giant Maersk, ZDNet, August 16, 2017

[67]Eric Newcomer, Uber Paid Hackers to Delete Stolen Data on 57 Million People, Bloomberg, November 21, 2017

only from their countries directly to the United States, but from Milan in Italy to JFK in New York, and are resented by their US rivals for receiving multibillion subsidies due to state ownership. In their turn, the Arab companies condemn their American competitors for using US bankruptcy laws to *"shed debt as well as unfavorable business and labor contracts through strategic bankruptcy filings. This is something that all three US legacy carriers have taken advantage of in the past 15 years"* [68].

Catering Errors (Worldwide) 3.7 billion passengers were transported by airlines worldwide in 2016 [69]. During all but the shortest flights, there is a catering service for passengers, food which is provided by outsourced caterers. In some cases, the negligence of caterers towards food safety can threaten the health of passengers. For instance, in 2004, Gate Gourmet, located at Honolulu airport, supplied meals contaminated by Shigella sonnei bacteria to air carriers. Investigation into why some passengers had become ill led to the facilities of Gate Gourmet, where violations of hygiene regulations were discovered in the storage of food, the cleanliness of production and in other areas: all these factors had contributed to the bacteria flourishing [70].

Environmental Restrictions and Changes in Legislation (Worldwide) The toughening of emissions and noise restrictions can damage the business of transportation companies. For instance, in the 2000s, the EU imposed stricter limits and increased taxes for flying the older Soviet-style aircraft over Europe because they were so much noisier than their more modern counterparts. Implementation of these new limits made operating the Soviet era planes unprofitable for airlines from the former Soviet Union, which were forced to buy Boeings and Airbuses.

In order to reduce CO_2 emissions, and appeal to the green inclinations of its predominantly German market, the AIDA cruise company decided to build the first cruise liner in the world fuelled solely by liqufed natural gas instead of diesel. The ship was called AIDAnova, and followed on from the success of two dual-fuel cruise liners, AIDAprima and AIDAperla. The shipbuilders also received orders from Costa Cruises and P&O Cruises for LNG liners; like AIDA, these cruise companies are part of the American Carnival Corporation group.

Downturn and Demand Reduction on Transport Services (Worldwide) The general reduction in both global and national business activity has an impact on global supply chain intensity and therefore on the business of transportation companies; over time, this negative impact will also work in the other direction.

---

[68]Benjamin Zhang, The Middle East's 3 best airlines have infuriated their US competitors, Business Insider, July 28, 2015

[69]Another Strong Year for Air Travel Demand in 2016, International Air Transport Association, February 2, 2017

[70]Gate Gourmet Shigella Outbreak Lawsuits - Hawaii, Nationwide (2004), Marler Clark, April 1, 2008, http://marlerclark.com/news_events/gate-gourmet-shigella-outbreak-hawaii-nationwide

> **Key Risk Mitigation Measures in Transportation**
>
> - Careful recruitment, ongoing training and retention of skilled personnel (pilots, technicians, attendants, etc.) in order to maintain safe and high-quality transportation services.
> - Proper investment in, and safe and sustainable use of, transportation infrastructure: careful loading, regular service checks, timely updates on maintenance, etc.
> - Close cooperation with suppliers of transportation equipment to exchange risk information about the condition of vehicles and their performance in all situations, ensuring that repairs and updates are carried out in time and thus preventing accidents.
> - Timely and accurate weather forecasting for safe transportation.
> - Ensuring adequate protection or military cooperation to convey people and goods safely in conflict or war zones, or adequate information to set a course to avoid conflict areas.
> - Close coordination with law enforcement for the prevention of terrorist attacks, acts of sabotage, cyber attacks and other criminal action by third parties.

3.3 Hotels and Restaurants

The subsector includes the following industries according to ISIC Rev.4:

- 55—Accommodation
- 56—Food and beverage service activities
- 79—Travel agency, tour operator, reservation service and related activities

General Description and Key Features of this Subsector

The HORECA (HOtel/REstaurant/CAfe) subsector is one of the most labor intensive subsectors, which requires the employment of large numbers of minimally-qualified staff to carry out routine and limited tasks. Because of the low wages and status of the work, there is a high staff turnover. Since businesses can be small—perhaps even family owned—and require relatively low capital investment, they are easier to start up and there is high competition in the subsector. On the other hand, there is a high business mortality rate: it is easy to misjudge customers' needs and hard to make a decent profit. According to researchers at Ohio State University, 26% of independent restaurants fail during their first year of operation, 45% in their second year and almost 60% within 3 years [71]. The industry tends to duplicate the same pattern of service through the development of extensive networks under one reputed brand. In some cases, franchising is used to enable the rapid development of a network over a wide geographical area.

Critical Success Factors for an Organization Within the Subsector

A good reputation for customer care is the key competitive advantage in the subsector, so finding and retaining staff who genuinely want to serve and look after customers is the most critical factor for success. Hotels and restaurants have to give their customers high-quality service at reasonable prices. For this reason—and to be able to deal with potential emergencies with customers—cooperation with reliable suppliers is essential. The maintenance of service infrastructure, in good shape and in safe condition, is essential to the success of any HORECA business. If the business is a chain, like McDonalds or Hilton Hotels, the final important success factor is its ability to provide a consistent quality of service at all its service points around the world.

Stakeholders in this Subsector [72]

- Customers—45%
- Employees—35%
- Suppliers—10%
- Other—10%

Typology of Common Risks, Main Features of Major Accidents and Risk Mitigation Measures Within the Subsector

In our discussion of the risks facing the hospitality subsector, we will not touch on business risks like the wrong choice of location, misjudgment of target audience or pricing, menu problems, unsuccessful design and branding, conflicts with franchise partners, or the impact of wider economic depression.

---

71 H. G. Parsa, John T. Self, David Njite, Tiffany King, Why Restaurants Fail, Cornell Hospitality Quarterly, August 1, 2005

72 Our informed appraisal of the influence of each audience on a typical organization within the subsector (100% = combined influence of all audiences)

Poor Client Services

Shoddy customer care or obvious negligence of the needs of customers creates most crisis situations in the subsector. Most of the ruling disputes stay undetected—not only by the public, but even by the management of service organizations—because angry and disappointed clients just vote with their feet and try to find better service elsewhere. In some cases, clients demand to speak with a manager; they may even publish their negative reviews on special forums, blogs and social media to attract the attention of other clients and the general public, and thus "*punish*" the negligent service provider. The majority of conflicts occur when the action or inaction of service staff is perceived as inadequate by customers; in many cases, the service breakdown is due to non-human factors, but the staff don't explain the situation to customers, or fail to deal with it effectively. Especially with the advent of social media, poor customer care at one service point can damage the whole network of the brand in question: reading a bad review, other potential clients will assume that, if such shoddy or disrespectful service was allowed to occur in one place, it is likely to be repeated elsewhere. Poor service could include: the failure to keep premises clean, staff not complying with hygiene requirements or being allowed to show up at work with an infectious disease; slow, inadequate or even abusive responses to customer requests; violation of cooking procedures, food storage guidelines or regulations on the serving and mixing of alcohol; short-changing of clients or stealing their valuables; failure to inform clients with allergies about food ingredients; inability to provide first aid in the event of a life threatening emergency; or strikes in working hours that impact on clients. There are numerous outrageous examples of such behavior from service staff.

Arson at Dupont Plaza Hotel (Puerto Rico, 1986) 98 people died and more than a hundred were injured when a labor-management conflict turned ugly: three employees with a grudge against hotel management deliberately started a fire in order to punish the hotel owners. However, the fire soon became uncontrollable, spreading from the first-floor ballroom to the foyer, lobby, entrance and casino and ultimately filling the whole building with toxic smoke. Most of the people who died were in the casino: unbelievably, the emergency doors had been locked to prevent gamblers walking away with the proceeds. Because the accident took place during the afternoon of December 31, 1986—when the 22-story hotel was almost fully occupied by nine hundred guests—the evacuation was a challenge, which lasted through until the following evening and even required helicopter pickups from the hotel roof. A similar event occurred in 1982 at the Dorothy Mae Apartment Hotel in Los Angeles. The young nephew of the hotel manager, after a quarrel with his uncle, ignited gasoline on the floor of the apartment. The fire killed 25 people, and the arsonist was sentenced to 625 years in jail [73].

---

[73]George Ramos, 'It was as if the devil himself lived there.', Los Angeles Times, March 22, 1988

E. coli Outbreak at Jack in the Box (1993) This fast food chain promoted the unfortunately-named Monster Burger across its network. Demand was so high that staff at some Jack in the Box outlets failed to cook the meat for the burger properly. Unfortunately, some of the meat patties supplied to the network were contaminated by E. coli O157:H7. If the chain had complied with stricter regulations on meat roasting, following earlier warnings from both state health regulators and employees, the outbreak would never have occurred [74]. Due to the contamination, four children died and more than 700, mostly under 10 years old, were infected. The risk of an E. coli outbreak from undercooked meat was well known to the industry and to regulators, because two outbreaks had occurred at McDonald's in 1982 and 1983 [75].

Shigella Outbreak at Subway Restaurant in Illinois (USA, 2010) The outbreak, which affected at least 116 customers, occurred because of *"generally poor hygienic conditions and workers carrying the Shigella bacteria"* [76].

Prank Video by Two Domino's Pizza Employees (USA, 2009) and Burger King's Lettuce-Stomping Employee (USA, 2012) In both cases kitchen workers at the restaurants filmed and posted online videos of themselves breaking hygiene regulations in an intentionally disgusting way: a Domino's employee was filmed picking his nose with a piece of food and spitting into dishes being prepared, and a worker at Burger King appeared standing in two containers of lettuce, which were then used in food destined for customers.

Footlong Scandal at Subway (Global, 2013) In its advertising, the fast food chain boasted that its Footlong sandwiches really were a foot long—i.e. 12 in. or around 30 cm—but in 2013 some Subway branches were selling sandwiches of 11 or 11.5 in. This fact came to the notice of one customer from Australia, who posted his discovery on Facebook and evoked an enthusiastic response from all around the world. Subway was ridiculed for—literally—shortchanging its customers. The chain pleaded guilty as charged, and guaranteed that they would serve genuine 12-in. sandwiches at all their branches from then on [77].

Sweets Tainted by Pesticides (Pakistan, 2016) Implausibly enough, a sweet bakery was storing pesticide during the renovation of a neighboring pesticide shop in Punjab province, Pakistan. During a bitter feud between the two brothers who owned the bakery, the younger brother decided to take vengeance on the elder (*"I wanted to teach him a lesson"*) and mixed one bottle from the pesticide stock

---

74 Bill Marler, Publisher's Platform: Private AG, Food Safety News, August 21, 2011, http://www.foodsafetynews.com/2011/08/publishers-platform-private-ag

75 Jeff Benedict, Poisoned: The True Story of the Deadly E. Coli Outbreak That Changed the Way Americans Eat, Inspire Books, 2011, p.52

76 Subway Restaurant Shigella Outbreak Lawsuits - Illinois (2010), Marler Clark, http://marlerclark.com/news_events/subway-restaurant-shigella-outbreak-chicago

77 Size matters: Subway promises its Footlong sandwiches will be... a foot long, Associated Press, March 1, 2016

into some pastry ready for baking ("*I was so angry that I mixed the pesticides bottle in the sweets being baked at that time*"). Five kilograms of the tainted sweets was bought by one customer, who was celebrating the birthday of his grandson. He distributed the sweets among his relatives and friends. Tragically, 30 people died of food poisoning [78].

Measures to prevent such disastrous breakdowns in customer service include: careful selection of field staff; development of handbooks and guidelines on service standards; ongoing training of service staff to improve professional skills and customer care skills; development of service tutors; the promotion of a corporate culture based on service excellence; competition between service teams to "*go the extra mile*" in providing the best service; bonuses for employees who have given exceptional client service; employing mystery shoppers to assess staff independently; and advanced claim management to ensure a fast response to every complaint.

Supply Failures

Supply breakdowns, whether unintentional or deliberate on the supplier's part, can severely damage the business of a HORECA company. Common examples include the provision of raw food that is contaminated or in poor condition, buildings that are poorly constructed or contain highly flammable materials, or sudden power cuts.

Hepatitis a Outbreak at Chi-Chi Restaurant (USA, 2003) The Chi-Chi network of Mexican restaurants faced an outbreak of Hepatitis A caused by a delivery of onions from Mexico to its restaurant in Beaver Valley Mall, Pennsylvania. Four people died, 650 were infected and 9000 clients and staff members received preventive immunoglobulin injections. Chi-Chi's business was already unstable—the company had filed for bankruptcy—but the outbreak was the last straw, and in 2004 the majority of Chi-Chi restaurants in the United States and Canada were closed [79].

Infection Outbreaks at Taco Bell (USA, 2006, 2010) 71 customers in several states became sick after eating food infected with E. coli O157. According to the CDC, "*Evaluation of all [the] data indicates that shredded lettuce consumed at Taco Bell restaurants in the northeastern United States was the most likely source of the outbreak*" [80]. Then in 2010, 155 customers went down with salmonella, which was again traced to consumption of Taco Bell lettuce.

Peanut Corporation of America Salmonella Outbreak (USA, 2009) This outbreak cost nine lives and affected 714 customers in 46 states. A FDA investigation

---

[78] Pakistani shopkeeper killed dozens with 'revenge' sweets, AFP, May 6, 2016

[79] Bill Marler, Publisher's Platform: Private AG, Food Safety News, August 21, 2011, http://www.foodsafetynews.com/2011/08/publishers-platform-private-ag

[80] Multistate Outbreak of E. coli O157 Infections Linked to Taco Bell, US Centers for Disease Control and Prevention, December 14, 2006

revealed appalling sanitation conditions at two plants of the Peanut Corporation of America (PCA) where there were cockroaches, rats, mold, dirt, accumulated grease, a leaky roof and bird droppings. Not surprisingly, further investigation of PCA's business practices followed, revealing that salmonella had been detected in PCA foods in a private lab test long before the outbreak—but that the PCA's chief executive had ordered employees to continue supplying the contaminated products. On hearing from the manager at one of the plants that a shipment of peanut butter had been put on hold awaiting tests, the CEO notoriously emailed: *"Just ship it"* [81]. This butter was not sold directly to retail customers, but distributed to industrial clients—food manufacturers, bakeries, restaurants, cafeterias and other service providers—to be used in the manufacture of ice cream, crackers, cookies, cereal, pet treats, and so on. In the end, more than 2833 peanut-containing products were affected and had to be recalled by the PCA [82]. The company went bankrupt, and the owner and CEO was convicted of 70 criminal charges and sentenced to 28 years in jail in September 2015.

Meat Adulteration Scandal at Burger King (Ireland and UK, 2012–2013)
Unknown to their corporate customers, several European slaughterhouses and meat processing plants were adding horsemeat and pork to their beef products. Horsemeat poses no threat to humans, but some veterinary drugs used to maintain the health and performance of racehorses could be dangerous to people who consumed the meat. Along with chains of food stores and supermarkets, the adulterated meat was supplied to Burger King without the knowledge of the fast food chain. As a result, Muslim and Jewish clients of Burger King consumed not only horsemeat, but also pork, which is prohibited by their faith.

Expired Meat Supplied to Fast Food Chains (China and Japan, 2014) Shanghai meat supplier Husi Food Co. delivered expired chicken and beef—repacked and relabeled as fresh meat—to its corporate customers in China and Japan, including Starbucks, Burger King, McDonald's, KFC, Pizza Hut and others [83,84]. A similar fraud was discovered in Taiwan in 2014, where cooking oil was produced from recycled waste oil—collected from restaurant fryers or discarded animal parts, fat and skin—by Chang Guann Co. and supplied to 1256 bakeries, restaurants, food manufacturers and other corporate customers [85].

---

[81] Moni Basu, For first time, company owner faces life sentence for food poisoning outbreak, CNN, September 21, 2015

[82] Multistate Outbreak of Salmonella Typhimurium Infections Linked to Peanut Butter, 2008–2009, US Centers for Disease Control and Prevention, May 11, 2009

[83] Joe McDonald, Starbucks, Burger King Affected by China Meat Scandal, Associated Press, July 22, 2014

[84] Lucy Hornby, McDonald's and KFC hit by China food safety scandal, Financial Times, July 21, 2014

[85] Stacy Hsu, Chang Guann fined NT$50 m over oil, The Taipei Times, September 10, 2014

Counterfeit Alcohol in Bars and Restaurants (Russia, 2010s) In the first 9 months of 2016, there were more than 9300 deaths and 36,000 cases of serious poisoning from the consumption of adulterated or low quality alcohol in Russia [86]. Some cases occurred in expensive and respectable restaurants and bars, the owners of which had received fake alcohol from their suppliers.

Power Outage at Paris Hotel & Casino in Las Vegas (USA, 2016) The vacation of around 3000 guests of the hotel was interrupted when the whole building was evacuated after a power cut caused unintentionally by a contractor working in the basement. The hotel and casino only fully recovered after 20 h.

Measures to tackle the risk of supply problems could include: companies taking direct control over the production process and delivery of raw material from their suppliers, in order to mitigate potential emergencies with customers; in-house testing of the quality of supplied materials; the development of wide networks of suppliers in order to mitigate the risk of a delivery blackout due to problems with one particular supplier; advanced client claim management to detect possible risks; and even, if it is economically feasible, the development of in-house production facilities to take full-scale control over the production process.

Infrastructure-Related Crises

Fires, slippery floors or stairs, sewage failures, poor ventilation or inadequate gas and smoke extraction from cooking areas could threaten the lives of customers or at least leave clients dissatisfied with their visit. Of course, the most challenging infrastructure-related risk is a fire at a service point.

The US National Fire Protection Association provides countrywide statistics on hotel fires for 2009–2013. During this period, there were 3520 fires in hotel buildings annually, with nine civilian deaths and 120 civilian injuries. Half of the events and 25% of the civilian deaths were connected with cooking equipment; 75% of deaths were from fires set off by smoking materials; 11% of fires were caused by electrical failure or the malfunction of electrical equipment [87]. The worst fire tragedies include: a skyscraper fire at a hotel in Seoul, South Korea in 1971—there were 164 deaths, mostly because the building was poorly designed for safety in emergencies; a fire at the MGM Grand in Las Vegas, USA in 1980 killed 85 people and injured more than 600, mostly through smoke inhalation, after a short circuit in the restaurant.

Statistics from the US Fire Administration show that, in this country alone, there are around 5900 restaurant fires every year, though with only 75 injuries, 59% of

---

[86]Concerning poisoning by alcohol-containing products, Russian Federal Service for Supervision of Consumer Rights Protection and Human Welfare, November 24, 2016, http://rospotrebnadzor.ru/about/info/news/news_details.php?ELEMENT_ID=7405

[87]Richard Campbell, Structure Fires in Hotels and Motels, National Fire Protection Association, September 2015

which are associated with cooking (91% of cooking fires are assessed as small) [88]. Fire accidents at nightclubs and discotheques tend to be far worse. Some examples of the consequences of malfunctions with electrical equipment: a fire at the Beverley Hills Supper Club in Kentucky, USA in 1977 led to 165 deaths; and in 1996 at the Ozone Disco Pub in Manila, Philippines, fire claimed at least 162 lives. Gas leaks and subsequent explosions are also very dangerous: 64 people died in a gas explosion at a restaurant in Taichung City, Taiwan in 1995; a gas cylinder exploded at a restaurant in the Indian state of Madhya Pradesh in 2015, causing 104 deaths; and in April 2018, the explosion of a carbonic acid gas cylinder at a Burger King in the Armenian capital Yerevan injured nine customers, mainly children.

Fires set off by fireworks or pyrotechnics going out of control during a performance include incidents at the Station Nightclub in Rhode Island, USA in 2003, when fire and toxic smoke from the polystyrene sound insulation caused 100 deaths; at the Republica Cromanon Nightclub in Buenos Aires, Argentina in 2004, when 194 died; at the Santika nightclub in Bangkok, Thailand in 2009; at the Lame Horse nightclub in Perm, Russia in 2009 (156 deaths); at the Kiss nightclub in the state of Rio Grande do Sul in Brazil in 2013 (242 deaths); and at the Colectiv nightclub in Bucharest, Romania in 2015 (64 deaths).

Measures to minimize fire risks include: proper design of service points; timely investment in basic equipment, such as fire extinguishing systems—well located fire extinguishers, fire blankets, smoke detectors and sprinkler systems; careful storage of dangerous materials like cooking oil, gas cylinders and cleaning agents; regular staff training in fire containment and evacuation procedures; close coordination with emergency services on mutual fire drills, communication and independent assessment of risks; strict prohibition of fireworks inside or outside service points; and clear fire instructions for clients.

Other Risks

Terrorist Acts against Hotels, Restaurants and Nightclubs In 1946, the British administration for Palestine, which was located at the King David Hotel in Jerusalem, was attacked by Zionist militants, who planted a bomb in the basement of one wing of the hotel; 91 people died including 28 British nationals. In 2002 more than 200 people, the majority of them Australian tourists, were killed after Islamist radicals made a suicide attack at a nightclub and planted a car bomb nearby at the resort of Kuta Beach in Bali, Indonesia. On November 26, 2008, during a Pakistan-sponsored attack by terrorists on the Indian city of Mumbai, the targets included places that were popular among foreigners such as the Leopold Cafe, the Taj Mahal Palace and the Oberoi Trident hotels; in the attacks a total of 165 people died

---

[88]Restaurant Building Fires, US Fire administration, Topical Fire report Series Volume 12, Issue 1, April 2011

including 26 foreigners, and 304 were injured [89]. On June 26, 2015, an Islamist radical shot 38 European tourists at the Riu Imperial Marhaba Hotel near Sousse in Tunisia. On November 20, 2015, 20 people died in an Islamist terrorist attack on the Radisson Blu hotel in Bamako, Mali. There have been several attacks on hotels in Afghanistan and Pakistan by the Taliban and other movements.

Prevention measures include: close collaboration with state security agencies for the assessment of possible risks; coordinated action in the event of early warnings or changes in the terrorist threat level; rapid response to attacks by special forces; and mutual drills and training for security agencies and the staff of service organizations, in order to enhance the survival skills of staff—and thus improve the survival chances of guests—in the event of a hostage crisis.

Natural Hazards Extreme weather events can pose a danger to guests, staff and infrastructure. A massive avalanche in Galtur, a popular ski resort in the Austrian Alps, claimed 31 lives on February 23, 1999. The resort was in an area which had never experienced such a destructive avalanche, so the death toll and the damage to civil and tourist infrastructure were tremendous. This incident prompted a reassessment of which zones should be considered "safe", and the erection of a special anti-avalanche dam in addition to the wider reconstruction of damaged buildings. While the region responded to the disaster, several nearby ski resorts were closed at a loss of around £5 million a day [90].

In December 2004, the Sumatra–Andaman earthquake and resulting tsunamis sent a knockout blow to the tourist industry on the west coast of Thailand: at least 1953 foreign tourists died, more than 250 hotels were affected, and there was a shortfall of more than US$1 billion in tourism revenues in the first quarter of 2005 [91]. It also harmed the tourist sectors of Malaysia, Sri Lanka and the Maldives, because tourism in these countries too is very much geared towards beach and water-sport vacations.

Climate change could damage the long-term prospects of the whole industry. For example, according to predictions by the Institute for Snow and Avalanche Research and the CRYOS Laboratory, if temperatures continue to rise at their current rate, the Alps could face a 70% reduction of snow cover by the end of the century: below 1200 m, snow will completely disappear [92]. Obviously, the business of many Alpine ski resorts will be severely affected by the absence of stable show cover during the winter months. In this changing picture, a well-informed assessment of

---

[89]Mumbai Terrorist Attacks (Nov. 26–29, 2008), Government of India, January 5, 2009, https://fas.org/irp/eprint/mumbai.pdf

[90]Case study: Galtür Austria, BBC, http://www.bbc.co.uk/schools/gcsebitesize/geography/glacial_landscapes/avalanches_rev2.shtml

[91]The Indian Ocean Tsunami of 26 December 2004: Mission Findings in Sri Lanka and Thailand, Institution of Structural Engineers, 2006, p.165

[92]Christoph Marty, Sebastian Schlögl, Mathias Bavay, and Michael Lehning, How much can we save? Impact of different emission scenarios on future snow cover in the Alps, The Cryosphere, February 16, 2017

nature-related risks must be accompanied by investment into adequate infrastructure and continuous training of service staff in dealing with such emergencies.

Strikes of Personnel Staff may give excellent service but still go on strike, or struggle for better wages, a less exploitative schedule or holiday pay. In recent years, current or former staff of the Hyatt hotel network, McDonald's, Burger King, Domino's, Wendy's, Papa John's, KFC, and Pizza Hut, all participated in strikes, boycotts or disputes with employers in the United States. Management should monitor the general sentiment of workers, and try to work together with them to achieve what they are campaigning for in a mutually beneficial way. The development of nonfinancial motivations through promoting a strong corporate culture, and a fulfilling working environment, is also recommended.

Outrageous Behavior of Guests Though guests are the most important audience for a HORECA company, some of them can at times cause serious problems through their behavior when drunk or affected by other drugs—disturbing other clients or even damaging infrastructure. In some cases, clients could become arsonists, as in the fires at the Stardust Disco in Ireland in 1981 (48 deaths) and at a discotheque in Gothenburg, Sweden in 1998 (63 deaths). The response should be immediate from security staff and local police: complete exclusion of the offending guest and appropriate penalty for the damage caused.

Key Risk Mitigation Measures in Horeca

- Reducing the rate of staff turnover through the recruitment and retention of people who take genuine pride in looking after others, and are willing to work in stressful conditions for a modest salary.
- Brand protection by ensuring that management respond immediately and generously if any conflict or dispute arises with clients, suppliers, authorities, etc.
- Careful selection of suppliers of food, equipment, and other services because substandard quality in anything supplied to guests threatens the reputation of the HORECA provider.
- Maintenance of service infrastructure in good shape and in safe condition.

3.4 Information Services

This subsector includes the following industries according to ISIC Rev. 4:

- 60—Programming activities
- 62—Computer programming, consultancy and related activities
- 63—Information service activities

General Description and Key Features of the Subsector

These industries provide services by programming and maintaining software—from SCADA (Supervisory Control And Data Acquisition) and ERP (Enterprise Resource Planning) to simple web-sites. The originality and quality of the content mainly depend on creative and highly skilled specialists who can conceive of and develop innovative and unique solutions. The development of software is generally a project–oriented activity, while the maintenance of existing programs implies regular updates and ongoing improvement of codes where vulnerabilities have been detected. The ease with which content can be copied and transmitted all around the world in seconds makes government authorities an important player in these industries as global controllers of copyrights. This is a very competitive environment: only a small minority of the services "produced" and of the start-ups launched will be recognized as unique and fundamental, and do well enough to dominate the market.

Critical Success Factors for an Organization Within the Subsector

The critical factor in this subsector is the ability to create unique content that can attract the attention of customers all around the world, content of such originality that people cannot imagine anyone else having created it. Conceiving and developing such groundbreaking ideas mainly depends on finding people of inventive genius, recruiting and retaining highly qualified and motivated staff. Securing adequate financial reward for the production of original content requires intensive collaboration with the authorities, who control the value-added distribution of content through copyright laws, judicial authority and law enforcement. In some cases, a development company will need to attract distributors, resellers or integrators to deliver highly specialized versions of a software solution to individual customers according to their unique needs; thus the careful selection of such intermediaries between the vendor and corporate users, and continuous training of their staff to keep them up to speed with new software as it emerges, are also important for the development of the vendor's business.

Stakeholders in this Subsector [93]

- Employees—40%
- Customers—25%

---

[93]Our informed appraisal of the influence of each audience on a typical organization within the subsector (100% = combined influence of all audiences)

- Authorities—20%
- Distributors—10%
- Other—5%

Typology of Common Risks, Main Features of Major Accidents and Risk Mitigation Measures Within the Subsector

"Product" Related Crises

The main risks for a software project are of overrunning on budget and/or schedule, or failing completely because the intended result simply can't be programmed, requires a complexity of code which makes it unviable, or cannot be delivered for other reasons. In a survey of IT projects by American large companies, the Standish Group found that only 9% of the projects were successful: 61% of them were challenged, 31% were cancelled before completion, 52% went over-budget by an average of 189%, and 42% did not have the originally-proposed features and functions [94]. Another study by McKinsey and the University of Oxford, assessing the progress of 5400 large enterprise software projects, found that 66% of them were over budget, 33% had overrun, 17% made less money than expected and 17% were so poorly developed that they wasted enough money to threaten the survival of the company. To make IT projects more successful, McKinsey recommended that they focus on strategy and stakeholders during the development stages rather than on the exact budget and strict timetable, "*mastering technology and project content by securing critical internal and external talent*", "*building effective teams*" and "*excelling at core project-management practices*" [95]. These conclusions support our contention that the staff of a software project is critical for its success in creating IT solutions.

Venus Mission Failure (USA, 1962)   The Mariner-1 spacecraft was ordered to self-destruct in the first few minutes of its flight because the rocket deviated from its planned course. In addition to a malfunction of the radio beacon equipment, the back-up guidance codes for the navigation system that should have kept the rocket on course contained a repeated error: when the programmer was converting his original handwritten equations into digital codes, he mis-transcribed one symbol (a superscript bar) which cropped up several times.

Massive Radiation Overdoses from the Therac-25 (USA-Canada, 1985–1987)
At least six cancer patients were affected, some of whom were almost certainly

---

[94]CHAOS, The Standish Group Report, 2014, https://www.projectsmart.co.uk/white-papers/chaos-report.pdf

[95]Michael Bloch, Sven Blumberg, Jurgen Laartz, Delivering large-scale IT projects on time, on budget, and on value, McKinsey on Finance, Number 45, Winter 2013, pp.29–30

killed, by radiation overdoses from a Canadian radiotherapy machine called the Therac-25. When the cases were investigated, two mistakes were discovered in the programming codes, which complicated the coordination between software, hardware and service staff in determining the proper radiation dose for each patient. The previous model, the Therac-20, had also had programming errors—but it was equipped with special hardware that prevented overdosing. Ultimately, the Therac-25 had to be recalled.

Explosion of the Ariane 5 Rocket (French Guyana, 1996) On June 4 1996, the inaugural launch of the European heavy satellite delivery rocket Ariane 5 from Kourou spaceport in French Guyana failed. The subsequent investigation found that, in programming the inertial reference system responsible for measuring the position and movements of the spacecraft, programmers had used codes from the previous Ariane 4 rocket, without taking into consideration the new flight conditions of the more powerful Ariane 5. During the first flight, there was an operand error or *"arithmetic overflow"* during conversion of the value for horizontal bias, which was much larger than that during the Ariane 4 launches, from a 64-bit floating point number into a 16-bit integer signed format. This caused a hardware exception, which shut down the active and backup inertial reference systems. The resulting failure to send ongoing guidance and up-to-date attitude readings to the on-board computer led to a catastrophic deviation of the rocket from its course; suddenly put under intolerable aerodynamic forces, it disintegrated [96].

Mars Climate Orbiter Failure (Mars, 1999) and Launch Failure of Soyuz/ Fregat Upper Stage (Russia, 2017) The Climate Orbiter's mission was to study the atmosphere of Mars, but when the time came for the craft to maneuver into its orbit over the Red Planet, the thrusters accelerated too strongly. Instead of starting its orbit, the craft disintegrated in the planet's dense atmospheric layers. According to the head of the Mars Climate Orbiter Mission Failure Investigation, *"The 'root cause' of the loss of the spacecraft was the failed translation of English units [pounds-force seconds] into metric units [Newton-seconds] in a segment of ground-based, navigation-related mission software... The failure review board has identified other significant factors that allowed this error to be born, and then let it linger and propagate to the point where it resulted in a major error in our understanding of the spacecraft's path as it approached Mars"*. There were also measurement and communication failures between NASA and the subcontractors responsible for different parts of the Orbiter [97].

A quite similar situation caused problems for ROSCOSMOS (the Russian State Corporation for Space Activities) when the corporation opened its Vostochny

---

[96]ARIANE 5, Flight 501 Failure, Report by the Inquiry Board, The Chairman of the Board: Prof. J. L. Lions, Paris, July 19, 1996

[97]Douglas Isbell/Don Savage, Mars Climate Orbiter Failure Board Releases Report, Numerous NASA Actions Underway In Response, NASA Headquarters, Washington, DC, November 10, 1999, https://mars.nasa.gov/msp98/news/mco991110.html

cosmodrome in the Far East of Russia in 2016, to extend the operations of the existing Baikonur cosmodrome in the former Soviet republic of Kazakhstan. Until the commissioning of Vostochny, Baikonur had been the main launch pad for Soviet and Russian rockets since the end of the 1950s, when Earth's first artificial satellite was launched. During the launch of the Soyuz/Fregat freight launch vehicle from Vostochny on November 28, 2017, the vehicle deorbited and fell into the Atlantic Ocean. Previously, all 61 Soyuz launches, mainly from Baikonur, had been successful. An investigation concluded that a small hidden piece of code in the software of the ill-fated vehicle was adapted for the coordinates of Baikonur, and the code had not been adjusted to the location of Vostochny: *"there was such a combination of the parameters of the launch pad of the cosmodrome, the azimuths of the launch vehicle and the upper stage, which had not been met before"* [98].

Mapping Error at St. Mary's Mercy Medical Center (USA, 2003) During an update of the center's software, there was a mistake in mapping the status of 8500 recently discharged patients: instead of code "01", they were marked as "20", which was the system's code for a deceased patient. To make matters worse, the center duly informed the "expired" patients' insurance companies of their supposed demise— including Medicare, which subsequently suspended bill payments for their still very much alive clients [99]. Quite a similar case occurred in 2005 when, due to a software error within the Michigan Department of Corrections, 23 prisoners were released earlier than the courts had decreed.

Delay in Updating the Refund Monitoring Software Within the IRS (USA, 2006) In 2006, the Internal Revenue Service, the US tax enforcement service, lost between US$200 million and $300 million in improper refunds. The program that monitored potentially fraudulent refunds was not updated by a contractor in time for the 2006 filing deadline, and there was no time even to reinstall the old version to monitor the 2006 returns [100].

Errors in Windows Genuine Anti-Piracy Program (Worldwide, 2007) Microsoft tried to detect illegal usage of its products by activating Windows Genuine Advantage in its operation systems. In August 2007, more than 1000 legally bought Microsoft operation systems were accidentally blocked, and the owners were accused of using illicit versions by Windows Genuine Advantage. The outage lasted 19 h. Microsoft subsequently explained the cause of the crisis: *"Nothing more than human error started it all. Pre-production code was sent to production servers. The production servers had not yet been upgraded with a recent change to enable stronger encryption/decryption of product keys during the activation and validation*

---

[98]Conclusions of the Emergency Commission, ROSKOSMOS, December 12, 2017, https://www.roscosmos.ru/24451

[99]Larry Barrett, Hospital Revives Its "Dead" Patients, Baseline, February 10, 2003, http://www.baselinemag.com/c/a/Projects-Networks-and-Storage/Hospital-Revives-Its-QTEDeadQTE-Patients#sthash.wu4IVo2d.dpuf

[100]Mary Dalrymple. IRS computer glitch costs US millions, Associated Press, July 15, 2006

processes. The result of this is that the production servers declined activation and validation requests that should have passed" [101].

Windows Vista (Worldwide, 2007) Microsoft released Vista to replace their widely popular XP operation system in 2007, but reviews by computer critics slated the new system: it was slower than XP and required very powerful hardware to function at all. The product could not command a dominant market share and was finally replaced by Windows 7 in 2009.

Poor Quality of New Apple Maps (USA, 2012) Until the release of the iOS 6 in 2012, Apple used mapping from Google for its iPhones and iPads. From this point, the company began to rely on its own mapping product. However, in the initial months after its release, the quality of Apple Maps was widely criticized: there were so many mapping mistakes and bugs that many users found themselves lost when trying to navigate by the new maps. The most infamous case occurred in Australia, when people trying to reach the town of Mildura were directed by the application to a spot 70 km away in the Murray-Sunset National Park, a deserted area of bush. Several had to be rescued by local police, who were concerned that it was *"potentially life-threatening"* driving through the park during the scorching summer heat [102]. Though Apple insisted that 99% of the content of the maps was correct, even a 1% mistake rate in a navigation app designed for driving could potentially put customers' lives at risk. Apple apologized to users, assuring them that it was making constant improvements to the new app, asking for their active participation in spotting incorrect information and recommending that they also use alternative apps until Apple Maps was 100% accurate.

Failure of HealthCare.gov (USA, 2013) President Obama's long-awaited Patient Protection and Affordable Care Act committed to the launch of HealthCare.gov by October 1, 2013. The site was supposed to facilitate the provision of compulsory health insurance, but at reasonable prices, for millions of Americans—and during the first 3 weeks after the launch, up to 20 million visited it. However, the web-site had so many glitches that only a fraction of insurance seekers were able to apply smoothly online: of the 0.5 million subscribers predicted before the launch to apply in the first month, there were only around 27,000 who successfully chose health insurance plans by the end of October [103].

At first, the technical mistakes were blamed on the immediate surge of visitors to the site; but later analysis showed that there were software design flaws, complexities in the programming code and other deficiencies in the roll out of the project—all of which had contributed to the technical failures, and to a serious budget overrun.

---

[101]So what happened?, MSDN Archive, August 29, 2007, https://blogs.msdn.microsoft.com/wga/2007/08/29/so-what-happened

[102]Nick Thompson, Apple Maps flaw could be deadly, warn Australian police, CNN, December 11, 2012

[103]Sam Baker, It's Official: Obamacare Enrollment Is Super Low, The Atlantic, November 13, 2013

According to Harvard professor David Cutler, staff in the Obama administration *"were running the biggest start-up in the world, and they didn't have anyone who had run a start-up, or even run a business. It's very hard to think of a situation where the people best at getting legislation passed are best at implementing it. They are a different set of skills"* [104]. For example, the government body responsible for the project—the Centers for Medicare & Medicaid Services or CMS—gave 60 separate contracts to 33 different companies for *"individual tasks to support the implementation [of the site], but there was no single point-of-contact with responsibility for integrating contractors' efforts and communicating the common project goal to all 33 companies"* [105]. The selection of contractors was flawed, in some cases omitting any assessment of the contractors' competence to develop complex technical issues [106]. Moreover, in some cases, CMS awarded contracts before issuing requirements for suppliers and *"undertook the development ... without effective planning or oversight practices"* [107].

As a result of this lack of coordinated oversight, the contractors wrote tremendously complex code for the site, eventually racking up 500 million lines—many times the number in operating systems like Windows 8 (80 million lines) or online banking systems (75–100 million) [108]. The conditions of some contracts implied cost reimbursement, stipulating for example that *"the government pays all of the contractor's allowable incurred costs to the extent prescribed in the contract"* [109]. In the end, expenses for the development of the federal site jumped from US $56 million in September 2011 to $209 million by February 2014, the cost of the data hub rose from $30 million to $85 million and CMS's total obligations for the development of HealthCare.gov and its supporting systems were estimated at $840 million by March 2014 [110]. Only 2 months after the launch, the site was debugged and working in a tolerable manner. But, in an already polarized political environment, the whole episode was a disaster for the administration; it was the main reason for the resignation of Kathleen Sebelius, US Secretary of Health and Human Services, in 2014.

---

[104] Amy Goldstein, Juliet Eilperin, HealthCare.gov: How political fear was pitted against technical needs, Washington Post, November 2, 2013

[105] Daniel R. Levinson, Federal marketplace: Inadequacies in contract planning and procurement, Department of Health and Human Services, Office of Inspector General, January 2015, p.12

[106] HEALTHCARE.GOV. Ineffective Planning and Oversight Practices Underscore the Need for Improved Contract Management, United States Government Accountability Office, July 2014, p.12

[107] HEALTHCARE.GOV. Ineffective Planning and Oversight Practices Underscore the Need for Improved Contract Management, United States Government Accountability Office, July 2014, p.11–13

[108] Julianne Pepitone, To fix Obamacare website, blow it up, start over, CNNMoney, October 23, 2013

[109] HEALTHCARE.GOV. Ineffective Planning and Oversight Practices Underscore the Need for Improved Contract Management, United States Government Accountability Office, July 2014, p.12

[110] HEALTHCARE.GOV. Ineffective Planning and Oversight Practices Underscore the Need for Improved Contract Management, United States Government Accountability Office, July 2014, p.9

The launch of the UK's £12 billion National Programme for IT in the National Health Service also yielded disappointing results, prompting stinging criticism from the parliamentary Committee of Public Accounts: *"This saga is one of the worst and most expensive contracting fiascos in the history of the public sector... The benefits flowing from the National Programme to date are extremely disappointing"* [111]. Researchers at the University of Cambridge concluded that this, one of biggest civil IT systems in the world, was poorly implemented for several reasons: haste to keep up with an unrealistic timetable; design problems, including the *"failure to recognise that the longer the project takes, the more likely it is to be overtaken by new technology"*; the fact that *"the project was too large for the leadership to manage competently"*, and that there was *"a lack of clear leadership"* and *"a lack of project management skills"* [112]. Earlier, also in the UK, the implementation of a highly complex IT system within the Child Support Agency had faced many challenges.

Amazon Simple Storage Service Outage (USA, 2017) On February 28 2017, Amazon's cloud service suffered a 3½ h outage because a programmer entered a command incorrectly. The error triggered many servers of supporting subsystems to restart, which led to the outage. Because the service was used by up to 150,000 websites, which loaded trillions of files, the interruption affected many business including App Store, Apple Music, ICloud, ITunes, Adobe, Lonely Planet, Microsoft's HockeyApp, the MIT Technology Review, Airbnb and even the website of the US Securities Exchange Commission [113,114,115]. Previously in 2015, Apple iCloud Services experienced several outages lasting from 7 to 11 h due to internal DNS errors. Dropbox had a worldwide downtime in 2015 due to *"routine internal maintenance"* [116]. And a Facebook outage in 2010 continued for about 2½ h because a cluster of databases that deal with error processing became overwhelmed by repeated queries, after a malfunction in an automated system for verifying configuration values [117].

[111]Dismantled National Programme for IT in NHS: report published, UK Parliament, September 18, 2013

[112]Oliver Campion-Awwad, Alexander Hayton, Leila Smith and Mark Vuaran, The National Programme for IT in the NHS. A Case History, University of Cambridge, February 2014

[113]Summary of the Amazon S3 Service Disruption in the Northern Virginia (US-EAST-1) Region, Amazon, February 28, 2017, https://aws.amazon.com/ru/message/41926/

[114]Elizabeth Weise, Amazon mystery solved: A typo took down a big chunk of the Internet, USATODAY, March 2, 2017

[115]Jeffrey Dastin, Disruption in Amazon's cloud service ripples through internet, Reuters, February 28, 2017

[116]Kelly Fiveash, Dropbox DROPS BOX as service GOES TITSUP worldwide, The Register, August 30, 2015

[117]Robert Johnson, More Details on Today's Outage, Facebook, September 23, 2010, https://www.facebook.com/notes/facebook-engineering/more-details-on-todays-outage/431441338919

Measures to avoid or mitigate crises with the development, launch or operation of software products include: a constant search for talented people with unique knowledge and practical experience; ongoing advanced education of highly skilled specialists; increased attention to the strategic planning of the services being developed—a vision of success, clear and achievable requirements, a manageable team of contractors, and the best approach to progress assessment; the implementation of advanced tools for project management; the independent assessment of content development; prolonged testing of content before release; and proper claim management, with a fast response to any defects discovered—developers should provide an immediate fix through patches or updates of that version of content, or if immediate repair is impossible or there is a wider systemic failure, recall the defective solution completely.

Crises Induced by Action from Authorities or Third Parties

All software is vulnerable to external interference by hackers, national and foreign intelligence, and so on.

Privacy of Facebook Users (USA, 2009, 2018) Facebook used Beacon to track the movements of its users, for advertising purposes and to enable a user's location to be checked by their Facebook friends. This helped advertisers to target their promotions, but violated the privacy of individual users because the Beacon did not give users the option to reject constant surveillance. Eventually, the program was recalled, and its launch was later acknowledged to have been a mistake. However, the most outrageous breach of the privacy of Facebook's users was caused not by programming but by the policy of the network, which allowed third-party data aggregators—like advertising firms and universities—to receive access to the private data of millions of Facebook users.

In 2013, Facebook allowed a scientist from Cambridge University to launch the thisisyourdigitallife app, which asked Facebook users to participate in an academic survey by allowing the app to collect their personal data and that of their friends. Personal private information was collected for about 87 million users. The Cambridge Analytica political consulting firm received access to the data collected by the app. This firm focuses on supporting political campaigns online by means of special algorithms, which allow campaigners to target political messages very accurately and directly on social media; for instance, the firm helped Ted Cruz and Donald Trump during the 2016 US Presidential elections, but denies that Facebook data was used in these campaigns. When the data breach went public, there was widespread condemnation of Facebook and its founder. Some individual users, organisations and companies even deleted their Facebook profiles to protest against the network's apparent disregard for the privacy of its customers: according to Tech.pinions, a technology research group, 17% of a sample of 1000 Americans deleted the Facebook app from their phones, 11% deleted from other devices, and 9% deleted their account altogether [118]. Ultimately, Facebook was forced to apologize, and changed its policy on working with third-party data aggregators. Later, it came to light that, since its launch in 2004, Facebook has granted access of user data to more than 60 major device manufacturers including Apple, Amazon, BlackBerry, Samsung, and Microsoft to facilitate the integration of specific messaging features into their devices. Facebook explains: *"These partnerships work very differently from the way in which app developers use our platform"* and emphasizes that there were no cases of impropriate usage of users' data [119].

Downtime of PlayStation Network and Sony Pictures Entertainment Data Theft (2011, 2014) Sony PlayStation's entire online service was unavailable to its 77 million subscribers for 23 days after a cyber attack in April 2011. There were fears that the account numbers of 24.6 million users, and the credit card details of 12,700, had been stolen in the attack [120]. The perpetrators were never found. Sony was fined for poor IT security by the British Information Commissioner's Office. Then in November 2014, Sony Pictures Entertainment was hacked; this time information about employees, their relatives' health records and the company's upcoming movies were stolen. There had been threats of terrorist attacks if Sony released the movie '*The Interview*', a comedy about an attempt by two American journalists to assassinate the North Korean leader Kim Jong-un. Assuming that the cyber attack was connected with these threats, US officials accused North Korea of involvement. However, no evidence was provided to back this up, and the hackers were never identified.

Yahoo! Accounts Info Leakage (2013, 2014) Yahoo! was hacked twice—first in 2013 when information about more than a billion user accounts was stolen, then again in 2014 when 500 million accounts were compromised. Information about the breaches was only released by Yahoo! in 2016, when 200 million accounts were put up on sale on the Internet. The US authorities blamed the 2014 attack on Russian state hackers, though Russia strongly objected to the allegation. There were similar intrusions into MySpace (more than 360 million user records), LinkedIn (more than 100 million), Dailymotion (85 million), LivingSocial (50 million), Adobe (at least 38 million), and others.

Accusations Against Telegram Messenger of Supporting ISIL (Since 2015)
Pavel Durov, former founder of VKontakte (the most popular social network in Russia), emigrated to the West and decided to develop Telegram: a free online messenger that was secure enough to prevent any unauthorized intrusion, even by intelligence services. In 2017, the number of Telegram users hit 100 million with

---

[118]Carolina Milanesi, US Consumers Want More Transparency from Facebook, Tech.pinions, April 11, 2018 https://techpinions.com/us-consumers-want-more-transparency-from-facebook/52653

[119]Gabriel J.X. Dance, Nicholas Confessore, Michael LaForgia, Facebook Gave Device Makers Deep Access to Data on Users and Friends, The New York Times, June 3, 2018

[120]Ben Gilbert, Sony Online loses 12,700 credit card account numbers, 24.6 million accounts compromised [update 2], Engadget, May 2, 2011

15 billion messages daily. The service became popular with those who had good reason to want to avoid state surveillance—including members of the Islamic State, who planned atrocities such as their attacks in France. However, Durov refused to collaborate with the authorities in the anti-terrorist fight. Rob Wainwright, Director of Europol, observed in May 2017: *"There are some that simply won't co-operate with us. One in particular causing major problems for us is Telegram... [the messenger gives] some co-operation but nowhere near what we are getting from Facebook, Twitter and some of the others"* [121]. Durov's lack of cooperation with Russian intelligence services in eliminating terrorist communication on the messenger led to the banning of Telegram in Russia in the spring of 2018.

Countless Zero-day Vulnerabilities and "Backdoors" in Popular Software from Microsoft, Adobe, Apple, Oracle, etc. (USA, Ongoing) In June 2012, it emerged that the US National Security Agency (NSA) had been developing a secret cyber military program since the administration of George W. Bush, which continued under the presidency of Barack Obama. Called "Olympic Games", this program was directed towards the launch of a remote cyber-attack on the Natanz underground nuclear site in Iran, where there were around 5000 centrifuges for the enrichment of uranium. According to David E. Sanger—chief Washington correspondent of The New York Times, who revealed the existence of the program in his book "Confront and Conceal: Obama's Secret Wars and Surprising Use of American Power"—the attack was intended to avert the possibility of an Israeli air attack on the site, which was considered increasingly likely if the Iranian nuclear program continued unchecked. Programmers from the American and Israeli military worked together to develop what was soon to gain worldwide notoriety as the Stuxnet virus.

The centrifuges at the Natanz site were controlled by Siemens Supervisory Control and Data Acquisition (SCADA) software, which is used to control and monitor many industrial processes. Like most software worldwide, the SCADA system used the Microsoft Windows platform, which had a number of "zero-day" vulnerabilities. These are bugs that are unknown to the software vendor—and therefore to users and the general public—so that, if hackers discover and exploit these bugs, both vendor and users will have "zero days" to prepare for the attack [122].

The Stuxnet virus got into the system through several of these vulnerabilities but then targeted the SCADA software, changing the speed control regime for the Natanz centrifuges so that they span much too fast, while giving feedback to the monitoring system that all was as normal. By the time the sabotage had been identified, 984 out of around 5000 centrifuges on the site had been destroyed [123,124]. According to Hillary Clinton, then US Secretary of State, the virus set

---

[121]Dominic Kennedy, Message app used by Isis refuses to fight jihadists, The Times, May 4, 2017

[122]Stefan Frei, Security Econometrics. The Dynamics of (In)Security, A dissertation submitted to the ETH ZURICH for the degree of Doctor of Science, 2009, p.94

[123]David E. Sanger, Confront and Conceal: Obama's Secret Wars and Surprising Use of American Power, Crown Publ. Group, 2012, pp.188–209

[124]David E. Sanger, Obama Order Sped Up Wave of Cyberattacks Against Iran, The New York Times, June 1, 2012

Iran's nuclear program back by several years [125]; thus, in terms of the damage it caused, the attack was a spectacular success. But there was an error in the programming: when the infected computers were connected online, the virus recognized the Internet as a local network. It spread to infect other computers worldwide, and was eventually detected by antivirus specialists from Belarus.

By May 2012, Sergey Ulasen, one of the key players in uncovering Stuxnet, was working for the Russian anti-virus company Kaspersky Lab. The company had been asked by the UN to investigate a new and far more sophisticated virus called Flame. Because the new virus contained code segments similar to Stuxnet, the suspicion was that the viruses had either been developed by the same programmers or that Flame's developers had helped the Stuxnet team [126,127]. There were computers in several Middle Eastern countries infected by the Flame virus, but most were at the Iranian Oil Ministry. Like Stuxnet, Flame used vulnerabilities in the Windows platform but, where Stuxnet had been introduced into the Natanz computer network through an infected USB stick, Flame was able to enter local networks as part of Microsoft's regular Windows Update service.

Then, in October of the same year, Kaspersky Lab announced that they had found a third virus exploiting vulnerabilities in Microsoft's Office and Excel platforms. The "Red October" virus cropped up—mainly in Russia and the former Soviet Republics—in the computer networks of military installations, energy providers, nuclear and other infrastructure, research firms and government embassies, and was designed to facilitate espionage. According to Kaspersky's Global Research & Analysis Team, Red October's structure *"rival[led] in complexity the infrastructure of the Flame malware"* [128].

The fact that the United States is actively planning to use cyber weapons as a key element in future warfare was clearly indicated in NSA documents leaked to the German magazine Spiegel by former NSA employee, and now famous whistleblower, Edward Snowden. NSA material ranging from a recruitment ad for new interns to more secret strategy documents shows that the agency plans to be able to *"remotely degrade or destroy opponent computers, routers, servers and network enabled devices by attacking the hardware"*, *"erase the BIOS on a brand of servers that act as a backbone to many rival governments"* and *"paralyze computer networks and, by doing so, potentially all the infrastructure they control, including power and water supplies, factories, airports or the flow of money"* [129].

---

[125]Con Coughlin, Stuxnet virus attack: Russia warns of 'Iranian Chernobyl', The Telegraph, January 16, 2011

[126]Ellen Nakashima, Greg Miller, Julie Tate, US, Israel developed Flame computer virus to slow Iranian nuclear efforts, officials say, The Washington Post, June 19, 2012

[127]Flame virus hit Iran's oil industry but officials say antidote found. Israeli vice-prime minister suggests Israel might have been behind cyberattack, The Associated Press, May 30, 2012

[128]The "Red October" Campaign - An Advanced Cyber Espionage Network Targeting Diplomatic and Government Agencies. Kaspersky Labs' Global Research & Analysis Team, Jan. 14, 2013

[129]Jacob Appelbaum, Aaron Gibson, Claudio Guarnieri, Andy Müller-Maguhn, Laura Poitras, Marcel Rosenbach, Leif Ryge, Hilmar Schmundt and Michael Sontheimer, The Digital Arms Race: NSA Preps America for Future Battle, Spiegel, January 17, 2015.

More than 40,000 people were already working for the US Cyber Command by 2013, engaged in digital surveillance but also in the preparation of destructive network attacks. The previous year, NSA and Cyber Command chief General Keith Alexander reasoned that *"Part of our defense has to consider offensive measures"* [130,131]. And as Barack Obama remarked In February 2015, *"We [Americans] have owned the Internet. Our companies have created it, expanded it, perfected it in ways that they [non-American vendors] can't compete"* [132]. Such clear declarations of military intent from a country with a near-monopoly in software and online solutions—after all the United States produce more than 90% of commercial software worldwide—make uncomfortable reading for other countries, many of which have a very different domestic and foreign policy outlook to that of Washington. From the point of view of these countries, the rest of the world is potentially at the mercy of an American hybrid attack—a combination of cyber attacks with conventional military measures. On the other hand, with both its civilian and military infrastructure more heavily IT-dependent than that of any other country, the US is itself very vulnerable to cyber attack.

By actively developing the kind of malware we have described, the US Cyber Command have also created a global black market: any new "zero-day" vulnerability discovered in widely used software could fetch a high price from the right—or wrong—buyer. Since it will not yet be tackled by any existing computer security programs, the new bug can be used for national defense purposes: as long as its existence is kept secret, a weapon exploiting the bug remains usable [133]. It is already standard practice for Microsoft and other software companies to inform the government of any security gap they discover in their products, some time before they issue updates to tackle them: an obvious suspicion is that this is to give the NSA time to exploit bugs before the wider tech community is aware of them. Since zero-day bugs stay undetected for an average of 312 days, this represents a generous window of opportunity [134].

The software giants who produce the vast majority of global digital software infrastructure—Microsoft, Oracle, Cisco, Apple, and Adobe—are all US-based companies, and all of their applications contain many defects and require continuous

---

[130] Jacob Appelbaum, Aaron Gibson, Claudio Guarnieri, Andy Müller-Maguhn, Laura Poitras, Marcel Rosenbach, Leif Ryge, Hilmar Schmundt and Michael Sontheimer, The Digital Arms Race: NSA Preps America for Future Battle, Spiegel, January 17, 2015

[131] Tom Simonite, Welcome to the Malware-Industrial Complex, The US government is developing new computer weapons and driving a black market in "zero-day" bugs. The result could be a more dangerous Web for everyone, MIT Technology Review, February 13, 2013, http://www.technologyreview.com/news/507971/welcome-to-the-malware-industrial-complex

[132] Kara Swisher, White House. Red Chair. Obama Meets Swisher, February 2015, ReCode, http://recode.net/2015/02/15/white-house-red-chair-obama-meets-swisher

[133] Hardin Tibbs, The Global Cyber Game: Achieving strategic resilience in the global knowledge society, The Defence Academy Cyber Inquiry Report, 2013, p.53, http://www.futurelens.com/wp-content/uploads/2014/04/The-Global-Cyber-Game.pdf

[134] Secunia Yearly Report 2011, The evolution of software security from a global enterprise and end-point perspective. Feb. 14, 2012 (secunia.com/resources); The Known Unknowns – Analysis of zero-days (http://techzoom.net/Publications/Papers/knownunknowns)

updating with new patches. The malware attacks on critical infrastructure we have described above all used vulnerabilities in Microsoft systems. Statistics from Net Applications in April 2016 show that more than 91% of desktop computers currently use the Windows platform, giving Microsoft a near-total monopoly of the world market [135]. Yet hundreds of bugs are revealed every year in the company's software! Other leading products such as Adobe Reader, Flash Player and Oracle Java are in the same situation: they install billions of copies worldwide every year [136], but discover hundreds of new vulnerabilities.

With the ever-increasing complexity of software, vulnerabilities are ever more frequent [137], and the market leader Microsoft is typically the world leader in this respect too. For both Microsoft and Apple, the annual incidence of Common Vulnerabilities and Exposures (CVE) fluctuated between 150 and 400 over the decade from 2005 to 2014. Google is catching up—their CVE rate was less then 10 per year before 2008, but rose to almost 300 in 2011. The rates for Cisco, IBM and Oracle are even higher, in the range of 350–450 per year from 2012 to 2014 [138]. The trend is not improving for most vendors: if anything, the overall number of CVEs is still increasing.

Since the vast majority of their potential targets use Microsoft software with the update function enabled, state cyber forces have every opportunity to make one-off attacks on millions of computers. If they can find an undiscovered zero-day vulnerability, such attacks can be disguised as legitimate Windows updates; this was the approach taken with the Flame virus. Among the revelations of whistleblower Edward Snowden was the evidence of Microsoft's close collaboration with the NSA's Prism program, which was able to access encrypted messages on Outlook, Skype, and Onedrive [139]. If Microsoft was willing to collaborate with the US intelligence through their flagship internet products, we should assume that they will also have given the NSA back door entry points into all of their software and enabled the agency to use their continuous update function—giving access to millions of computers worldwide. Thus, in addition to its conventional capability, the US Army almost certainly has the capacity to launch a full-fledged cyber attack on any adversary.

The vulnerabilities in American software revealed by Snowden's whistleblowing worried many customers all around the world, and threatened the long-term prospects of software vendors. This was obviously the motivation for a coalition of tech firms including Apple, Facebook, Google and Microsoft, along with security experts and a number of civil society organizations, to write an open letter to President

---

[135]Desktop Operating System Market Share, April 2017, http://www.netmarketshare.com/

[136]The market share of Windows Desktop Programs (2011 Data) are, according to the Secunia Yearly Report 2012: Adobe Flash Player −98%, Oracle/Sun Java - 86%, Adobe Reader - 82%, Apple Quicktime - 52%.

[137]National Vulnerability Database (NVD), http://nvd.nist.gov

[138]Data provided by Dr. Stefan Frei, based on the NVD vulnerability database for the last 10 years for all vulnerabilities, and vulnerabilities with CVSS>7.

[139]Glenn Greenwald, Ewen MacAskill, Laura Poitras, Spencer Ackerman and Dominic Rushe, Microsoft handed the NSA access to encrypted messages, The Guardian, July 12, 2013

Obama on May 19, 2015: *"We urge you to reject any proposal that US companies deliberately weaken the security of their products. We request that the White House instead focus on developing policies that will promote rather than undermine the wide adoption of strong encryption technology. Such policies will in turn help to promote and protect cybersecurity, economic growth, and human rights, both here and abroad. Strong encryption is the cornerstone of the modern information economy's security. Encryption protects billions of people every day against countless threats—be they street criminals trying to steal our phones and laptops, computer criminals trying to defraud us, corporate spies trying to obtain our companies' most valuable trade secrets, repressive governments trying to stifle dissent, or foreign intelligence agencies trying to compromise our and our allies' most sensitive national security secrets. Encryption thereby protects us from innumerable criminal and national security threats. This protection would be undermined by the mandatory insertion of any new vulnerabilities into encrypted devices and services. Whether you call them 'front doors' or 'back doors', introducing intentional vulnerabilities into secure products for the government's use will make those products less secure against other attackers. Every computer security expert that has spoken publicly on this issue agrees on this point, including the government's own experts. In addition to undermining cybersecurity, any kind of vulnerability mandate would also seriously undermine our economic security. US companies are already struggling to maintain international trust in the wake of revelations about the National Security Agency's surveillance programs. Introducing mandatory vulnerabilities into American products would further push many customers—be they domestic or international, individual or institutional—to turn away from those compromised products and services"* [140].

"Back Door" Entries in Hard Drives (Worldwide, 2015)   As if it was not enough for the NSA to have substantial access to any computer using one or all of the leading softwares on the market, it is also likely that back door entries have been left open to allow the agency access to computer hard drives. In February 2015, Kaspersky Lab reported on their investigations into malware by the Equation group, which was found to be deeply embedded in hard drives produced by 12 top manufacturers including Toshiba, Seagate and Western Digital. The Equation malware turned up in personal computers used in business and administration: banks, telecom companies, energy companies, nuclear researchers, even government and military institutions. It was discovered in 30 countries including Russia, China, Iran, Pakistan, Syria, Afghanistan, Mali, Yemen and Algeria. Kaspersky's investigation found that the malware gave its developers total remote control over any infected computer, enabling them to access data, intervene in operations and even re-format the infected hard disk. Reuters correspondent Joseph Menn reported: *"A former NSA employee commented... that Kaspersky's analysis was correct, and that people still in the intelligence agency valued these spying programs as highly as Stuxnet [see above*

---

[140]Letter to President Obama, May 19, 2015, https://static.newamerica.org/attachments/3138%2D%2D113/Encryption_Letter_to_Obama_final_051915.pdf

and footnotes [141,142]. *Another former intelligence operative confirmed that the NSA had developed the prized technique of concealing spyware in hard drives"* [143]. The existence of such hidden software, or even the suspicion that it may be present, is likely to put potential customers off and provoke public outrage—especially in the regions that have been targeted. Individuals will resent the possible violation of their privacy, and neither commercial companies nor government organizations will wish to lay themselves open to interruption or sabotage of their systems in a geopolitical dispute.

In a crisis caused by third party action, a company becomes the victim in the eyes of users, state officials and the media, and can recruit the help of national law-enforcement and cyber security agencies to find and prosecute the malefactors. Response measures include additional attention to the cyber resilience of services and programs, the continuous search for vulnerabilities, immediate release of patches if new problems are discovered, ongoing collaboration with national cyber security agencies, monitoring of the latest actions of hackers, sharing experience among company personnel, and the recruitment of *"white hackers"* to test weak spots within services and products.

In the event of pressure from cyber security bodies to install "zero-day" vulnerabilities to enable surveillance, or grant access to the private data of people considered a security risk, a vendor could lobby for national laws to be changed or request a court decision on the legality of such demands. Nevertheless, it will be better for the vendor not to invite conflict with state representatives, which may simply motivate governments to seek a new legal footing for greater powers of surveillance.

Key Risk Mitigation Measures in Information Services

- Continuous search for errors in programming code and prompt updates for any bug revealed.
- Ongoing recruitment of people of inventive genius, and the retention of highly qualified and motivated staff.
- Ability to build constructive relationships with national and international security authorities to reach a consensus on a workable balance between national security and the privacy of customers.
- Close cooperation with law enforcement agencies worldwide to protect copyright and fight hackers.

---

[141] David E. Sanger, Confront and Conceal: Obama's Secret Wars and Surprising Use of American Power, Crown Publ. Group, 2012, pp. 188–209

[142] David E. Sanger, Obama Order Sped Up Wave of Cyberattacks Against Iran, The New York Times, June 1, 2012

[143] Joseph Menn, UPDATE 3-Russian researchers expose breakthrough US spying program, Reuters, February 16, 2015; Equation: The Death Star of Malware Galaxy, Kaspersky Labs' Global Research & Analysis Team, Feb. 16, 2015, http://securelist.com/blog/research/68750/equation-the-death-star-of-malware-galaxy and https://securelist.com/files/2015/02/Equation_group_questions_and_answers.pdf

3.5   Telecommunications

The subsector includes the following industries according to ISIC Rev. 4:

* 60—Broadcasting activities
* 61—Telecommunications

General Description and Key Features of the Subsector

Telecom and broadcasting have many similarities with utilities: for companies in all three subsectors, maintaining a non-stop supply to customers 24-7-365 with fast and innovative services is the main priority. The high importance of the telecommunication and broadcasting industries for national security—and for the smooth function of society—makes these industries highly dependent on national authorities. There is a general framework for regulation of telecoms companies; governments demanding access to telecoms networks for surveillance over the masses; restrictions on international investment in the industries; and in countries less affected by the global trend towards deregulation, limits on the procurement of telecommunication equipment and even on tariffs for services. More moderate deregulation of telecom—such as the issue of a limited number, sufficient for healthy competition, of licenses to different providers, or the principle of number portability to enable customers to switch providers if they so wish, or a robust framework of antitrust/ competition law—all these allow for competition within the subsector, keep prices down for customers, and increase the quality of services and the reliability of national telecommunication infrastructure, since if one operator's network goes down others will still be available. A competitive environment also motivates operators and manufacturers to invest in research, innovation and the next wave of new technology in order to seize the competitive advantage for a while.

The industry has a strong propensity for intra-industrial benchmarking of technologies, prices, marketing tools and client service innovations because of the nature of telecommunication services. Clients consume minutes of calls and megabytes of data with little real understanding of the technology and methods of service delivery; thus, when choosing a provider, they compare the prices of calls and data first—only assessing coverage and reliability in practice, after they have made their choice. Most services are provided not through single purchase orders, but on a subscription basis. The role of personnel in a modern telecom company is more limited than elsewhere in the service sector because of the automation of many operations; earlier in its history, the subsector was heavily dependent on an enormous field staff team, who was responsible for connecting every local or long-distance call. Nowadays, the staff of a telecom company is mainly occupied with the resolution of clients' issues in call centers, the maintenance of telecommunication infrastructure and with call-outs to respond to network failures and service breakdowns.

Critical Success Factors for an Organization Within the Subsector

For a telecom company, the critical factors are the ability to provide customers with a stable and fast connection at a reasonable price, to provide an uninterrupted supply to customers 24-7-365 and, in the event of an interruption, to ensure a faster emergency repair response than competitors. It is worth mentioning here what is known as the service recovery paradox: customers who have lost services, but had a fast and effective response by the service company, are more loyal than those who have had uninterrupted service. Thus, what may sometimes be an inevitable network and service breakdown is a valuable opportunity for a telecommunication company to demonstrate that it can solve difficult problems effectively, and a fast and skilled repair call-out team will foster customer loyalty and ensure customer retention. Finding people with a passion to serve clients, and innovative people with top technology talent, is therefore of high importance for all telecom companies. Other important factors are the ability to obtain state licenses to provide modern telecommunication services, to adapt to the continuous development of regulation as it responds to new markets, and to work with suppliers on the implementation of innovative and integrated technologies along with the expansion of network coverage.

The fast pace of technological change and the relentless pressure of competition represent a huge challenge. For example, the decades-long golden age of the landline business was interrupted in the 1990s and 2000s by the new technologies of mobile voice and message transmissions. Between them, these technologies also destroyed the business of pager operators. The full-fledged development of Iridium, a global satellite phone operator with 66 satellites, was limited by price and technology competition from ground mobile networks. Unable to pay its colossal development debts with the income from an insignificant number of subscribers, the company nearly went bankrupt—but then found a far more lucrative client for its satellite communication system in the form of the US Department of Defense. As a result, Iridium has survived, even though 850,000 subscribers in 2016 seems modest in comparison with 4.8 billion unique mobile subscribers, and 7.9 billion SIM connections, worldwide [144,145]. And in their turn, mobile operators already face the end of their dominance as data transmission through broadband and WiFi networks competes with them in providing access to the Internet, and Skype, Viber, WhatsApp and other online messaging apps offer customers cheap or free voice/video calls and messaging. Ultimately, mobile operators could simply become a *"data tube"* in the incoming digital era.

---

[144]Iridium Announces Fourth-Quarter and Full-Year 2016 Results; Company Issues 2017 Outlook, Iridium Communications Inc., February 23, 2017

[145]Number Of Global Mobile Subscribers To Surpass Five Billion This Year, Finds New GSMA Study, GSMA February 27, 2017

Stakeholders in this Subsector [146]

- Customers—40%
- Authorities—20%
- Employees—20%
- Investors—10%
- Suppliers—5%
- Other—5%

Typology of Common Risks, Main Features of Major Accidents and Risk Mitigation Measures Within the Subsector

Crises Induced by Technical Factors

The main factor determining customer satisfaction with a telecommunication service is the reliability, stability and quality of the connection in any geographical location where the company has promised to provide one and within the usage limits agreed in their contract with the customer. Therefore outages, or lapses in the service quality expected and promised, are the main risk for any telecommunication company. Network failures could occur for a number of technical reasons, including physical malfunctions of network devices, computers, or cables, bugs in the software, or staff mistakes in operating or servicing network equipment or software. According to Heavy Reading research, cellular operators alone lose up to US$20 billion a year worldwide due to outages, which on average occur five times a year for most mobile operators [147,148].

Fire at Telephone Exchange (USA, 1988) On May 8, 1988 after a fire in the switching room at the Hinsdale Central Office of the telephone company Illinois Bell, more than 1.5 million customers were affected, including hospitals and the airports of O'Hare and Midway. The accident influenced emergency service lines and local, long distance, and cellular calls. Because it was an electrical fire, dry chemical extinguishers were used, but these proved ineffective—so firefighters had to wait until the power had been switched off to the whole exchange before it was

---

[146]Our informed appraisal of the influence of each audience on a typical organization within the subsector (100% = combined influence of all audiences)

[147]Mobile Broadband Brings High-Profile Outages, Heavy Reading Finds, Light Reading, October 22, 2013, http://www.lightreading.com/services-apps/broadband-services/mobile-broadband-brings-high-profile-outages-heavy-reading-finds/d/d-id/706202

[148]Mobile Network Outages & Service Degradations, Heavy Reading Survey, February 2016

safe to use water. It took around 4 weeks to fully restore service, especially to the private addresses of 38,000 local customers [149].

AT&T Network Outages (USA, 1990, 1998)
In December 1989, AT&T uploaded a software system to all of its 114 toll switching centers, which processed over 115 million telephone calls daily, in order to increase productivity and ensure faster recovery from inevitable failures. All the centers were then connected with each other and constantly sending messages about their status: in case of a local failure, the center in question could send message to the others, informing them that it was unable to receive calls and asking to be recognized as unavailable. On January 15, 1990, a center in Bedminster, N.J. experienced a technical malfunction and rebooted. However, due to a coding error in the recently uploaded software, the center started rebooting again and again and sending misleading messages to other switchers; since they were governed by the same defective code, these centers in their turn began to reboot and send the same status updates. It took 21 h for AT&T to completely solve this nationwide breakdown [150]. A quite similar event occurred on April 13, 1998 when one of the switches, produced by Cisco Systems Inc., failed to upgrade software and again sent error messages to 145 other centers, overloading the AT&T network for 26 h [151].

America Online Downtime (USA, 1996) Routine maintenance of "*high capacity switches within the local area network*" led to a 19-h connection outage from the biggest Internet provider in the world at the time. 6.3 million customers were affected.

Ostankino TV Tower Fire (Russia, 2000) At 540.1 m, this TV tower is the tallest structure in Europe. It was completed in 1967, and broadcasts television channels to more than 15 million people located in Moscow and the surrounding area. On August 27, 2000, fire broke out at a height of 460 m because old cables—constructed in the 1960s and never upgraded—could no longer withstand the increased loads of modern television. The plastic insulation of the cables began to melt, setting off new fires at lower levels of the tower. In the intense heat, 120 of the 149 steel cables that stabilized the concrete structure of the tower were ruptured. Fortunately, the tower remained standing and later the cables were restored. The accident cost the lives of three people. It took more than 10 days to fully resume TV broadcasting in the Moscow region and more than a year to complete reconstruction works.

Diversion of the World Internet Through Turkish Telecom (Turkey, 2004) The Border Gateway Protocol (BGP) is a routing protocol that provides information about the current connectivity and availability of all Internet networks. The protocol

---

[149]Hinsdale Central Office Fire, Final report, Executive summary, March 1989, Forensic Technologies International Corporation

[150]Dennis Burke, All Circuits are Busy Now: The 1990 AT&T Long Distance Network Collapse, California Polytechnic State University, November, 1995

[151]Matthew A. Telles, Yuan Hsieh, The Science of Debugging, Coriolis, 2001, pp.40–50

works on the principle of trust between connected networks, and perceives all information transmitted from any network as true. One of the functions of the BGP is to create a global routing table. On December 24, 2004, due to an error in the operation of BGP, Turkish Telecom sent messages to all the other networks on the global routing table, advertising over 100,000 routes—which if true would have made it the best routing path for the entire global Internet. Internet traffic responded to this apparent increase of bandwidth by moving to Turkey from sites such as Amazon, Microsoft, Yahoo and CNN. TTNET was immediately overwhelmed with traffic and sites all over the world went down for several hours [152,153].

911 Service Outage and Rogers Wireless Network (Canada, 2013) More than nine million customers of the Rogers Wireless network, the largest wireless carrier in the country, suffered from interruptions of voice and text messaging caused by software problems on October 9, 2013. The most dangerous consequence of this downtime was the failure of the provider to supply emergency calls to 911. The elimination of landlines and payphones in Canada made the situation even worse for people who needed emergency help. In a similar incident in August 2014, T-Mobile US failed to provide a connection to the 911 system in the United States for 3 h. Eventually, the operator was fined US$17.5 million by the US Federal Communications Commission.

Multi-Carrier Outage Due to Reliance on a Single Network in the Southeast (USA, 2015) Probably as a result of a fiber cable being cut on the AT&T network, Verizon Wireless, T-Mobile US and Sprint faced outages in Tennessee, Alabama, Kentucky, Indiana and Georgia in 2015. The main cause of this multi-carrier outage was the dependence on AT&T as the sole backhaul provider in the countryside. The backhaul part of the network connects the core network to the smaller networks of individual carriers; in the rural Southeast of the US, there is little economic advantage for wireless carriers in constructing a full-fledged network, so they constructed their own cell towers but used AT&T as backhaul [154].

Disruption Due to Congestion (Worldwide Problem during Holidays and Disasters) Outages are frequent during peak loading of networks during holidays and disasters, when people are getting in touch with friends and family.

Measures to prevent such crises include faster restoration of service outages, timely investment in infrastructure, continuous monitoring of weak spots and bottlenecks in existing infrastructure, duplication of critical elements of the network, automatic chemical systems to rapidly extinguish electrical fires caused by

---

[152]Internet-Wide Catastrophe—Last Year, Dyn Guest Blogs, December 24, 2005, http://dyn.com/blog/internetwide-nearcatastrophela/

[153]Andrey Robachevsky, Five obstacles to security of the global routing system, Russian Institute for Public Networks, 2012, http://www.ripn.net/articles/secrout2012/

[154]Jon Brodkin, An AT&T problem allegedly caused outage on Verizon, Sprint, and T-Mobile, ARStechnica, August 5, 2015

short circuits or overloading, and advance preparation for network congestion at peak periods.

Poor Customer Service

Customers can be dissatisfied with the service they receive for a range of non-technical reasons: a poorly functioning call-center or operator, making clients wait for ages before their question is dealt with; unskilled support personnel; negligence by staff towards the customer's particular needs; billing errors which overcharge customers—who will perceive overbilling as stealing their money; inflexible tariff plans, which do not suit the needs of a customer; operators who are not given the authority to give a customer a few days' grace when the customer cannot pay on time; companies actively promoting new services when their networks are clearly not ready to transmit increased traffic; operators being unable to facilitate online access to a customer's account in order to reduce client-staff interactions; or badly-laid cables into a customer's home which are untidy, inconvenient or cause damage to the house.

To tackle these kinds of problems, a company should keep track of customer preferences and requests, train service staff at frontline stores and call-centers to improve their advanced customer care skills, and develop advanced customer relationship management (CRM) and billing systems.

Crises Caused by External Events or Actions of Third Parties

Service cuts and network crashes can also be provoked by natural disasters, power blackouts or by the actions of terrorists, saboteurs, hackers, foreign military forces, and so on.

Congestion Due to Natural Hazards and Terrorist Acts (Worldwide) Landline and mobile networks were overloaded during the Loma Prieta earthquake in San Francisco in 1989, during the Northridge earthquake in Los Angeles in 1994, during the Kobe earthquakes in Japan in 1995, in New York on the day of the 9/11 attacks in 2001, on the coast of Thailand during the Sumatra–Andaman earthquake in 2004, during the London Underground bombings in 2005... and following many other events when huge numbers of people were asking for emergency help, or calling relatives and friends to check that they were alive and well.

NATO Aggression Against Yugoslavia (1999) For more than 70 days during the bombardment of Belgrade, NATO was also attacking critical infrastructure—including communication centers, which were eliminated in the early stages of the conflict. However, when it came to bringing down Internet services, results were modest because of the existence of decentralized IP networks, and backup channels like satellite links, cellular networks, and even amateur packet radio [155].

Telecom Outage During North American Blackout (USA and Canada, 2003) This widespread power blackout, which affected up to 50 million people and lasted from 2 to 4 days, also had an impact on mobile networks and Internet connections through WiFi networks and broadband. Although the majority of cell towers were equipped with backup batteries to cover short-term power breaks (from 4 to 6 h), they could not provide stable services during a blackout lasting days. Landline services appeared to be the most reliable form of communication during the blackouts, because most switchers had backup diesel generators. Similar events on a smaller scale occurred in San Francisco and in Texas in 2007, and in New Zealand in 2016. In July 2007, the 365 Main datacenters in San Francisco, which hosted several popular sites, lost power from Pacific Gas and Electric because of transformer problems. The datacenter was unable to restore the service because 6 backup generators failed to launch properly. And in November the same year, the Rackspace datacenter in Texas failed to re-start its backup generators after a truck struck a utility pole and damaged a power transformer. The blackout lasted more than 4 h. In February 2016, due to a power outage at one of Vodafone's operations centers, customers in New Zealand lost access to landline and broadband services for at least 5 h.

Hurricane Katrina (USA, 2005) During the hurricane, power grid infrastructure was severely damaged. For instance, 6 weeks after the hurricane, nearly 50% of New Orleans was still without power. The blackout also damaged telecommunications infrastructure in the affected areas, because some backup batteries and generators were flooded, and even undamaged sites faced fuel supply disruptions because of flooded roads and security issues. Quite apart from the power supply problem, the hurricane physically destroyed 2.475 million lines and nine central offices. Emergency services were unreachable. Only a few AM stations were able to continue broadcasting weather forecasts and news of recovery action after the hurricane: the most famous example was the continuing operation of WWL Radio from a closet when the studio was inoperable after the disaster. In addition, there were more than 3000 cell towers in the hurricane zone, most of which were destroyed or partially damaged. The rapid response of mobile operators to install hundreds of transportable "*cell-on-wheels*" towers with generators allowed cellular communications to be restored within a week of the disaster, which struck the Gulf Coast region and particularly New Orleans [156]. Flooding during Hurricane Sandy in 2012 caused power blackouts, interrupted the operation of several data centres and some

---

155 Anthony M. Townsend, Mitchell L. Moss, Telecommunications infrastructure in disasters, Center for Catastrophe Preparedness and Response, New York University, April 2005, p.8

156 Alexis Kwasinski, Wayne W. Weaver, Patrick L. Chapman and Philip T. Krein, Telecommunications Power Plant Damage, Assessment Caused by Hurricane Katrina – Site Survey and Follow-Up Results, EEE Systems Journal, Volume: 3, Issue: 3, September 2009, pp.277–287

911 emergency call centres, and destroyed around 25% of wireless cell towers in the affected areas [157].

Hengchun Earthquake (Taiwan, 2006) The epicenter of the earthquake was off the coast of the southern part of Taiwan, which led to the disruption of several submarine communications cables. This affected Internet traffic between Taiwan, Singapore, Hong Kong, South Korea and Japan for over a month [158].

Internet Submarine Communications Cable Sabotage (2008, 2011) Countries in the Middle East and South Asia were affected after several suspicious breakages of submarine Internet cables. Officially, the disruptions were attributed to ships' anchors and natural deterioration of cables due to salt water and damage from rocks on the sea floor, but conspiracy theorists soon spread the rumor that the cables had been cut by Israeli agents [159,160].

Entire Country Cut off from Internet by Scrap Metal Collector (Georgia and Armenia, 2011) On March 28 2011, a 75-year-old woman in Georgia, who earned a little money to augment her modest pension by collecting copper and other metal, accidentally cut a fiber optic cable while digging for scrap metal near a railway line. The cable in question was of strategic importance, supplying Internet to Georgia, Armenia and Azerbaijan from Europe along Georgian railways. The accident led to 5 h' downtime for more than three million Internet users in Armenia, and several local outages in the Georgian capital Tbilisi.

Terrorist Attacks on Telecommunication Facilities (Nigeria, Afghanistan, Iraq) Boko Haram, the West African wing of the Islamic State since 2015, has attacked security agencies, politicians, Christian churches, markets, public schools, and hospitals. Since they accuse the Nigerian telecom companies of assisting the government in tracking their members, they also targeted the infrastructure of mobile operators in five cities in Nigeria in September 2012, damaging 150 base stations. Similar accusations of intelligence assistance were put forward by the Taliban against cell networks in Afghanistan in 2008, resulting in the destruction of 30 cell towers over 3 weeks. Afghan operators were forced to suspend coverage of the networks during the night, as requested by the Taliban, in order to prevent new attacks on their infrastructure [161]. A similar attack was implemented by ISIS at the

---

[157]Sinead Carew, Hurricane Sandy disrupts Northeast US telecom networks, Reuters, October 30, 2012

[158]Choe Sang-Hun and Wayne Arnold, Asian Quake Disrupts Data Traffic, New York Times, December 28, 2006

[159]Mike Whitney, Three Internet Cables Slashed in a Week: Has Iran lost all Internet Connectivity?, Global Research, February 3, 2008

[160]Richard Sauder, Connecting The Many Undersea Cut Cable Dots - 9 Or More?, Rense, February 5, 2008

[161]Freedom Onuoha, The Costs of Boko Haram Attacks on Critical Telecommunication Infrastructure in Nigeria, E-International Relations, November 3, 2013

Al-Tanf border crossing between Iraq and Syria in May 2017, when two communication towers were demolished [162].

Cyber Attacks on Telecommunication Infrastructure (Worldwide) The risk of such attacks has grown progressively in recent years, and they are expected to become ever more frequent in the future. A severe attack on the global Internet occurred on October 21, 2016, through the Internet performance management company Dyn. The company is responsible for mapping the IP addresses of all domain names through the Domain Name System (DNS). Dyn was exposed to the largest denial of service (DDoS) attack in history—up to 100,000 malicious endpoints with a strength of 1200 gigabytes per second [163]—resulting in problems for end-users trying to open web-sites in North America and Europe. Similar but smaller attacks on root name servers—the core servers for the DNS—took place in 2002, 2007 and 2015. T-Hrvatski Telekom, a subsidiary of Deutsche Telekom, faced outage because of a DDoS attack in September 2015. In November 2016, hackers tried to access the private information of six million customers of Three Mobile, a UK mobile phone operator, by means of employee logins. Fortunately, they were not able to reach information about payments, cards or bank accounts [164]. TalkTalk, another British company providing pay television, telecommunications, Internet access and mobile services, were not so lucky. In October the previous year, they had experienced a hacker attack that breached information about more than 150,000 customers—and in this case 10% of the leakage included customer financial details.

Of course, the responsibility for these kinds of telecommunication failure lies either with the third parties who caused them—whether intentionally or not—or in the case of natural disasters, with nobody... but operators still bear some of the blame if they have failed to prepare for natural disasters and blackouts, or to anticipate the possible harmful actions of evil-doers.

Crises Induced by Authorities and Security Agencies

There are two main types of crises with relation to authorities. The first stems from the possible impact of national regulation and deregulation of services, and the licensing processes that obtain in most countries. The second type occurs when there is perceived to be a question of national security: when governments or their national security agencies try to control telecommunication infrastructure during social upheavals, to impose censorship or to monitor the activity of Internet users

---

[162]Chris Tomson, In pictures: ISIS blows up two communication towers in raid on Iraqi border post, Al-Masdar Al-'Arabi, May 20, 2017, https://www.almasdarnews.com/article/pictures-isis-blows-two-communication-towers-raid-iraqi-border-post/

[163]Nicky Woolf, DDoS attack that disrupted internet was largest of its kind in history, experts say, October 26, 2016

[164]Subrat Patnaik, Paul Sandle, Richard Chang, Elaine Hardcastle, Cyber hack of UK's Three puts customer information at risk, Reuters, November 18, 2016

and voice and messaging conversations. The most sophisticated systems of such state surveillance or control are currently in the United States and China. America's capability of widespread surveillance—at home and abroad—over phone, message and internet traffic became known to the global public through the whistle blowing of Edward Snowden. The Chinese surveillance and censorship program, known as the Golden Shield Project or the Great Firewall of China, may be less well-leaked but must be assumed to be extensive.

Blocking of YouTube and Facebook in Pakistan (2008, 2010) Pakistan Telecom and other national internet providers were forced to start blocking access to YouTube and Facebook for local users, after the publication on both platforms of material deemed to be anti-Islamic. In 2008, this national blocking resulted in users worldwide being unable to visit YouTube for 2 h.

Shutdown of Telecommunication Services During Anti-Government Protests (Worldwide Practice) During several days of upheavals on Tahrir Square, the Egyptian authorities tried to block Internet access—especially to social networks—by means of the state-owned Telecom Egypt, which operated the majority of fiber-optic cables in the country. Similar action was taken by the Libyan government during the early stages of anti-Gaddafi protests. Attempts or threats to block social network and Internet access to protestors were also made in Moldova (2009), in Iran (2009), in the UK (2011), in Cameroon (2017), and during the Catalonia referendum (October 1st, 2017) and have doubtless occurred elsewhere.

Cancellations of 2G Licenses Due to Corruption (India, 2012) In February 2012, the Indian Supreme Court canceled 122 regional mobile licenses, which had been issued on the authority of the then telecom minister in 2008. The problem was that certain companies had been favoured in the decision-making process, and these companies had been offered extremely low prices on the licenses—which had been sold at 2001 prices, at a loss of up to US$35 billion of missing revenue to the Indian government. There were accusations that the Minister had received around US$665 million in bribes from the participants. Recall of the licenses affected up to 5% of around 900 million mobile users, who had only 4 months to change their operator. The crisis also damaged the Indian business of the Norwegian Telenor and Russian MTS, both of whom had acquired local companies with the problematic licenses with no knowledge of the corruption that had tainted the issuing process. The total investment of the companies with revoked licenses in the Indian market was estimated at between US$25 and $30 billion. The cancelled licenses were offered again for auction in November 2012 [165,166].

Companies from this subsector have to follow laws, regulation and license requirements in order to minimize the probability of conflict with regulators. They

---

[165]Vikas Bajaj, Indian Court Cancels Contentious Wireless Licenses, The New York Times, February 2, 2012

[166]Dhananjay Mahapatra, 'A Raja made Rs 3000cr in bribes', The Times of India, February 11, 2011

must also maintain a careful and delicate balance between national security and the privacy of customers when collaborating with authorities.

Key Risk Mitigation Measures in Telecommunications

- Securing a non-stop supply to customers 24-7-365 with fast, innovative and high quality services.
- Maintaining constructive relations with national authorities, because control over and protection of telecommunication infrastructure is a matter of national security for many states. Governments also determine many spheres of activity of the telecommunication business, and can help to protect infrastructure from hackers, vandals, terrorists, etc.
- Brand protection by responding immediately to unsatisfied customers.
- Recruitment and retention of technically skilled personnel with passion to serve clients.
- Continuous research and development of innovative solutions.

3.6   Financial and Insurance Services

The subsector includes the following industries according to ISIC Rev. 4:

- 64—Financial service activities, except insurance and pension funding
- 65—Insurance, reinsurance and pension funding, except compulsory social security
- 66—Activities auxiliary to financial service and insurance activities

General Description and Key Features of this Subsector

Customers will invest their money with a given financial institution if it has a reputation as being safe, financially sound and well informed—in short, if it is a trusted brand. The subsector is highly dependent on national and international financial authorities: central banks, finance ministries, and regulators. It is these authorities that determine monetary and economic policy on the deregulation of the finance industry. Moreover, adequate oversight over the subsector is a question of national security. Since it involves a tremendous number of manual operations, the subsector relies heavily on employees and on their honesty and skill in managing a client's money or serving a customer. Automation has increased the speed of data handling, reduced staff mistakes and given customers remote access to financial services, but these efforts have an unwanted side effect: the whole financial system now relies on IT infrastructure, which is dangerously vulnerable to cyberattacks and technical failures.

Critical Success Factors for an Organization Within the Subsector

Trust: no transaction is possible without mechanisms that ensure (some form of) trust between contracting parties that their deal will be upheld and secure. Ability to provide customers with reasonably priced services or lucrative profits, without violating the regulatory framework or the solvency and soundness of the institution. Good relations with government and regulatory authorities: lobbying for the interests of the subsector, access to liquidity, influence on state monetary, economic and legislation policies. Adequate control over highly qualified staff to prevent unauthorized decision-making and white-collar crime. Motivating frontline service staff to serve customers politely, kindly and accurately, and maintain united corporate standards. Development of advanced, reliable and protected IT systems to ensure high quality data processing and give customers flexible access to financial services.

Stakeholders in this Subsector [167]

- Customers—35%
- Authorities—30%
- Employees—25%
- Other—10%

Typology of Common Risks, Main Features of Major Accidents and Risk Mitigation Measures Within the Subsector

Trust and its Role in the 2008 Financial Crisis: Lessons for the Future

The credit crisis that started in 2007 is a perfect illustration of the deep fact that financial price and economic value are based fundamentally on trust—not on fancy mathematical formulas, not on subtle self-consistent efficient economic equilibrium—but on trust in the future, trust in sustainable economic growth, trust in the ability of debtors to face their liabilities, trust in financial institutions to play their role as multipliers of economic value, trust that your money in a bank account can be redeemed at any time you choose. Usually, all of us take these facts for granted, as an athlete takes for granted that her heart will continue to pump blood and oxygen to her muscles. But what if she suffers a sudden heart attack? What if normal people

---

[167]Our informed appraisal of the influence of each audience on a typical organization within the subsector (100% = combined influence of all audiences)

happen to doubt the banks? Then, the implicit processes of a working economy—all we take for granted—start to dysfunction and spiral into a global collapse.

Observers and pundits have attributed the 2007–2009 financial crisis to the mortgage-backed securities linked to the bursting of the house price bubble; the irresponsible lending, overly complex financial instruments and conflicts of interest; too many transactions between parties with "*asymmetric information*" (i.e. one side knew more than the other) but where even the "*experts*" were confused, until in the end the blind were leading the blind into market illiquidity; the spreading of risks by packaging and reselling imagined valuations to unsuspecting investors. In fact, to get the full story of the gradual erosion of trust, we need to go back at least to the late 1990s and to understand the interplay of three successive bubbles. In the late 1990s, the ITC "*new economy*" bubble was expanding at full steam. By the end of the decade, the GDP growth rate of the US was roughly double that of Europe, a bonanza attributed to a comparable gain in productivity that people assumed would be permanent. However, it turned out to be a bubble that burst in the first half of 2000, triggering two dismal years for the stock market: between 2000 and 2003, shares lost more than 60% and there was a (mild) recession.

To fight the recession and the negative economic effects of the collapsing stock market, the Fed engaged in a proactive monetary policy, slashing their interest rate from 6.5% in 2000 to 1% in 2003 and 2004; along with expansive Congressional real-estate initiatives, this fueled what can now be acknowledged as one of the most extraordinary real-estate bubbles in history, with excesses on a par with those during the famous real-estate bubble in Japan in the late 1980s. As Alan Greenspan himself documented in a scholarly paper researched at that time—during which he was Federal Research Chairman—the years from 2003 to 2006 witnessed an extraordinary acceleration of the amount of wealth extracted by Americans from their houses [168,169]. The negative impact on consumption and income of the collapse of the first ITC bubble were happily forgotten in the wake of a renewed enthusiasm and a sense of abundance that permeated the very structure of US society. With wanton abandon, both the public and Wall Street were the (often unconscious) creators of a third bubble, this time inflated by subprime mortgage-backed securities (MBS) and complex packages of associated financial derivatives.

But let us be clear. These financial instruments were great innovations which, in normal times, would indeed have provided a win-win situation: more people have access to loans, which become cheaper because banks can sell their risks to the supposedly bottomless reservoirs of investors worldwide with varying appetites for different risk-adjusted returns. However, the excesses culminating in the third

---

[168] Alan Greenspan and James Kennedy, Estimates of Home Mortgage Originations, Repayments, and Debt On One-to-Four-Family Residences, Finance and Economics Discussion Series, Divisions of Research & Statistics and Monetary Affairs, Federal Reserve Board, Washington, D.C., 2005–41.

[169] Alan Greenspan and James Kennedy, Sources and Uses of Equity Extracted from Homes, Finance and Economics Discussion Series, Divisions of Research & Statistics and Monetary Affairs, Federal Reserve Board, Washington, D.C., 2007–20.

bubble were so enormous that, as has been argued by many astute observers, too many of the MBS were "*fragile*": they were linked to two key unstable processes, the value of houses and the loan rates. The "*castles in the air*" of bubbling house prices promoted a veritable eruption of investment in MBS, and this investment pushed up the demand for and therefore the prices of houses—until it became apparent that these mutually as well as self-reinforcing processes were unsustainable.

In March 2007, the first visible signs of a worrisome increase in default rates appeared, followed in the summer of 2007 by the startled realization of the financial community that the problem would become much more serious. In parallel with an acceleration in defaults by homeowners and a deteriorating economic outlook, late 2007 and early 2008 saw a rapid increase—punctuated by dramatic bankruptcies—in the estimates being made of the cumulative losses facing banks and other investment institutions. Initial guesses were in the range of tens of billions, but soon this became hundreds of billions, then more than a trillion. The Federal Reserve and Treasury stepped up their response in proportion to the ever-increasing severity of the uncovered failures—but were either unaware of the enormity of the underlying imbalances, or unwilling to take the measures that would address the full extent of the problem. To be blunt, governments were blissfully unaware of what should have been the entirely predictable cumulative effect of these three bubbles, because each one appeared to "*solve*" the problems left by the previous one. Monetary easing, the injection of liquidity, successive bailouts: all these addressed the symptoms, but ignored the sickness.

The sickness was and remains the cumulative excess liability present in all sectors of the US economy: US household debt as a percentage of disposable income at around 130%, total bank debt as a percentage of GDP currently around 110%, US Government debt at 65% of GDP, corporate debts at 90% GDP, state and local government debts at 20% GDP, unfunded liabilities of Medicare, Medicaid, and Social Security in the range of 3–4 times GDP. Such levels of underlying liability, along with the recurring bubbles of fake expansion we have described, have produced a highly reactive and unstable situation in which sound economic valuation becomes unreliable, further destabilizing the system. This sickness has only been worsened by measures that disguise or deny it—like the misapplied innovations of the financial sector with their flawed incentive structures, and the de facto support of an all-too willing population, eager to believe that wealth extraction could be a permanent phenomenon. To be sustainable in the long term, the growth of wealth has to be equal to the actual productivity growth; in developed countries, this has been about 2–3% in real value over the long term.

Because of this failure of good governance, the crisis has accelerated to such an extent that it has forced coordinated action among all major governments. While their involvement has finally restored confidence, the crisis has revealed the depth of the problem. At one level, the loss of trust between financial institutions stemmed from the asymmetric information—the gaps in knowledge or understanding between different parties involved—on their suddenly escalating counter-party risks, making even the most solid bank seem capable of default and leading to credit paralysis. This could be addressed by financial measures and market forces. But at a deeper level, people in the street had lost confidence, as they saw a succession of problems, timidly addressed by decision makers proposing ad hoc solutions to extinguish the fire of the day (Bear Stearns, Fannie, Freddie, AIG, Washington Mutual, Wachovia), with no significant result but only more deterioration.

Nothing can resist loss of trust, since trust is the very foundation of society and economy. That people did not make a run on the banks was not, given today's insurance policies against the catastrophes of the past, sufficient indication to the contrary. In fact, there has been an *"invisible run on the banks"*, as electronic and wire transfers have been accelerating in favor of government-backed Treasury bills and other *"safe-haven"* assets [170,171]. A significant additional impediment to the restoration of public trust in the aftermath of the crisis was that the Fed, Treasury and concerted government actions are perceived as supporting a status quo where *"gains are private while losses are socialized"*.

How to restore and ensure continuity of trust in the financial sector? A number of actions must be considered. While guarantees and direct equity investments seem better tools than acquisition of bad debts at above-market prices, one needs also to examine the actions that go beyond the short-term and which have been mostly absent in the debates. First, governing bodies should play to the intelligence of the crowd. By explaining truthfully (is this possible nowadays?) the genuine cause of the problems: more than a decade of excesses and three successive and inter-related bubbles, the fact that the liabilities are enormous and that the budget has in the end to be balanced, that accelerating borrowing into the future cannot be a sustainable strategy. As humans, we are more inspired to trust when failures are acknowledged than when blame is shifted to someone else. This is the core reason why going to the fundamental source of the problems may be part of the solution in restoring confidence that a solution exists at minimal cost.

Secondly, the issue of fairness is essential for restoring confidence and support. Banks have acted incompetently in the recent bubble by accepting package risks, by violating their fiduciary duties to the stockholders [172], by letting the compensation/incentive schemes run out of control. There is an absolute need to rebuild and ensure the continuing support of public confidence, and this applies equally to the regulators. The banking system needs a constant re-assessment of the role of regulations to deal with conflicts of interest, address cases of moral hazard (where someone will take far greater risks with someone else's money than they would with their own), and enforce the basic principle of well-functioning markets, that investors who took risks to earn profits must also bear the losses. For instance, to fight the rampant moral hazard that has fueled and continues to create new bubbles, one should revisit

---

170D. Sornette, Ein Schweizer Souveränitätsfonds, Politik & Wirtschaft, Schweizer Monat, October 2015, 1030, pp.26–31

171Richard Senner and Didier Sornette, The 'New Normal' of the Swiss Balance of Payments in a Global Perspective: Central Bank Intervention, Global Imbalances and the Rise of Sovereign Wealth Funds, http://ssrn.com/abstract=2990512

172John C. Bogle, The fiduciary principle: no man serve two masters, The Journal of Portfolio Management, 2009, 36.1, pp.15–25

regularly the possibility and nature of actions by shareholders to be given "*claw-back*" permission: the legal right to recover senior executive bonus and incentive pay, if it proves to be ill-founded. In addition, many of the advisors and actors of the 2008 drama had vested interests and strong conflicts of interest, including the Fed, the Treasury and the major banks acting on behalf or with the approval of the Treasury. An independent elected body could be one way to address this problem, ensuring separation of interest and power.

The three bubbles mentioned above provide vivid support for the Minsky hypothesis that, by leveraging opportunities in speculative euphoria, financial markets produce endogenous instabilities. Contrary to standard economic theory that views crises as only caused by exogenous shocks which are inherently unpredictable, endogenous crises are predictable, which makes the existence of the turmoil of the last decade all the more upsetting as it could probably have been mitigated. Regulations are the natural response taken by governments in times of crises, but there are many problems with them. Regulations are necessary to ensure fairness and structural stability—even though short-term swings are not necessarily nefarious and should not be combated by all means, since profit in investment is the compensation for taking risks. The problem is that most regulations are either too simple or too complex to provide the intended benefits. The main difficulty with regulation lies in the "*law of unintended consequences*" or "*illusion of control*": all regulations have negative secondary effects, which are almost never foreseen. The Sarbanes-Oxley Act of 2002, enforcing enhanced standards for all US public company boards, management, and public accounting firms, is a case in point: the cost of being a publicly held company in the US increased by 130%, forcing many small businesses and foreign firms to deregister from US stock markets, to the advantage of centers elsewhere. In addition, the complexity of the new accounting reports made the information even less transparent, undermining the very goal of the Act. As another relatively recent example, consider the mortgage rates that soared from 5.87% to 6.38% during the week of 13 Oct. 2008 in an unexpected reaction to the Treasury financial rescue plan at that time: in response to its announcement that it would take equity stakes in banks and guarantee new bank debt, investors were prompted to buy bank debt and sell bonds backed by home loans, making access to loans even more expensive for households.

There is however a domain where, in hindsight, regulations would have had great benefits in defusing the present crisis: financial derivatives. Again going back the 1990s, Alan Greenspan, supported successively by then Treasury Secretaries Robert Rubin and Laurence Summers, convinced the US Congress to make the fateful decision not to pass any legislation that would have supervised the development and use of financial derivatives—notwithstanding various attempts by legislators, and the call from expert financiers like Warren Buffet and Georges Soros who warned years before the present crisis about these "*weapons of financial mass destruction*". Having been one of the most vocal supporters of the self-regulatory efficiency of financial markets, Greenspan has been writing in his memoirs that the villains were the bankers whose self-interest he had once bet upon for self-regulation. This brings us back full circle to the key issue: regulations should aim at ensuring a culture of integrity and ethical behavior. The most advanced research in behavioral and experimental economics in particular informs us that proportionate punishments and the promotion of cultural norms in society are key mechanisms for building morality and cooperation.

Fundamentally, the present problems reflect the fact that the "*financial economy*" has outgrown the "*real economy*". Mechanisms designed to control the overgrowth of the financial economy would provide an intrinsic foundation for stability. The financial world provides many services such as efficient access to funding for firms, governments and private people. Furthermore, it works as an effective storage of value, which should reflect the "*real economy*". But the extraordinary growth of the component of wealth associated with the financial world has been artificial because based on multipliers amplifying the virtual and more fragile components of wealth. A vivid example is the fact that, during bubbles, the market valuation of funds investing in bricks-and-mortar companies often far exceeds the sum of the value of the companies in which they have invested. Objective measures and indicators can be developed to quantify the ratio of wealth resulting from finance compared with the total economy. For instance, when it is measured that, on average, 40% of the income of major US firms is coming from financial investments, this is clearly a sign that the US economy is "*building castles in the air*". Guidelines could be drawn to flag warning signals to central banks and governments when the ratio of financial wealth to the value of the "*real economy*" goes above an acceptable range defined by a consensus among economists, and actions could be taken to moderate the growth of this ratio. These indicators should be the key targets of modern central banks.

Insolvency of Financial Institutions: Due to Mistakes in Management or White-Collar Crime

The main threat to the business of any financial institution originates from the potential failure of its staff to manage clients' money profitably. Such failures can have one or a number of causes: unintentional mistakes in portfolio management; macroeconomic challenges like economic downturns, which could force clients to withdraw their money and provoke liquidity problems, or market crashes; deregulation measures on highly complex financial instruments, which can make management and oversight a very difficult task; extending loans to borrowers with a questionable credit history, leading to credit defaults; and deliberate criminal action by staff in money management, such as corrupt investment advice, shortchanging, rough handling, and so on.

Barings Bank Collapse Due to Unauthorized Trading (Singapore/UK, 1995)

Barings PLC was the oldest bank in Britain and one of the oldest in the world, its name a byword for sober financial reliability. But in February 1995, Barings collapsed, sending tremors through the sector worldwide. A Singapore-based trader at the bank called Nick Leeson had single-handedly lost over US$1.4 billion (£827 million) in unauthorized transactions.

After the deregulation of the British financial sector under Margaret Thatcher's Conservative government in the mid 1980s, traditional commercial banks were free to provide services such as securities brokerage and underwriting, which had until then been the preserve of the investment banks. The intention was to give British banks an advantage in the international markets, and consolidate London's position as a leading financial center. Following the deregulation, Barings had started working in the futures market, and made more than half of its profits from securities. To ensure that he always received money from London headquarters to fund the speculations he was making between the stock exchanges in Japan and Singapore, Leeson started manipulating the accounts, leading Barings management to believe that the Singapore subsidiary was highly profitable. By the time the true scale of his losses came to light, they were colossal enough to bring Barings to bankruptcy.

How could Leeson have been able to conceal his true position from supervisors, external auditors, and regulators until the final collapse? [173]. Barings' top management did not have sufficient knowledge of the futures market to spot the impending disaster at the Singapore branch in time; and this most traditional of financial institutions relied too much on the integrity of its employees, lacking stringent internal controls. Leeson's bosses simply took the reports from Singapore on trust, passing them on without scrutiny to the Bank of England. As for Leeson himself, he later admitted that what had driven him to conceal his spiralling losses and keep claiming profits was his *"fear of failure"*. We are all influenced by our surroundings and, both on the Singapore exchange and within Barings, he was immersed in an environment that valued only success and profit, where failure and loss could not be countenanced. So, he dug himself deeper and deeper into debt, unable to bear his *"incompetence, negligence and failure"* being exposed [174]. Also, although he knew that his deception would have dramatic consequences, Leeson did not foresee that it would bring catastrophe to the entire Barings business. After the fraud came to light he was sentenced to $6^1/_2$ years in prison in Singapore, but the sentence was reduced for good conduct and due to serious illness and he was released in 1999. He has since become a consultant and conference speaker in the perception and control of risk.

Unauthorized trading scandals also occurred at Société Générale in 2008, when Jerome Kerviel caused losses equivalent to $7.2 billion, and at UBS in 2011 when Kweku Adoboli lost over $2.3 billion.

LTCM collapse (USA, 1998) The unfortunately titled Long-Term Capital Management (LTCM) was a hedge fund, which collapsed in 1998 [175]. The circumstances of the collapse could be seen with hindsight as an early warning of the subprime mortgage crisis of 2007–2008: it was brought on by the excessive leverage of LTCM's positions, and it showed how the demise of a single firm could impact

---

[173]Report to the Board of Banking Supervision Inquiry into the Circumstances of the Collapse of Barings, Bank of England, July 18, 1995, Conclusion chapter, subsection "Outline"

[174]Documentary "Going Rogue", Journeyman Pictures, December 2011

[175]See Donald Mackenzie, An Engine, Not a Camera: How Financial Models Shape Markets, The MIT Press, August 29, 2008 for a detailed account.

the financial system all over the world. LTCM's distinguished board included Myron S. Scholes and Robert C. Merton, who won the Nobel Memorial Prize for Economics in 1997. The fund developed arbitrage positions: looking for bonds, equities and traded options whose prices temporarily appeared to be too widely spread in two different markets and effectively betting that they would converge. This strategy reaped remarkable returns in the first few years; but, as markets became more efficient, investment opportunities decreased, and with an ever-growing capital base, LTCM became very highly leveraged [176]. Thus, in early 1998, the firm had assets of around US$129 billion—but only $4.72 billion of this was equity and over $124.5 billion was borrowed—a staggering debt to equity ratio of over 25 to 1 [177]. At such high leverage, a small loss of say 2% on LTCM's positions would have translated into a 50% loss on LTCM equity, which would be quite disastrous for LTCM itself and their creditors, including most of the biggest banks on Wall Street. But already in 1997, there had been a financial crisis in the East Asian markets, and the following year the Russian government defaulted on its (internal) bond payments. This was LTCM's nightmare scenario: the convergence arbitrages on which they had gambled so heavily actually diverged, bringing huge losses. Since the firm's capital was funded by the same financial institutions with which it traded, there was concern on Wall Street that, if LTCM had to liquidate its securities to pay off its debts, prices would be pushed further down: a positive feedback loop that could set off a chain reaction of catastrophic losses for the entire financial system. Thus, in September 1998, executives from 16 financial institutions met at the New York Federal Reserve to agree on a bail out of LTCM: a US$3.6 billion recapitalization of the fund to avoid further liquidation and thus prevent a global market meltdown. As we have observed, the LTCM debacle had important lessons to offer. But unfortunately, the Federal Reserve, the US treasury and financial regulators worldwide either failed to learn those lessons or chose to ignore them: essentially similar high-leverage practices have been allowed to develop across the industry, and their catastrophic consequences continue to unfold.

Collapse of Lehman Brothers and the American Subprime Mortgage Crisis (2008) The largest bankruptcy in US history was in September 2008, when the investment bank Lehman Brothers collapsed, losing assets of more than US$600 billion [178]. It was the clearest sign so far that the American real estate bubble, built on the widespread lending of mortgages to "subprime" customers, had burst. Ultimately, the banks foreclosed on more than eight million American households, who

---

[176]Leverage refers to any investment position using borrowed funds for the purchase of an asset, with the expectation that the after-tax income resulting from the asset price appreciation will exceed the borrowing cost.

[177]P. Jorion, Risk Management Lessons from Long Term Capital Management, European financial management, 2000, 6.3, pp.277–300

[178]The loss estimation for this bankruptcy assumes that the asset valuations before the bankruptcy were true. Since extraordinary inflation of asset valuations has been one of the causes for the bankruptcies, this considerable loss figure must be taken with a grain of salt.

were forced to leave their homes. Over the following 21 months, more than US$17 trillion of household wealth was destroyed. Worldwide financial and economic crisis inevitably followed during 2008 and 2009 [179], leading to the worst recession for half a century. The world's stock markets lost more than US$30 trillion [180]. By 2008, there had been decades of deregulation in the financial sector worldwide, and US financial regulation had become fragmented and ineffectual. Fundamentally, the crisis stemmed from the seductive but irresponsible idea that the growth of income and the economy can be sustained indefinitely on credit and financialisation rather than on genuine real-economy-based productivity growth; this illusion became practically an article of faith amongst governments and financial institutions all over the world [181].

In 1933, following the Great Depression, a key piece of financial regulation called the Glass–Steagall Act came into force. The act made it illegal for any single financial institution to work in investment banking, commercial banking and insurance. But by the late 1970s, economists and politicians had forgotten the hard lessons of the Great Depression. The finance industry successfully lobbied for significant weakening of regulation in the sector: deregulation of the American economy would increase competition, prices of goods and services would go down, and economic growth would accelerate. Over the next 20 years, there was significant deregulation of the activity of commercial banks, savings and loan associations, and the mutual funds established by investment banks. Financial institutions could now invest in risky assets including debt securities or derivatives (allowed from 1983), interest and currency exchange rates (from 1988), stock indices (from 1988), precious metals such as gold and silver (from 1991), and equity stocks (from 1994), and provide higher-risk loans with higher interest payments [182].

In the geopolitical context at the time, the US government had an ideological point to prove in taking a more *"hands-off"* approach to the economy: the Soviet Union, the ultimate experiment in total government control, had collapsed. And the financial lobby also turned to the academic world, funding leading professors and researchers to study the possible advantages of deregulated markets [183]. Their research gave lobbyists intellectual credibility, and enabled them to argue a legal basis for deregulation. It took the financial sector an estimated US$2.7 billion on lobbying the federal government; but ultimately, they convinced politicians to take apart the legal framework set up in the wake of the Great Depression. Political

---

[179]Markus K. Brunnermeier, Deciphering the Liquidity and Credit Crunch 2007–2008, Journal of Economic Perspectives, 2009, 23(1), pp.77–100

[180]Justin Yifu Lin, Policy Responses to the Global Economic Crisis, Development Outreach, World Bank Institute, Volume 11, Issue 3, December 2009, pp.29–33

[181]Didier Sornette, Peter Cauwels, 1980–2008: The Illusion of the Perpetual Money Machine and what it bodes for the future, Risks 2, 2014, pp.103–131

[182]The Financial Crisis Inquiry Report, The Financial Crisis Inquiry Commission, Washington, D.C., January 2011, p.35

[183]Documentary film "Inside Job", Director Charles Ferguson, 2010

"*investment*" from the sector was not confined to direct lobbying during this period: they also put more than US$1 billion into the political campaigns of sympathizing candidates [184]. In 1999, Congress passed the Gramm-Leach-Bliley Act—essentially a repeal of the Glass–Steagall Act—removing all restrictions against financial institutions being involved simultaneously in banking, securities and insurance operations.

Already a year earlier in 1998, with the approval of the Federal Reserve, Citibank had merged with Travelers Insurance Group to establish the largest financial institution in the world, Citigroup Inc. But because of the legal framework in place before deregulation, banking, securities and insurance were still covered by separate regulators. This, of course, also applied to the new merged financial institutions: no single government regulator had yet been set up to oversee their activities across all three sectors. Thus, no single person—neither on the regulatory side nor on the senior management of the new breed of financial institutions—fully understood the risks involved across their whole range of activities, within which different sections of the same institution may have conflicting interests.

The most bewildering area was the ever-expanding range of new derivatives. The creation and calculation of derivatives was already highly complex and, with the financial institutions continuing to lobby against them, government regulators had little chance of maintaining proper control over innovative financial instruments. According to then Federal Reserve chairman Alan Greenspan, there was no need for such oversight of over-the-counter (OTC) derivatives in any case. In 1998, he declared: "*Regulation of derivatives transactions that are privately negotiated by professionals is unnecessary*" [185]. Thus, unrestricted by effective regulation, the worldwide derivatives market—90% of which was OTC derivatives—increased 60-fold over just two decades. From its 1990 value of US$10 trillion, the market reached US$605 trillion (of underlying face value) by 2009 [186]; by comparison, the world GDP in 2010 was around US$65 trillion [187]. After the mortgage crisis in autumn 2008 and confidence in the private transactions of professionals had shattered, Greenspan admitted: "*Those of us who have looked to the self-interest of lending institutions to protect shareholders' equity (myself especially) are in a state of shocked disbelief*" [188].

The Clinton and George W. Bush administrations were both keen to increase home ownership, and set ambitious targets to do so. This would create millions of new jobs in construction and stimulate economic growth. In December 2001 the

---

[184]The Financial Crisis Inquiry Report, The Financial Crisis Inquiry Commission, Washington, D.C., January 2011, p.xviii

[185]Testimony of Chairman Alan Greenspan, The regulation of OTC derivatives, Before the Committee on Banking and Financial Services, US House of Representatives, July 24, 1998

[186]Top 10 Challenges for Investment Banks 2011, Accenture, 2010, Chapter "Challenge 2: Dealing with OTC Derivatives Reform"

[187]In search of growth, The Economist online, May 25, 2011

[188]Kara Scannell, Sudeep Reddy, Greenspan Admits Errors to Hostile House Panel, The Wall Street Journal, October 24, 2008

Federal Reserve cut their interest rate to 1.75%, the lowest for 40 years, and the Fed rate reached and remained at 1% for most of 2003 and 2004. With the mushrooming range of new financial products, potential borrowers had easy access to cheap credit. People would remortgage their houses to raise the money to pay medical bills, get the kids through college, or take time out to start a new business. In the brave new financial world, the house you lived in was just another asset. With no increase in wages, home refinancing shot up from US$460 billion in 2000 to US$2.8 trillion 3 years later [189].

Traditionally, mortgage lenders had always selected borrowers carefully: if borrowers proved unable to keep up fixed-rate payments for 30 years, the lenders too would face serious consequences. The financial institutions depended on the reliability of their debtors for their own stability. Even into the 1990s, only *"prime"* borrowers who could meet tough lending conditions could expect to get a mortgage. First-time home buyers, for example, were expected to make a 20% down payment. But, in the deregulated market of the later 1990s, lenders could set lower and lower standards for borrowers. People with no credit history or proof of income— *"subprime"* borrowers—found themselves being offered credit. Even with toothless regulation, how was this in the interest of lenders? The financial sleight of hand that made it possible was known as the *"securitization pipeline"*. First, multiple loans were packaged by lenders into residential mortgage–backed securities. Then, investment banks like Goldman Sachs, Merrill Lynch, Bear Stearns and Lehman Brothers repackaged these securities into collateralized debt obligations (CDOs). Finally, CDOs were promoted to traditionally more conservative American investors such as hospitals, endowment funds and retirement schemes, and global investors such as pension funds and sovereign funds: they were claimed to be as safe as US Treasuries, and had the same AAA-rating, but investors could expect a higher yield. In fact, CDOs were *"tranches"* of mortgage–backed securities from debtors ranging from the prime to the distinctly subprime [190,191].

Moreover, the banks were insured against potential default by credit default swaps (CDSs) provided by leading insurance companies like American International Group (AIG). AIG was the largest insurance company in the world, and had issued CDSs for a total insured value of at least $379 billion by 2007 [192]. Crucial to the confidence of investors further along the pipeline was the CDOs' AAA rating. Rating agencies like Moody's, Standard & Poor's and Fitch were respected as the arbiters of financial credibility, but investment banks *"paid handsome fees to the*

---

[189]The Financial Crisis Inquiry Report, The Financial Crisis Inquiry Commission, Washington, D.C., January 2011, p.5

[190]Paul Muolo and Mathew Padilla, Chain of Blame. How Wall Street Caused the Mortgage and Credit Crisis, John Wiley & Sons, pp.185, 219

[191]The Financial Crisis Inquiry Report, The Financial Crisis Inquiry Commission, Washington, D.C., Jan. 2011, pp. 117, 119, 170, 278, 339, 393

[192]Matthew Richardson, Why the Volcker Rule Is a Useful Tool for Managing Systemic Risk, NYU Stern School of Business, 2012, p.8

rating agencies to obtain the desired ratings" [193]. The agencies expected between US$0.5 million and $0.85 million of fees for each mortgage-related security.

When CDOs were first created, their security rested on a broad range of debts and would have justified the rating; but as lending grew, keeping track of the quality of borrowers was more difficult. Nevertheless mortgage-related securities continued to receive the highest rating: Moody's, for example, gave the AAA stamp of approval to nearly 45,000 such securities between 2000 and 2007. In 2006, the agencies made US$887 million on mortgage ratings, accounting for 44% of all corporate revenue that year. Moody's alone rated 30 mortgage-related securities as AAA every working day. Ultimately, during the crash in 2007–2008, 83% of the mortgage securities rated AAA during 2006 would be downgraded [194].

Having bought a mortgage securities package, the owners of the package would receive all payments from borrowers—which meant that the original lenders need no longer concern themselves with the financial situation of debtors. Lenders were thus free to make riskier loans with impunity, and subprime mortgages shot up from 7.4% of the US mortgage market in 2002 to 23.5% in 2007 [195]. Obviously, the total mortgage market increased as a result, bringing ever higher profits to the financial institutions. The more new CDOs were issued, the greater the cut for the investment banks—so they encouraged lenders to offer new credit to anyone who could be persuaded to take it.

The model caused, but also suited, a constantly expanding housing market. Borrowers took advantage of the easy credit and got the market moving, and lenders had scant concern about the security of their money because they were passing the risks to investors through CDOs and other mortgage securities; the investors in their turn felt they had a safe bet, because the risks were insured through CDSs. At the full spate of this expansion in late 2000, Alan Greenspan *"argued that the financial system had achieved unprecedented resilience"* [196]. Mortgage securities, with their AAA ratings, seemed to be among the safest investments on the market.

By trading in derivatives over the counter, the five major investment banks (Bear Stearns, Goldman Sachs, Lehman Brothers, Merrill Lynch, and Morgan Stanley) were able to operate with very high leverage on their capital. Sometimes their leverage ratios were as high as 40 to 1: in other words, for every US$40 in assets, there was in fact only US$1 in capital to cover losses. But such high ratios represented a huge gamble on assets continuing to rise in value—even a 3% drop

---

[193]The Financial Crisis Inquiry Report, The Financial Crisis Inquiry Commission, Washington, D.C., January 2011, p.44

[194]The Financial Crisis Inquiry Report, The Financial Crisis Inquiry Commission, Washington, D.C., January 2011, p.xxv

[195]The Financial Crisis Inquiry Report, The Financial Crisis Inquiry Commission, Washington, D.C., January 2011, p.70

[196]The Financial Crisis Inquiry Report, The Financial Crisis Inquiry Commission, Washington, D.C., January 2011, p.83

would be enough to bankrupt a major investment bank [197]. The investment banks' response to this apparently reckless risk will not by now surprise the reader: through credit default swaps, their brokers were simultaneously trading for and against the housing boom.

In June 2006, Richard Bowen, chief business underwriter of the largest financial multinational Citi, discovered that up to *"60% of the loans that [were bought] and packaged into obligations were defective"*. He alerted Robert Rubin, chairman of the Executive Committee of the Board of Directors of Citi and a former US treasury secretary. But there was no reply from Rubin on this early warning—instead, Richard Bowen's *"bonus was reduced, and he was downgraded in his performance review"* [198]. This active discouragement of any dissent was common practice among the major *"players"* of the securitization pipeline. Whether they came from an insightful researcher or the managing director of a department, warnings of imminent disaster were ignored. At around this time, Richard Fuld—CEO of Lehman Brothers—was purging the bank of anyone who not only realized, but had also the courage to say, that they were heading for serious trouble. Most investment bank executives simply failed to understand the corrosive influence of OTC derivatives on their business. Even when asset values started to fall, rapidly depleting the capital they had over-leveraged to cover their liabilities, they continued to reassure investors, competitors, partners and the authorities of the financial stability of their organizations. At a meeting of Lehman Brothers shareholders in April 2008—just after the collapse of competitors Bear Stearns—Richard Fuld assured those present that *"the worst . . . [is] . . . behind us"* [199]. Some sources would argue that he had insufficient grasp of the new financial instruments—the CDOs and CDSs—since his expertise was as a bond trader. Even after he had voted to file for bankruptcy for Lehman, Fuld would maintain: *"There was no capital hole at Lehman Brothers. At the end of Lehman's third quarter [of 2008], we had US $28.4 billion of equity capital"* [200].

At the butt end of the securitization pipeline, among the insurance giants, the situation was no better. Senior executives at AIG later admitted to the FCIC commission that *"they did not even know about these terms of the [credit default] swaps until [July 2007 when] the collateral calls started rolling"* [201]. AIG were supervised by the unfortunately-named Thrift Supervision—but even they failed to

---

[197]The Financial Crisis Inquiry Report, The Financial Crisis Inquiry Commission, Washington, D.C., January 2011, p.xix

[198]The Financial Crisis Inquiry Report, The Financial Crisis Inquiry Commission, Washington, D.C., January 2011, p.19

[199]Lorraine Woellert, Yalman Onaran, Fuld Targeted by Lawmakers as Surrogate for Wall Street Excess, Bloomberg, Oct. 6, 2008

[200]The Financial Crisis Inquiry Report, The Financial Crisis Inquiry Commission, Washington, D.C., January 2011, p.235

[201]The Financial Crisis Inquiry Report, The Financial Crisis Inquiry Commission, Washington, D.C., January 2011, p.243

uncover the true level of risk the company was underwriting [202]. When the stricken banks and investors started turning to AIG to cash in their credit default swaps, the insurers did not have the liquidity to bail them out. Instead, in September 2008, they themselves were bailed out by the US government for a colossal US$85 billion. The whole house of cards had finally collapsed.

Bernard Madoff Ponzi Scheme (USA, 2008) Madoff Investment Securities was one of the most reputable investment firms on Wall Street for decades. The founder was well connected with the financial, cultural and political elite of the United States and attracted private investors, charities and hedge funds at home and abroad with the exceptionally stable return rate he offered: a guaranteed 10–12% per year, regardless of the volatility of the market. He came across as a solid and reliable person, the key market maker on Wall Street. Then, when some hedge funds tried to withdraw their money during the subprime crisis of 2008 which we have discussed above, the fraud was disclosed. Involving an estimated US$64.8 billion, it was the largest Ponzi scheme in history: the stable returns for the older investors in the scheme were not profits from the shrewd investment of their money, but came straight out of the accounts of the latest tier of new investors.

The fact that fraud on such a tremendous scale could go undetected for decades was due in no small part to Bernard Madoff's shrewd mastery of manipulation, but it was also made possible by the deregulation of the US financial sector in the 1990s and 2000s. Madoff had a close relationship with some high-level executives at the Securities and Exchange Commission [203]; in any case, the commission had limited resources for digging into the business of Madoff Investment Securities—even though a number of companies and investigators warned the commission of a possible Ponzi scheme almost a decade before the fraud was finally discovered. In the end, Bernard Madoff was sentenced to 150 years of prison for this white-collar crime against thousands of investors.

Poaching of Clients by Two Vice Presidents of Goldman Sachs (USA, 2017)
The investment bank accused two former vice presidents of stealing its customers, by persuading clients to transfer their money into the management of a new independent investment firm set up by the two before they resigned from Goldman Sachs. The bank maintained that such behaviour from its former staff violated employment agreements, and took legal action against them [204].

Avoiding disaster in the complex world of finance requires both sophisticated and common-sense solutions, including: careful selection when hiring staff or making internal promotions; adequate control over highly qualified and ambitious

---

[202]The Financial Crisis Inquiry Report, The Financial Crisis Inquiry Commission, Washington, D.C., January 2011, p.243

[203]Christoph Luetge, Johanna Jauernig, Boudewijn be Bruin, Business Ethics and Risk Management, Springer Science & Business Media, 2013, p.27

[204]Bob Van Voris, Neil Weinberg, Goldman Sachs Sues Two Ex-Executives Over Client Poaching, Bloomberg, May 23, 2017

employees to prevent unauthorized decisions and white-collar crime; advanced risk management software which limits access, enables multi-level decision-making, and accurately models finances to reliably predict solvency; a corporate culture that encourages and motivates whistleblowers; advanced security and surveillance systems to monitor the actions of employees; careful and financially realistic selection of borrowers; and advanced monitoring of the economic situation and the liquidity of other market players.

But, of course, when the whole system, from banks to underwriters, rating agencies, regulatory agencies, investors, government officials, the central bank and the public are all together taken in by the frenzy and the illusion of the *"perpetual money machine"* [205], there is not much to do, except surf the opportunities and prepare for the crash and exploit it as a number of contrarian investors did [206].

Poor Customer Service

Good customer service in a potentially fraught area like finance can be challenging: staff must serve clients with a good attitude, have a genuine desire to solve the customer's needs, be patient even with difficult clients, scrupulously work in the customer's best interest, have a good command of IT systems and internal procedures, and in some areas an understanding of very complex financial tools. Failing in any of these respects could leave the customer disappointed, frustrated and in some cases furious—and in a highly competitive subsector, the institution will have lost at least one person's business, and probably more if the aggrieved client publicizes their negative experience in forums, or on social and traditional media. Bad customer care does not cause prominent large-scale crises, but it does cause millions of micro-crises, which chip away at the reputation of financial institutions. As with other service network business, an incident at one branch of a financial institution threatens the whole network, because customers tend to perceive such crises as corporate problems of the whole institution.

The ongoing prevention of poor customer service requires: careful selection of service staff; motivation of frontline staff to serve customers politely, kindly and accurately according to unified corporate standards; training of staff on the frontline and in call centers to develop advanced customer care skills along with the professional expertise they need; skilled operation of IT systems and business processes; advanced claim management; and ongoing analysis of consumer preference and behavior, informing the development of advanced customer relationship management (CRM) systems. The automation of some services could reduce servicing time and minimize potential conflict situations in the interaction of clients with staff.

---

[205]D. Sornette and P. Cauwels, 1980–2008: The Illusion of the Perpetual Money Machine and what it bodes for the future, Risks 2, 2014, pp.103–131

[206]Michael Lewis, The Big Short: Inside the Doomsday Machine, W. W. Norton & Company, February 1, 2011

Risks Through Information Technology Breakdowns

Sweeping automation of services, ever faster operations and increasing numbers of transactions all have a positive influence on the efficiency of a financial organization. However, the risks associated with this trend bring new challenges, such as the need to constantly maintain the resilience of IT systems against technical failures anywhere on a global network, and against attacks by hackers who want to obtain access to customers' data and ultimately their money.

Black Monday Crash and Knight Capital Collapse (USA, 1987, 2012) On Monday October 19 1987, stock indexes in the US fell by more in a single day than on any previous day in the history of the market: the Dow Jones Industrial Average and S&P 500 dropped by 22.6% and 20.4% respectively. During the first months of 1987, American stocks had grown significantly, but several days before the crash there was news about a larger-than-expected US trade deficit and subsequent devaluation of the American dollar. Investors all around the world began to cut their losses by liquidating their positions, and by the opening of the Monday trading session in the United States, there was massive bear trend pressure on stocks caused by futures. The plunge was exacerbated by a widespread "*portfolio insurance*" option in the software that already governed trading, which was programmed to sell accumulated securities immediately if their prices dropped below a predetermined level. The prevalence of such commuter models, all programmed according to the same principle, caused a cascade of stock selling. It was the first time in the modern history of trading when computer networks designed to protect against losses brought such serious damage to investors and investment banks. This crisis was not initiated by a bug in the software, but the collapse of Knight Capital Group in 2012 was. The investment company faced losses of $440 million when a high-frequency trading algorithm, recently introduced but not adequately tested, bought at high and sold at low on approximately 150 different stocks during less than hour.

JPMorgan Chase Data Breach (USA, 2014) The largest bank in the country was attacked by hackers in 2014. They got into 90 servers at JPMorgan Chase and stole the personal information of about 76 million households and seven million small businesses: names, addresses, phone numbers and emails of the clients. Fortunately, financial data was not compromised [207].

IT Outage of PayPal (Worldwide, 2015) On October 302,015, PayPal was unable to provide transactions globally for several hours after a power outage of one of their data centers. At that time, the payment platform arranged up to 12.5 million transactions daily and had more than 170 million accounts, with a total volume of $235 billion annually. So even 2-h outage represented more than $50 million losses for merchants [208].

---

[207] Jessica Silver-Greenberg, Matthew Goldstein, JPMorgan Chase Hacking Affects 76 Million Households, The New York Times, October 2, 2014

[208] Leena Rao, PayPal outage may have cost merchants millions, Fortune, October 30, 2015

For a financial institution, responding to the risk of serious IT problems includes substantial investment into computer systems and defense software, and constant training of IT staff to manage this kind of risk. Collaboration with government bodies responsible for fighting cyber crime is also vital.

Other Risks

Other risks for financial organizations include: scams, especially directed at clients with medical and car insurance, which can have a serious financial impact on insurance companies; physical robbery of an individual bank branch, endangering clients and staff; and pressure from regulators on a bank to disclose information about their clients to tax authorities.

Key Risk Mitigation Measures in Financial and Insurance Services

- Maintaining an excellent relationship with other participants in economic activity so that they trust the financial institution at all times.
- Development of cooperative relations with the authorities, because the activity of financial institutions influences the national and even international economy and should therefore be regulated properly.
- Careful selection and retention of honest-minded and highly-qualified staff, and complex and detailed oversight over those staff.
- Careful selection of borrowers and clients.

3.7   Professional Services

This subsector includes the following industries according to ISIC Rev. 4:

- 69—Legal and accounting activities
- 70—Activities of head offices; management consultancy activities
- 71—Architectural and engineering activities; technical testing and analysis
- 73—Advertising and market research
- 74—Other professional, scientific and technical activities

General Description and Key Features of the Subsector

Professional services organizations support the business operations of other companies by assisting in their internal business processes, or by providing unique solutions in management consulting, legal support, architecture, marketing or public relations. The majority of professional services work with their clients on one or a series of specific projects. The caliber of personnel has a higher influence on business in this subsector than in any other in the global economy, because organizations sell unique ideas and solutions developed by highly qualified employees with vast experience in the field. Many organizations in these industries are named after their founders, who have created a brand by offering unique professional solutions and now guarantee professional excellence by their name and reputation. A partnership structure gives stability to the business, eliminates conflicts and helps retain the most qualified staff within an organization.

Critical Success Factors for an Organization Within the Subsector

Success for a professional services provider rests on the ability to attract and retain the best minds in a field, and motivate them to develop unique and unbeatable ideas and solutions. The other key factor is how the organization relates to its potential market: it must convince customers that it offers better solutions than its competitors, and develop a long-term relationship with them as a trusted outsource of many internal corporate functions.

Stakeholders in this Subsector [209]

- Employees—50%
- Customers—40%
- Partners/suppliers—5%
- Other—5%

Typology of Common Risks, Main Features of Major Accidents and Risk Mitigation Measures Within the Subsector

Poorly Judged Professional Advice

The majority of accidents occur because even highly qualified and experienced staff can make errors of judgment in assessing what their client needs. Public disclosure of such mistakes can damage the reputation of the provider, because other clients will question the professional compliance of its staff.

---

[209]Our informed appraisal of the influence of each audience on a typical organization within the subsector (100% = combined influence of all audiences)

McKinsey & Co. and Mistaken Advice to Swissair (Switzerland, 1990s) In the 1970s and 1980s, Swissair's business was stable and the airline thrived. Then in 1992, Switzerland chose not to join the forthcoming European Economic Area (EEA). After this, the national flag-carrier Swissair struggled to expand its operations in Europe against the competitive pressure caused by the liberalization of air transportation within the EEA. Moreover, the airline was making little progress in forging alliances in the wider global market. Frustrated in its desire to grow and compete, Swissair hired management consultants McKinsey & Company, who created a so-called "*hunter*" strategy for the airline. They would make a billion-dollar expansion into the European market by buying stakes in regional air carriers like the Belgian Sabena, several small French airlines, the charter airlines LTU and Air Europe, the Polish LOT and others. In return for Swissair's financial involvement, the regional companies agreed to use the back-office services of their new stakeholder (catering, booking, technical, logistics and cargo services) and to comply with Swissair's management position in many aspects of the business.

However, after several years of implementation, the "*hunter*" strategy failed, for two major reasons. Firstly, some of the acquired air carriers did not perform well financially, which generated debts and losses for Swissair; and secondly several unions took an uncompromising position towards the proposed changes, which made implementation difficult [210]. Finally, in October 2001, the financial pressure became intolerable and all Swissair's planes were grounded, affecting up to 40,000 customers. While the consultants were not responsible for the implementation of the strategy they had recommended, it was nevertheless their misjudged advice that led to the bankruptcy of the airline in 2002: there is information that the former head of the Swiss branch of McKinsey & Co, who joined the board of Swissair, insisted on the implementation of the "*hunter*" strategy [211].

According to Duff McDonald, who wrote the investigative book "*The Firm: The Story of McKinsey and Its Secret Influence on American Business*" [212], the leading management consulting firm in the world made several huge mistakes in advising corporate clients. Thus, the infamous Enron (see below) was consulted by McKinsey & Co—but the firm did not allow its staff to make recommendations on the financial issues that ultimately led to the collapse of Enron. McKinsey proposed the US$350 billion merger between America Online and Time Warner, which was later described in the New York Times as "*one of the biggest disasters that have occurred in our country*" [213]. In 1980, the firm had been wildly off the mark in its forecast for AT&T of the number of people who would be using cellphones by 2000: in reality,

---

[210]A scary Swiss meltdown, The Economist, July 19, 2001

[211]Ulrich Steger, Helga Krapf, The Grounding: did corporate governance fail at Swissair?, IMD, 2002, p.5

[212]Duff McDonald, The Firm: The Story of McKinsey and Its Secret Influence on American Business, Simon & Schuster, 2014

[213]Tim Arango, How the AOL-Time Warner Merger Went So Wrong, The New York Times, January 10, 2010

this number reached 108 million, while only 0.9 million was predicted
[214]. McKinsey & Co. also advised Railtrack, the company that took over railway
infrastructure after the privatization of British Rail, proposing that they should
radically change the maintenance strategy to reduce investment in infrastructure:
*"Railtrack should 'sweat' its assets. This meant replacing its cyclical system of
rail maintenance with a programme where infrastructure was mended on an as-
and-when basis"* [215]. The changes led to so many infrastructure and safety-
related accidents that Railtrack's assets were eventually transferred to the state-
controlled Network Rail. These examples are just the most notorious among many
others in the consultancy's checkered career. However, this should not obfuscate
the fact that McKinsey & Co has also contributed to some major business suc-
cesses, and helped develop many ideas that have become accepted norms of
modern business.

Mistakes in Architectural Design (Worldwide) Architectural mistakes and
design shortcomings can lead to the collapse of buildings or huge additional
spending on rework. We have described several cases in the "Construction"
subchapter above: the misjudgment of seismic risks in the Soviet republic of
Armenia before a major phase of high-rise construction there; errors in calculating
the weight of the proposed Hotel New World in Singapore, which led to its collapse;
the miscalculation of the load-bearing capacity of a walkway at the Hyatt Regency
Kansas City; or the vulnerability of the Citigroup Center in New York to hurricane
winds. Some other noteworthy mistakes in design include these: the collapse of the
Hartford Civic Center roof (Connecticut, USA, 1978); the reconstruction of the Ray
and Maria Stata Center at MIT when the innovative design of the buildings led in
practice to problems that were at best inconvenient and at worst dangerous (Massachusetts, USA, 2000s); the collapse of the roof of Transvaal Park into a crowded
swimming pool due to obvious mistakes in its design (Moscow, Russia, 2004–28
died); the concave glass face of the Vdara Hotel & Spa building, which reflected and
focused the sun's rays into the hotel swimming pool area below, severely burning
some guests (Las Vegas, USA, since 2009); or similar sun reflection problems at the
Walt Disney Concert Hall in Los Angeles and the Walkie-Talkie skyscraper in
London.

   Response measures include: thorough analysis of the competence and integrity
of potential employees; vigilant internal control—including independent internal
assessment of the solutions being offered to customers; active collaboration
with external control, by inviting and working with regulators and advanced
industrial experts; and claim management under the control of the owners of the
business.

---

214 Andrew Ross Sorkin, McKinsey & Co. Isn't All Roses in a New Book, The New York Times,
September 2, 2013
215 Clayton Hirst, The might of the McKinsey mob, The Independent, January 20, 2002

White-Collar Crime

In some cases, employees in professional services may choose to use their skill and experience not for the prosperity of clients and society but for their own benefit, through deliberate deception of clients, insider trading and other criminal action.

Role of Arthur Andersen in the Rise and Fall of Enron (USA, until 2001)

Usually, scams against the interests of clients lead to total destruction of a professional services company. The most infamous case in the history of the subsector is the collapse of the Arthur Andersen audit firm, after it came to light that they had played an active part in falsifying the accounts of the American company Enron. After a gradual descent into a corrupt and mercilessly competitive corporate culture, Enron went bankrupt in December 2001 with losses of US$63.4 billion. For a short time, this was the worst bankruptcy in US corporate history, though this record did not stand for long: a similar accounting scandal brought down global giant WorldCom in 2002, with losses of $107 billion, and in 2008 Lehman Brothers collapsed losing more than $600 billion [216].

Even before its unsustainable expansion, Enron was a large and broadly-based concern: it had approximately 3500 domestic and foreign subsidiaries and affiliates and worked in the wholesale and commodity markets, the operation of gas transmission systems, and the management of retail energy services, energy-related assets and broadband services. Over nearly a decade leading up to the bankruptcy, Enron executives were routinely cooking the books to increase the company's perceived revenue. The higher the company's value rose in the market, the higher their own income became—in the short term—and the internal climate of the company put relentless pressure on its employees to meet short-term targets. By 2000, Enron was declaring revenues of over $101 billion, giving the impression that the company was the seventh largest in the USA [217]. Most of this huge apparent revenue was through trading operations, especially the trading of energy derivatives. The human cost of the bankruptcy was colossal. The company's 59,000 shareholders—who included university endowments and pension funds—lost more than $60 billion between them; up to 25,000 employees lost a total of $2 billion in stock options and pension funds; and as for Enron's 20,000 creditors, they received between 14 and 25 cents on every dollar lent to the company [218]. The other major casualty of Enron's collapse was the business of their auditors.

---

[216]These loss estimations for these three bankruptcies assume that the asset valuations before the bankruptcy were true. Since extraordinary inflation of asset valuations has been one of the causes for the bankruptcies, these considerable loss figures must be taken with a grain of salt.

[217]Report of investigation of Enron Corporation and related entities regarding federal tax and compensation issues, and policy recommendations. Volume I: Report, US Joint Committee on Taxation, Feb. 2003, p.5

[218]Dick Carozza, Interview with Sherron Watkins. Constant Warning, Fraud Magazine, January/February 2007

Where many others could be seen as innocent victims of the disaster, this was not the case for Arthur Andersen, who went to the wall the following year through irreparable damage to their reputation. They had been one of the "Big Five" accounting firms in the world, along with Ernst & Young, Deloitte & Touche, Price Waterhouse Coopers and KPMG. In the 1990s, Andersen diversified beyond their main auditing practice to move into accounting consultancy. The two arms of the company were supposed to be in a cooperative partnership, but they competed with each other, failing to communicate when their clients were in difficulty. Furthermore, the continuous growth of revenue became the only priority as Andersen sought to win and keep the big accounts, regardless of the quality of those clients or even the legality of their own recommendations [219]. Enron became Andersen's second largest client; the largest was WorldCom, which as we have mentioned filed for bankruptcy in 2002.

The accountants' Houston office could hardly have had a closer symbiosis with Enron, who were their main client [220]. Enron employed Arthur Andersen to provide both audit and consultancy services. In addition, they did not have their own internal audit department [221]; instead they outsourced their "*internal audit*" to Andersen and even their internal accountants and controllers were often former Andersen executives [222]. Thus, between internal and external auditing fees, Andersen were earning $1 million a week from Enron [223]. On the advice of Andersen consultants, Enron employed more aggressive accounting, including the use of "*special purpose entities*": separate companies set up solely to keep some transactions off Enron's balance sheet. Andersen's chairman Joseph Berardino later testified that "*in the previous year (2000), Andersen had received $52 million in fees from Enron, of which only $25 million could be directly attributed to the audit. Of those fees, $13 million were clearly for consulting work and the remaining $14 million is arguably related to the audit because it is work that can 'only be done by auditors'*" [224]. Staff bonuses at Andersen's Houston office depended on the stable growth of Enron; and as one case study put it, "*[l]ured by promises of undreamt-of-wealth, many Andersen employees aspired to work for Enron and were therefore*

---

[219]Jennifer Sawayda, Arthur Andersen: An Accounting Confidence Crisis, Daniels Fund Ethics Initiative, University of New Mexico, pp. 2,6,

[220]Gary M. Cunningham, Jean E. Harris, Enron And Arthur Andersen: The case of the crooked E and the fallen A, Global Perspectives on Accounting Education, Volume 3, 2006, p.31

[221]Dick Carozza, Interview with Sherron Watkins. Constant Warning, Fraud Magazine, January/February 2007

[222]C. William Thomas, The Rise and Fall of Enron. When a company looks too good to be true, it usually is, Journal of Accountancy, April 2002

[223]E. Banks, The Insider's View on Corporate Governance: The Role of the Company Secretary, Springer, November 2003, p.177

[224]Peter C. Fusaro, Ross M. Miller, What Went Wrong at Enron: Everyone's Guide to the Largest Bankruptcy in US History, John Wiley & Sons, 2002, pp.127–128

very reluctant to 'rock the boat' with the company" [225]. So closely enmeshed with the fortunes of their client as they were, it is perhaps no surprise that Andersen auditors were willing to approve falsified accounts in order to earn more. Clearly, Arthur Andersen's headquarters should have had an eye on their Houston office. But the company lacked strong internal control over its regional units, and its top management were delighted with the Houston office's continuous growth; thus, they refrained from awkward questions about exactly what was going on there. The final act of the Houston office in this disastrous episode saved them from detailed exposure of their misconduct, but ultimately cost Arthur Andersen its reputation: when Enron's false accounting came to light, the Houston office shredded thousands of documents detailing their work for Enron from 1997 to 2001 [226]; Andersen were only found guilty of obstruction of justice, and were fined just $0.5 million [227]. But investors worldwide now questioned the veracity of the firm's accounting reports for any of their clients. Arthur Andersen never recovered from the Enron case: auditing depends on trust, and an auditing company stands or falls with its reputation.

Galleon Group Insider Trading Scandal (USA, 2010) Rajat Gupta was the managing director of McKinsey & Co. from 1994 to 2003. After resigning from this executive position, he became a senior partner emeritus of the company and was invited to be an independent member of the boards of Goldman Sachs, Procter&Gamble and other American corporate giants and non-profit organizations. Gupta was a close friend of Anil Kumar, a senior partner at McKinsey & Co., and the billionaire Raj Rajaratnam, founder of the Galleon Group hedge fund. In 2009, officials accused the three friends of insider trading. For several years, state investigators had been recording telephone calls to Rajaratnam from Gupta, Kumar and many other well-informed sources, giving him confidential information about the plans of large American companies. Rajaratnam used this inside information to play the stock market, and gave cash to several of his sources in return for the tip-offs. Rajat Gupta insisted that he did not benefit from leaking boardroom business; but ultimately, he was sentenced to 2 years for this white-collar crime. Moreover, the crisis put in question the moral values of senior executives of the leading management consulting firm, because it was the first time that partners of the firm had been involved in violating financial security laws through such callous disclosure of their clients' secrets. McKinsey & Co. was forced to draw a boundary between Gupta, Kumar and the firm. In addition, new stricter rules were established for employees of the firm and members of their families limiting the trading of securities [228].

---

[225] Paul H. Dembinski, Carole Lager, Andrew Cornford and Jean-Michel Bonvin, Enron and World Finance. A Case Study in Ethics. pp.196–197

[226] Jennifer Sawayda, Arthur Andersen: An Accounting Confidence Crisis, Daniels Fund Ethics Initiative, University of New Mexico, pp.5,8

[227] Elizabeth K. Ainslie, Indicting Corporations Revisited: Lessons of the Arthur Andersen Prosecution, American Criminal Law Review, Vol. 43:107, p.107

[228] Anita Raghavan, In Scandal's Wake, McKinsey Seeks Culture Shift, The New York Times, January 11, 2014

According to Duff McDonald, author of The Firm, "*McKinsey does what all its competitors do, which act as de facto industrial spies. The firm would surely take umbrage at the suggestion, but the whole notion of 'competitive benchmarking' is just a fancy way of telling one client what the other clients are up to, with the implicit — and somewhat dubious — promise that their most sensitive secrets will not be revealed*" [229].

Tampering of Samples/Forging of Documents in Thousands of Drug-Related Cases (USA, 2003–2012) A juridical disaster emerged when it came to light that Annie Dookhan, a chemist at the Hinton State Laboratory Institute in Boston, had falsified narcotics-related tests: for instance, she had added cocaine to drug-clear samples, or forged documents proving that tests had been positive or negative. Over her time at the institute, she had conducted more than 40,000 tests—sometimes up to 500(!) tests per month. By maintaining such a tremendous work rate—three times the average productivity in the field—Dookhan perhaps felt she had something to prove to her colleagues: she had in fact started this nefarious career by forging her Masters diploma in chemistry [230]. The forged tests were the pretext for exculpating and releasing around 24,000 convicted persons from prison. In the end, Dookhan was herself sentenced to 3 years for obstruction of justice, tampering evidence, perjury and falsification of records [231].

Plagiarism of Logo Design for the 2020 Olympic Games in Tokyo (Japan, 2015) The 2020 Olympic committee was forced to hold a new round of selection for the logo of the Games after allegations that Kenjiro Sano, designer of the proposed logo, had taken his design from the logo of the Théâtre de Liège in Belgium, previously created by designer Olivier Debie [232]. Accusations of plagiarism of advertising and marketing ideas are common in this subsector.

Measures to tackle white-collar crime include: careful selection of staff; enhanced internal control over the decisions and actions of key staff members; developing a corporate culture that promotes long-term sustainability of business over short-term profit; advanced compensation systems, including a clear pathway to obtaining partner status, to motivate staff to build the long-term sustainable development of the firm; a corporate culture that encourages whistleblowers; and close collaboration with regulators, state investigators and prosecutors in the event of suspicious staff activity.

---

[229]Duff McDonald, The Firm. The Story of McKinsey and Its Secret Influence on American Business, Simon & Schuster, 2013, p.191

[230]Elaine Quijano, Massachusetts lab tech arrested for alleged improper handling of drug tests, CBS News, September 28, 2012

[231]Shawn Musgrave, DAs say Dookhan drug-tampering case nearing an end, Boston Globe, March 25, 2017

[232]Justin McCurry, Tokyo 2020 Olympics logo scrapped after allegations of plagiarism, The Guardian, September 1, 2015

Key Risk Mitigation Measures in Professional Services
- Ongoing assessment of the quality and adequacy of advice and solutions provided by professional services staff towards corporate clients.
- The ability to attract and retain the best minds in a field, and motivate them to develop unique and unbeatable ideas and solutions.

3.8   Scientific Research and Education

This subsector includes the following industries according to ISIC Rev. 4:
- 72—Scientific research and development
- 85—Education

General Description and Key Features of the Subsector

Scientists are responsible for the accumulation, systematization and distribution of knowledge about all spheres of our civilization, the Earth and the universe. In spite of the rise of e-learning, real teachers remain the key source for delivering high quality education. The quality of teaching staff and of the education they deliver, the comparative rating of an educational organization and the public comments and achievements of alumni determine potential students' choice of educational institution. The influence of staff on the business process in this subsector is one of the highest within the service sector.

Critical Success Factors for an Organization Within the Subsector

It is above all the ability to recruit and retain brilliant researchers and teachers that determines the success of scientific and educational organizations. Other key factors are: a consistent record of high-quality education and high student performance over decades/centuries, which maintains a good reputation and brand; the ability to satisfy the expectation of students, their parents and society; the foresight to anticipate the trends of future development and teach not what is now in demand, but what will be in demand; and the maintenance of a safe environment for the educational process with respect to transportation, condition of buildings, fire and food safety, and entertainment.

Stakeholders in this Subsector [233]

- Employees—60%
- Customers—25%
- Regulators—10%
- Other—5%

Typology of Common Risks, Main Features of Major Accidents and Risk Mitigation Measures Within the Subsector

Poor Quality Education and Staff Failures

The subsector has relatively few large-scale crises, but countless micro-crises—poor quality of research or teaching, or rough or unfair handling of an individual student by teachers—which occur frequently. Mediocre teaching and poor academic performance are usually the result of a downward spiral: the inability to retain the best teachers and researchers leads to a gradual deterioration in the caliber of lecturers, diminishing standards of teaching and therefore of learning; students feel they are being short-changed and give negative reviews; prospective employers are skeptical of graduates from the institution. All of this damages the reputation of an educational establishment in the long term, leading to low student enrolment and making it less attractive as an employer to the brightest potential academic staff... and so the vicious circle continues.

The reputation of an educational institution can also be damaged by faculty misconduct, such as fraudulent research, sexual harassment of students, tyrannical behavior in the workplace, professors bullying colleagues and students [234] and vice-versa [235], and abuse of power [236], any kind of corruption around invigilating or marking exams or assessments, poor occupational safety during practical laboratory work, and so on. Here are several examples.

Scientific Misconduct There are thousands of cases where researchers have falsi-fied their results, or have been caught plagiarising ideas or even passages from

---

[233]Our informed appraisal of the influence of each audience on a typical organization within the subsector (100% = combined influence of all audiences)

[234]Workplace bullying in academia (https://en.wikipedia.org/wiki/Workplace_bullying_in_academia)

[235]Amy May and Kelly E. Tenzek, Bullying in the academy: understanding the student bully and the targeted 'stupid, fat, mother fucker' professor, Teaching in Higher Education 23 (3), 275–290 (2018).

[236]Hannah Devlin and Sarah Marsh, Hundreds of academics at top UK universities accused of bullying, https://www.theguardian.com/education/2018/sep/28/academics-uk-universities-accused-bullying-students-colleagues (28 Sep 2018).

scientific publications by other authors. A researcher at the University of Edinburgh calculated that *"1.97%... of scientists admitted to have fabricated, falsified or modified data or results at least once – a serious form of misconduct by any standard—and up to 33.7% admitted other questionable research practices... However, it is likely that, if on average 2% of scientists admit to have falsified research at least once and up to 34% admit other questionable research practices, the actual frequencies of misconduct could be higher than this"* [237]. According to the Medline database at the US National Library of Medicine, between 1966 and 2001 around 300 sets of research results in the biomedical field alone were later retracted for the above reasons [238]. Exposure of such fraudulent practice could not only destroy the career of the scientists involved, but damage the reputation of the institutions where it took place.

Staff-to-Student Sexual Harassment An investigation by the Associated Press found that, of the three million school teachers working in state education in the US, a total of 2570 were accused of sexual misconduct between 2001 and 2005. In 1801 cases, there was sexual harassment of young people, 80% of whom were students [239]. And according to the US Conference of Catholic Bishops, 4450 out of 110,000 Roman Catholic priests were involved in molesting minors between 1950 and 2002 [240]. And 1200 cases of suspected harassment or abuse were recorded within the Boy Scouts of America from the early 1960s through to 1985 [241]. The disclosure of even one such outrageous case casts a shadow on the reputation of a whole educational institution, because parents naturally worry that their own child may be the next victim. And sexual abuse horrifies people at such a visceral level that such cases usually hit the national headlines.

Corruption in Knowledge Assessment In 2009, 178 teachers at 44 Atlanta public schools were involved in falsifying the test results of their pupils to improve the statistics for their own and their schools' academic performance. Teachers erased and corrected wrong answers, seated their pupils during tests so that lower performing children could cheat off their higher-scoring classmates, hinted at correct answers during the tests, and allowed some children to change incorrect answers on the following day [242]. Finally, in 2015, eleven people were convicted of racketeering and making false statements, and jailed for between 5 and 20 years.

---

[237]Daniele Fanelli, How Many Scientists Fabricate and Falsify Research? A Systematic Review and Meta-Analysis of Survey Data, PLOS ONE, May 29, 2009

[238]Bridget Murray, Research fraud needn't happen at all, American Psychological Association, February 2002, Vol 33, No. 2

[239]Martha Irvine, Robert Tanner, Sexual Misconduct Plagues US Schools, Associated Press, October 21, 2007

[240]Draft survey: 4450 priests accused of sex abuse, CNN, February 17, 2004

[241]Jason Felch, Kim Christensen, Release of Scouts' files reveals decades of abuse, Los Angeles Times, October 19, 2012

[242]Robert E. Wilson, Michael J. Bowers, Richard L. Hyde, Georgia investigation into cheating in Atlanta Public Schools, Office of the Governor of Georgia, July 6, 2011, p. 18

In 2011, the University of North Carolina (UNC) found itself at the center of a scandal after it was revealed that 3100 student-athletes did not attend all their "*mandatory*" courses between 1993 and 2011, but still successfully passed tests and exams. How could a system of "*paper classes*"—initially under the umbrella of "*independent study*" but later including "*seminar courses*" for which the lectures never took place—have been allowed to develop over nearly 20 years, under the nose of the professor of the faculty in question and with the knowledge of academic counselors and most of the student body? The main motives of the academic support tutor who developed the scheme, and the faculty which tacitly supported it, were a genuine compassionate desire to help student-athletes from poor backgrounds to graduate the UNC, the considerable income generated by the university's sporting activity (particularly in football and basketball) and the promotion of the university's brand nationwide through sport. The "*paper classes*" scheme allowed the best athletes, in return for their exceptional sports performance, to be excused from many of the obligations of their classmates: in some cases, they even defended plagiarized papers, yet they ultimately received higher grades than the average UNC student [243].

In 2017, the Russian Federal Air Transport Agency decided to annul the licenses of pilots discovered to have forged documents or lied about their education. The majority of these fake licenses had been issued by private flying schools after the collapse of the Soviet Union, where the only requirement for successful graduation as a pilot was money, irrespective of academic or practical progress during the course. In response to the licensing scandal, the leading Russian air carrier Aeroflot agreed with demands that all private flying schools in Russia should be closed due to the unreliable quality of their pilot training for civil aviation. Only state-owned establishments, under no financial pressure to cut corners, should be allowed to work in this sphere: "*Almost all pilots whose certificates have been annulled had graduated from private aviation training centers. Some institutions of this type ... issued ... falsified documents about the health [of pilots], [their] training progress, [and their] knowledge of English*" [244].

Poor Occupational Safety During Practical Laboratory Work The negligence of teaching staff towards safety in supervising laboratory work can lead to grim results: injury or even death of students through fire, gas explosions, poisoning, leakage of dangerous bacteria, and so on. For example, 13 students in Australia died in an explosion in 1974 when their teacher was demonstrating how a rocket engine works [245]. In 1982, a hydrogen sulphide explosion at the Colorado School of Mines

---

[243] Kenneth L. Wainstein, A. Joseph Jay III, Colleen Depman Kukowski, Investigation of Irregular Classes in the Department of African and Afro-American Studies at the University of North Carolina at Chapel Hill, October 16, 2014

[244] Aeroflot supports the decision of Rosaviatsiya to cancel falsified pilot certificates, Aeroflot, May 5, 2017

[245] Data of the Laboratory Safety Institute, http://www.labsafetyinstitute.org/MemorialWall.html

in the United States resulted in the death of a student [246]. And in 2010, a microbiology lab at Clark College in Washington State was responsible for a salmonella typhimurium outbreak, which infected 109 people in 38 states and caused one death [247,248].

To avoid or minimize the risk of deteriorating standards or faculty misconduct, an academic institution must select carefully when hiring staff, pay teachers and researchers the salary they deserve, keep a vigilant eye on the quality of education being delivered, develop a system of student claim management (including whistleblowing) to detect any early warning signs, and look after the ongoing development of staff—both in their professional fields and in the areas of scientific/educational ethics and occupational safety.

Educational Infrastructure-Related Accidents

Whatever the quality of teaching, an unsafe school or college environment can endanger students. Such threats could originate from accidents on the way to or from school, unprotected electrical outlets, broken toys, loose shelves, uncovered or wide open classroom windows, dirty or slippery floors, poorly cooked food in the student cafeteria, weak infection control, the leakage of dangerous materials, fire hazards, poorly constructed buildings, and so on. Here are some of the most tragic cases from different parts of the world.

Natural Gas Leak at New London School (USA, 1937)   295 people died and more than 300 were injured in an explosion of leaked natural gas from the heating system of this school in Texas. In order to reduce heating bills, the school had bought raw mixed gas from nearby oil wells, which had not been treated by the addition of methanethiol, which usually gives domestic gas its familiar odor to enable leaks to be detected promptly. It was only a matter of time before somebody switched on some electrical equipment, causing a spark that ignited the leaked gas.

Elbarusovo School Fire (USSR, 1961)   106 children and 4 teachers died in a fire when a physics teacher was repairing a benzene electric generator. There was a benzene leak which ignited; fire soon engulfed the whole school building, which eventually collapsed, and the corpses of victims were so badly burnt that most of them could not be identified.

ABC Day Care Center Fire (Mexico, 2009)   49 children died in a fire at the ABC Day Care Center in the Mexican city of Hermosillo.

---

246 Ibid

247 Human Salmonella Typhimurium Infections Associated with Exposure to Clinical and Teaching Microbiology Laboratories (Final Update), CDC, January 17, 2012

248 Bill Marler, Salmonella Escapes for Clark College Lab, June 26, 2012, www.marlerblog.com/legal-cases/salmonella-escapes-for-clark-lab

Yuba City High School Choir Bus Disaster (USA, 1976) 28 students died due to the negligence of a school bus driver, who failed to recognize low air pressure in the hydraulic brakes. The bus fell over 20 feet from a slip-road and landed on its roof. According to the NHTSA's National Center for Statistics and Analysis, there were 1191 school-transport-related crashes between 2005 and 2014 in the United States, in which 1332 people died. 8% of the fatalities were passengers on school transportation vehicles, 21% were pedestrians, cyclists, and others and 71% were occupants of other vehicles involved. Over the same period, 111 school-age pedestrians died in school transportation-related crashes [249].

Typhoid Outbreak at Madrid School (Spain, 1991) 54 students and a cook at a public school in Móstoles, a suburb of Madrid, were hospitalized after an outbreak of salmonella typhi.

Bihar School Meal Insecticide Poisoning (India, 2013) 23 children died after eating a school lunch cooked in oil from a canister previously used for storing insecticide.

Stage Collapse at Indiana Westfield High School (USA, 2015) Tens of students fell through a stage floor into a room below the stage during a musical performance. 13 participants were hospitalized.

Regular staff training on safety issues, and evacuation drills with students, are mandatory. To prevent these kinds of accident, management should also make regular inspections with engineers and representatives of relevant governmental bodies. These should include an assessment of existing infrastructure and the agreement on a road map to deal with any risks that have come to light. Moreover, there should be continuous investment in upgrading infrastructure, vehicle fleet and any other areas in need of maintenance and improvement.

Threats from Natural Disasters and Third Parties

Any educational institution could become the victim of an external event, whether through natural disaster, terrorist action, arson or a shooting attack by a current or former student.

Aberfan Coal-Waste Slide (UK, 1966) 116 children at Pantglas Junior School and 28 adult residents of the town of Aberfan perished in a landslide of coal waste from the nearby Merthyr Vale colliery. For years, the mine had been accumulating huge slag heaps, one of which collapsed on the town after a period of heavy rain on October 21, 1966 [250].

---

[249]Traffic Safety Facts 2005–2014 Data, NHTSA's National Center for Statistics and Analysis, May 2016, https://crashstats.nhtsa.dot.gov/Api/Public/Publication/812272

[250]Aberfan. The mistake that cost a village its children, BBC, http://www.bbc.co.uk/news/resources/idt-150d11df-c541-44a9-9332-560a19828c47

Beslan School Hostage Crisis (Russia, 2004) On September 1 2004, 34 Chechen terrorists with ties to Al Qaeda captured 1128 schoolchildren, parents and teachers at School #1 in the city of Beslan in North Ossetia, a North Caucasian republic of the Russian Federation. Most of the hostages were forced together into the school's gymnasium, where they spent more than 50 h. Anti-terrorist special forces stormed the school after the terrorists set off two large explosions in the gymnasium, wounding most of the hostages. Between earlier hostage executions, these later explosions and the ensuing battle to rescue the remaining hostages, 334 people including 186 children were killed and more than 700 injured.

Destruction of Enterprise High School by a Tornado (USA, 2007) Eight students died and 50 were hospitalized when a tornado destroyed the Enterprise High School in Alabama in February 2007.

Nepal Earthquake and Destruction of Local Schools (2015) After the most destructive earthquake to hit the country for 80 years, more than 9000 people died in Central Nepal. The quake and the aftershocks that followed completely destroyed more than 5000 primary and secondary schools and damaged 16,000 of them [251].

School Shooting Incidents (USA) The ease of accessing firearms in the United States has led to a tremendous number of shooting incidents in schools, colleges and universities. The most tragic of these were in 1999 at Columbine High School in Colorado (13 victims), in 2007 at Virginia Tech (32 victims), and in 2012 at Sandy Hook Elementary School in Connecticut (26 victims).

We would recommend making an assessment of the probability of potential external risks according to national statistics for such events, and regional and local statistics for natural hazards. The assessment should be discussed with relevant government agencies, and with any nearby organizations that could pose a threat to the educational organization. Response measures could include improving the safety of educational buildings, enhancing infrastructure, continuous monitoring of potential natural hazards, evacuation training for staff and students and the exchange of risk information with other organizations.

Key Risk Mitigation Measures in Scientific Research and Education

- Ongoing assessment of the quality of scientific research undertaken and education provided by staff.
- The ability to recruit and retain brilliant researchers and teachers.
- Maintaining scientific and educational infrastructure in a safe and clean condition.

---

[251]Ewan Watt, Nepal earthquake: '5000 schools destroyed and thousands more damaged', Theirworld, April 29, 2015, http://theirworld.org/news/nepal-earthquake-8216-5000-schools-destroyed-and-thousands-more-damaged-8217

3.9   Health Care

This subsector includes the following industries according to ISIC Rev. 4:

- 86—Human health activities
- 87—Residential care activities
- 88—Social work activities without accommodation
- 75—Veterinary activities

General Description and Key Features of this Subsector

Medical care is one of the most labor intensive subsectors, which requires the employment of large numbers of highly educated, qualified medical staff. The automation of modern medical services is mainly confined to monitoring equipment and patient health databases: diagnosis, the determination of appropriate medical treatment and even surgical operation are, and will be for decades, the responsibility of real medical staff. The quality of medical care is unavoidably variable, because of the uniqueness of each patient's health condition and the need therefore to provide a unique treatment. Correspondingly, 70–80% of health care breakdowns occur through human error, usually involving interpersonal interactions within medical teams [252]. One should also note that the sector is severely dependent on the pharmaceutical industry and its lobbying efforts to promote particular drugs, vaccines, and treatments, and to influence regulation.

Critical Success Factors for an Organization Within the Subsector

A successful medical institution must be able to recruit and retain well qualified medical staff, provide customized health care for individual clients, and implement advanced technological and medical innovations in its treatment of patients, effectively and in time. To maintain a good reputation, a medical care organization needs not just to keep its customers happy during the treatment process, but to carry out effective treatment. Infrastructure and equipment must be used safely, and cleanliness and hygiene maintained. Comprehensive patient claim management is an essential element of quality management and improvement in medical services. Finally, patients must be offered a proper assessment of the treatments and drugs available to them, independent of the lobbying and promotion of the pharmaceutical companies.

---

[252]Schaefer H.G., Helmreich R.L., Scheidegger D., Human factors and safety in emergency medicine. Resuscitation, 1994, 28(3), pp.221–225

Stakeholders in this Subsector [253]

- Employees—40%
- Customers—25%
- Pharmaceutical and insurance industries—15%
- Regulators—15%
- Other—5%

Typology of Common Risks, Main Features of Major Accidents and Risk Mitigation Measures Within the Subsector

Crises Caused by Poor Health Care

The principal causes of incidents within the subsector are inadequate qualification or experience of medical staff, mistakes by doctors in the treatment of patients, and neglect of patients or excessive delay in responding to their needs. Because each patient is a unique customer, the subsector does not have large-scale crises, but medical mistakes generate micro-crises on a daily basis, which raise questions about the functioning of medical institutions.

In researching this subchapter, it has been difficult to select representative cases, because the majority have been "minor" incidents, which have had a disastrous impact on the condition of individual patients, but not on larger groups. But in the United States alone—where there is highly expensive and professional medical care—the annual death rate from such "minor" incidents is equivalent to 62 (!) Chernobyl accidents. Thus, according to researchers at Johns Hopkins University School of Medicine, health care errors are the third leading cause of death in the United States with 251,454 deaths annually (Chernobyl claimed around 4000 lives), after heart disease (614,348) and cancer (591,699). Michael Daniel summed up the research conclusions: "*[M]ost medical errors aren't due to inherently bad doctors ... [M]ost errors represent systemic problems, including poorly coordinated care, fragmented insurance networks, the absence or underuse of safety nets and other protocols, in addition to unwarranted variation in physician practice patterns that lack accountability*" [254,255]. Figures for the UK paint a similar picture: 3.6% of

[253]Our informed appraisal of the influence of each audience on a typical organization within the subsector (100% = combined influence of all audiences)

[254]Martin A Makary, Michael Daniel, Medical error-the third leading cause of death in the US, British Medical Journal, May 2016, 353

[255]Vanessa McMains, Johns Hopkins study suggests medical errors are third-leading cause of death in US, Johns Hopkins University, Office of Communications, May 4, 2016, https://hub.jhu.edu/2016/05/03/medical-errors-third-leading-cause-of-death/

deaths in hospitals in England between 2009 and 2013 were avoidable [256]. And according to the Institute of Medicine, *"Current estimates of the incidence of medication errors are undoubtedly low because many errors go undocumented and unreported"* [257]. According to a study by the Harvard Medical School, for every 1000 patient-days for critically ill patients, there are 80.5 adverse events, 36.2 preventable adverse events and 149.7 serious errors [258]. More than 1.5 million Americans are harmed every year by medical mistakes, including 400,000 preventable drug-related injuries [259]. In this country, 7000 people die annually through medical error due to the sloppy handwriting of doctors [260].

On a global scale, the number of tragic medical mistakes exceeds a million deaths annually (equivalent to 250 Chernobyl accidents), but the focus of the global risk management community is mainly on heavy industry rather than essential services. Thus, in the case of health care, we are faced with a cognitive bias, where the grim reality revealed by casualty statistics is eclipsed by more irrational fears of radioactivity, natural disasters, terrorism and so on.

Common medical errors affecting an individual patient include: ambulance mistakes in locating the address of patients or directing serious cases to the right hospital; team communication mistakes in the care of a given patient because of inaccurate record-keeping; mistakes during laboratory tests; misdiagnosis leading to detrimental courses of treatment such as unnecessary hospitalization, operations, drug treatment or avoidable infection with new diseases; mistakes in the type or dosage of drug treatment, including the development of unnoticed diseases or allergic reactions; errors in the dosage or placement of anesthesia; mismanagement during operations where patients are treated incorrectly—for instance the surgical removal of the wrong breast or limb; leaving tools or swabs in the patient's body after an operation; poor theatre hygiene such as badly disinfected tools or inadequate preparation of patients for an operation; equipment failure; and mistakes in post-operative recovery such as prescribing the wrong recovery regimen, incorrect use of traction, allowing bedsores to develop, giving the wrong drug or dosage, or incorrect connection and disconnection of intravenous drips or catheters. Outpatients can also suffer from ordering and delivery errors in pharmacies if an urgently needed drug is allowed to go out of stock [261].

---

[256]Ian Johnston, The third highest cause of death in the United States is mistakes by medical staff, The Independent, May 3, 2016

[257]Linda T. Kohn, Janet M. Corrigan, Molla S. Donaldson, To Err Is Human: Building A Safer Health System, Institute of Medicine, November 1999

[258]Rothschild J.M., Landrigan C.P., Cronin J.W., Kaushal R., Lockley S.W., Burdick E., Stone P. H., Lilly C.M., Katz J.T., Czeisler C.A., Bates D.W., The Critical Care Safety Study: The incidence and nature of adverse events and serious medical errors in intensive care, Critical Care Medicine, August, 2005, 33(8), pp.1694–1700

[259]Philip Aspden, Julie Wolcott, J. Lyle Bootman, Linda R. Cronenwett, Preventing Medication Errors, Institute of Medicine, 2007

[260]Jeremy Caplan, Cause of Death: Sloppy Doctors, Time, January 15, 2007

[261]Linda T. Kohn, Janet M. Corrigan, Molla S. Donaldson, To Err Is Human: Building A Safer Health System, Institute of Medicine, November 1999

However, although the vast majority of patient deaths are from breakdowns on an individual scale, there have been numerous cases where poor health care has affected larger numbers of patients. Thus, more than 250 infants and several mothers were infected by HIV because syringes were reused at a children's hospital in Elista (USSR) in 1988–1989. Single-use syringes were in short supply—so during pregnancy and after accouchement, the victims had received essential drugs by means of syringes shared with other patients, with only the needles being changed. The syringes were washed, but in solution that was rarely replaced. Because there had been hardly any cases of HIV in the USSR at that time, the infected children were not identified as HIV positive—and some were later hospitalized in other medical institutions in the region, where the reuse of syringes was also common. A similar case occurred among children in Romania in 1989 due to transfusions of unscreened blood and injections with improperly sterilized equipment [262]. In 2005–2006, more than a hundred children and several mothers in Southern Kazakhstan were infected with HIV during blood transfusions; the investigation concluded that staff did not adequately check the health condition of donors, that single-use syringes were reused and that there was under the counter smuggling of unscreened donor blood. And in 2010, it became known that more than 1800 patients at the St. Louis Veterans Affairs Medical Center in Missouri, USA, could have been exposed to HIV, hepatitis B or hepatitis C because technical staff were not observing the protocol for cleaning and sterilizing dental equipment [263]. Recently, it became known that Niels Hoegel, a German nurse who worked at two hospitals in Oldenburg and Delmenhorst, had killed at least 84 patients between 1998 and 2005 by administering dangerously high doses of a cardiac drug [264]. Hoegel explained that he had done so because he intended to *create medical emergencies so that he could glean gratitude and admiration by saving people from the brink of death*" [265]. He was sentenced to life in prison.

Another serious issue in the sector is the problem of health care-associated infections. In advanced countries, 7 patients out of every 100 acquire such infections, and in emerging countries 10 in every 100. According to the US Center for Disease Control and Prevention, in the United States around 1.7 million hospitalized patients every year get health care-associated infections and around 6% of them— almost 99 thousand people—die from these diseases. Many pathogens, which cause health care-associated infections are, or become, multidrug-resistant due to the intensive antimicrobial therapy which has been standard in hospitals over recent decades. Nevertheless, the simplest and still the most effective infection-control

[262]VA hospital may have infected 1800 veterans with HIV, CNN, July 1, 2010

[263]F.F. Hamers, A.M. Downs, HIV in Central and Eastern Europe, Lancet, 2003, #361, pp. 1035–1044

[264]David Rising, Nurse Convicted of 2 Murders Might Have Killed Another 84 Patients, Time, August 29, 2017

[265]German nurse 'admits to 30 patient killings', DW, Jan 8, 2015, http://www.dw.com/en/german-nurse-admits-to-30-patient-killings/a-18177310

procedure is for health care professionals and visitors to clean their hands with alcohol-based detergents [266].

Risk management in the medical services subsector should be focused on the reduction of medical errors by a range of complex solutions to improve the quality of care. These could include: careful selection when hiring staff; ongoing supplementary training of medical staff covering new technologies and methods of treatment, and sharing best practice experience with other hospitals; ongoing training of service staff to improve team communication and customer care skills; internal assessment of proposed treatments along with advanced patient claim management; maintaining high level of hygiene among staff and within hospitals; and proper and timely investment in infrastructure and medical equipment.

Infrastructure-Related Crises

Multiple casualties can be caused by fires in hospitals, road accidents involving ambulances, destruction of buildings by earthquakes, power cuts, and so on.

Hospital Fires (Different Countries) In 1960, 225 patients perished and more than 300 were injured in a fire at a mental asylum in Guatemala. The main cause of the disaster, which occurred during the night, was a candle left burning in front of a religious statue by one of the patients [267]. 55 patients were killed in 1980 in a fire at the Gorna Grupa psychiatric hospital in Poland. And in 1985, a blaze at the St. Emilienne Neuro-Psychiatric Institute in Argentina killed 79 people and injured 247 [268].

Ambulance Crashes (Worldwide) In the United States alone, an average of 4500 road accidents and 33 fatalities occur annually through ambulance crashes, according to statistics from the US National Highway Traffic Safety Administration over the 20 years between 1992 and 2011 [269].

Hospitals in Earthquakes and Tsunamis (Seismic Activity Zones) During an earthquake in 1971 in the San Fernando Valley in California, the Olive View medical center and Veterans hospital were severely damaged and some blocks collapsed. In 1994, the Northridge earthquake in California made 11 hospitals inoperable and caused US$3 billion damage to the state healthcare system [270]. In 2004, the Sumatra–Andaman earthquake and tsunami destroyed 30 hospitals in

---

[266]Mainul Haque, Massimo Sartelli, Judy McKimm, Muhamad Abu Bakar, Health care-associated infections—an overview, Nov 2018, Infect Drug Resist, pp. 2321–2333

[267]Jay Robert Nash, Darkest Hours, Rowman & Littlefield, 1976, p.223

[268]Lydia Chavez, Fire In Clinic Kills 79 In Buenos Aires, The New York Times, April 28, 1985

[269]Ground Ambulance Crashes, US National Highway Traffic Safety Administration, April 2014, https://www.ems.gov/pdf/GroundAmbulanceCrashesPresentation.pdf

[270]Hospital Seismic Safety, California HealthCare Foundation, January 2009

Indonesia, and left 77 severely damaged by flooding [271]. During the Haiti earthquake in 2010, which claimed more than 230,000 lives, only one hospital was in operation in the capital Port-au-Prince—a city of more than 800,000 people—because the others had all been destroyed or damaged [272].

Sewage Floods in Hospitals Many hospitals have been affected by sewage flooding, with the obvious risks that entails to seriously ill patients. Recent examples include Colchester General Hospital (UK, 2015), Venice Regional Bayfront Health (USA, 2015), and Cramlington Hospital (UK, 2016).

Poor Patient Conditions at Walter Reed Medical Hospital (USA, 2004–2007)
The institution was supposed to provide *"world class"* health care to American soldiers wounded in Iraq and Afghanistan. However, the condition of some of the buildings was so poor that the story became a nationwide scandal, even attracting the attention of George W. Bush who was invited by the patients to visit the hospital. In some patients' rooms, there were holes in the walls, mould and even infestation by mice and insects. Moreover, the quality of healthcare and efficiency of the health administration within the institution were under question.

Measures to try and avert such disasters include constant staff training on occupational safety issues; the establishment of contingency plans with other emergency agencies; routinely and repeatedly informing patients about what they should do in case of emergency; a thorough assessment of possible environment-related risks during the construction of medical institutions, and a good awareness of such risks during their subsequent use; and investment in fire alarms and extinguishers, other safety equipment and general infrastructure.

Crises Induced by Third Party Action

Promotion of Drugs by Pharmaceutical Companies (Worldwide) An important challenge to the quality of health care is the pressure put on doctors by pharmaceuticals companies to recommend their products to patients over others [273,274]. The greatest risk in the subsector is losing customer and government confidence that a healthcare system really heals patients, rather than just treating them for as long as possible in order to increase bills—for instance, millions of unnecessary operations are performed annually—or get more commission from pharmaceuticals companies for prescribing unnecessary drugs.

---

271 Indian Ocean Earthquake & Tsunami Emergency Update, Center of Excellence in Disaster Management and Humanitarian Assistance, May 25, 2005

272 Only one hospital open in Haiti's quake-hit capital, AFP, January 14, 2010

273 Engelberg, Joseph and Parsons, Christopher A. and Tefft, Nathan, Financial Conflicts of Interest in Medicine, January 26, 2014, https://ssrn.com/abstract=2297094

274 Philipp Aeby, Robert C. Brears, Wim Leereveld, Silvia Mancini, Nicole Neghaiwi, Robert Pojasek, Leesa Soulodre, Santiago Villa Chiappe, ESG risk in the corporate world, Issue #6 Pharmaceuticals, Reprisk, July 2014

For instance, evidence has emerged suggesting that industry-sponsored courses skew training material in favour of commercial interests [275]. This raises the question whether the pharmaceutical industry can be trusted to fund doctors' compulsory education without introducing bias.

In a Bloomberg Market document of 2009, David Evans reports that "*Pfizer and Lilly lead a parade of US companies that have paid US $7 billion in penalties after promoting drugs for uses not approved by the FDA*" [276]. According to Lon Schneider, a professor at the University of Southern California's Keck School of Medicine in Los Angeles, "*[pharmaceutical] companies regard the risk of multimillion dollar penalties as just another cost of doing business… There's an unwritten business plan. They're drivers that knowingly speed. If stopped, they pay the fine, and then they do it again*" [277]. Federal prosecutor Michael Loucks says: "*As drugmakers repeatedly plead guilty, they've shown they're willing to pay hundreds of millions of dollars in fines as a cost of generating billions in revenues*" [278].

In recent decades, there have been many Congressional inquiries on price gouging in the American pharmaceutical industry, which stands accused of "*gouging the American public with outrageous price increases, driven by greed on a massive scale*" [279]. A recent study of the Institute for New Economic Thinking shows that "*high drug prices restrict access to medicines and undermine medical innovation. An innovative enterprise seeks to develop a high-quality product that it can sell to the largest possible market at the most affordable price. In sharp contrast, the MSV [maximizing shareholder value]-obsessed companies that dominate the US drug industry have become monopolies that restrict output and raise price. These companies need to be regulated*" [280].

Measures to tackle this corruption and wrong-doing could include state regulation of the interaction of doctors with the pharma industry, public exposure and criminal prosecution of the worst cases, and ensuring that different pharma companies have equal access to doctors to explain their products. To address the allegation quoted above that "*drugmakers… [have] shown they're willing to pay hundreds of millions of dollars in fines as a cost of generating billions in revenues*", the logical response

---

[275]Jim Giles, Drug firms accused of biasing doctors' training, Nature, 2007, 450 (7169), pp.457–584

[276]David Evans, Big Pharma's crime spree, Bloomberg Markets, December 2009, pp.73–86

[277]David Evans, Big Pharma's crime spree, Bloomberg Markets, December 2009, pp.73–86

[278]David Evans, Big Pharma's crime spree, Bloomberg Markets, December 2009, pp.73–86

[279]William Lazonick, Matt Hopkins, Ken Jacobson, Mustafa Erdem Sakinç and Öner Tulum, US Pharma's Financialized Business Model, Institute for New Economic Thinking, July 13, 2017, No. 60, https://www.ineteconomics.org/uploads/papers/WP_60-Lazonick-et-al-US-Pharma-Business-Model.pdf

[280]William Lazonick, Matt Hopkins, Ken Jacobson, Mustafa Erdem Sakinç and Öner Tulum, US Pharma's Financialized Business Model, Institute for New Economic Thinking, July 13, 2017, No. 60, https://www.ineteconomics.org/uploads/papers/WP_60-Lazonick-et-al-US-Pharma-Business-Model.pdf

would be to scale up the size of the fines to make them really dissuasive. This would likely lead to the bankruptcy of the fined firm, which would then become an example to other would-be offenders.

Supply of Contaminated Cardiac Medicines to Hospital (Pakistan, 2011–2012)   More than 200 patients at the Punjab Institute of Cardiology in Lahore died, and more than 1000 were badly affected, after taking a drug supplied by a local producer under the brand name Isotab, which was contaminated with the antiparasitic pyrimethamine. The drug was dispensed free of charge to more than 46,000 people. The producer of the contaminated drug was responsible not only for manufacturing cardiac medicines, but also for antimalarial medications that contained pyrimethamine. Unfortunately, the defective batch of Isotab contained very high levels of pyrimethamine—14 times the normal dosage for a malaria patient, and enough to cause a very serious and potentially fatal reaction. However, once the cause of the problem was identified, it was reversible with a simple antidote, and the Institute had a record of all those supplied with Isotab on its IT system; thus, many potential casualties were warned and given the antidote in time [281].

To minimize the incidence of this kind of crisis, governments should have strict control over critical manufacturing; drug manufacturers could develop advanced customer claim systems; and hospitals should—as they generally do—closely monitor the condition of patients and react immediately in the event of an unexpectedly widespread adverse reaction to any drug.

Hospitals During Terrorist Attacks and Wars   In June 1995, Chechen terrorists kept more than 1648 hostages for several days at a local hospital at Budennovsk, a small town in southern Russia. Russian special forces made an unsuccessful attempt to storm the hospital, and during the crisis 129 people died and more than 400 were injured [282]. Terrorist or insurgent forces often establish their bases in local hospitals in order to discourage ground or air assault: if anti-terrorist forces attempt to dislodge them, the insurgents will blame the government for attacking a civilian target and accuse them of war crimes [283,284]. If a hospital finds itself at risk of being turned into a conflict zone in this way, management could ask for additional security from government agencies, and invest in appropriate surveillance and alarm systems; they should also consider further staff training on how to respond to terrorist attacks.

NSA Malware and British Hospitals (2017)   The strategy of massive automation of many services including health care has its Achilles heel: the problem of cyber

---

[281]Deadly medicines contamination in Pakistan, WHO, March 2013, http://www.who.int/features/2013/pakistan_medicine_safety/en/

[282]Prosecutor General Office reported on the investigation of the raid on Budennovsk by anniversary of the terrorist attack, Lenta, June 14, 2005

[283]Hamas uses hospitals and ambulances for military-terrorist purposes, Israel Ministry of Foreign Affairs, July 28, 2014

[284]Louis R. Mizell, E. Reed Smith, How terrorists target and attack health care facilities and personnel, EMS1, July 6, 2016

security. This was clearly demonstrated in the Wanna Decryptor virus cyber attacks in May 2017, which was based on leaked NSA zero-day vulnerabilities (we have described the collection of software vulnerabilities by NSA in the subchapter "Information services"). One of the organizations most severely affected by the attack was Britain's National Health Service, when several hospitals across the country were brought to a standstill due to virus contamination of the IT systems on which healthcare is now hugely dependent. The virus was able to penetrate into computers with outdated updates of Microsoft software [285].

Because hospitals are part of any country's critical infrastructure, government cyber security agencies urgently need to give special attention to vulnerability and adequate protection when installing new IT solutions, and develop better training for hospital IT staff to prepare them for such emergencies.

Key Risk Mitigation Measures in Health Care

- Comprehensive and unstinting focus on the reduction of medical mistakes.
- The continuous recruitment and retention of well-qualified medical staff, able to provide customized health care to individual clients and implement advanced technological and medical innovations in the treatment of patients, effectively and in time.
- Maintaining health care infrastructure in a safe and clean condition.

3.10 Entertainment and Sports Activities

This subsector includes the following industries according to ISIC Rev. 4:

- 90—Creative, arts and entertainment activities
- 93—Sports activities and amusement and recreation activities

General Description and Key Features of the Subsector

The subsector focuses on the organization of artistic performances by actors, singers, musicians, dancers, comedians, clowns, magicians, or acrobats through the set-up and promotion of concerts, operas, circuses, and other shows, and the running of the venues in which they take place; the operation of indoor and outdoor sports facilities and the organization of competitive sporting events; and the administration and operation of amusement parks. The interest of clients in the performance of artists

---

285Craig Timberg, Griff Witte and Ellen Nakashima, Malware, described in NSA documents, cripples computers worldwide, The Washington Post, May 12, 2017

or sports people mainly depends on the skills of these people and the service staff who work behind the scenes to deliver the unique quality of an artistic or sporting performance. Generally, such activity requires the gathering of a large number of people at an exact place and time. Most work within the subsector is project–related activity.

Critical Success Factors for an Organization Within the Subsector

The key determinant of success is the ability to find interesting, striking and popular acts (singers, magicians, sportsmen, and so on) who can attract the attention of potential customers and entertain them well enough to earn positive customer feedback. Having recruited the right caliber of performers, the other critical factor is the ability to manage large numbers of invited guests safely and maintain infrastructure properly.

Stakeholders in this Subsector [286]

- Employees—40%
- Customers—30%
- Security agencies—20%
- Other—10%

Typology of Common Risks, Main Features of Major Accidents and Risk Mitigation Measures Within the Subsector

Poor Quality of Performance

The major risk in the subsector is that of a bad artistic or sports performance: a theatrical flop, a lifeless and unfunny comedy show, the crushing defeat of a sport team, half-hearted sport matches that only happen to fulfil a contractual obligation, doping fraud by an individual sportsman or even a whole team, or negligence and disrespect from service staff towards clients during climbing expeditions, skiing holidays and so on. For example, the exposure of legendary cyclist Lance Armstrong as a drug cheat destroyed not only his own reputation, but the lifelong enthusiasm of

---

[286]Our informed appraisal of the influence of each audience on a typical organization within the subsector (100% = combined influence of all audiences)

many fans for the sport. The re-test of doping samples from the Olympic Games in 2008 and 2012 revealed more than 100 doping violations by athletes from 19 countries, including 15 gold medallists [287].

One year before the 2015 Great Nepal Earthquake, a group of sherpas—the local porters who carry baggage for Everest climbing expeditions and hang ropes for commercial clients up to the summit—staged a strike at Everest base camp, and even threatened some clients, in pursuit of better pay and compensation in the event of injury or death during climbing. The strike was provoked by the miserly compensation that had recently been paid by the Nepali government to the families of 16 sherpas who had perished in an avalanche on the way up the highest mountain on Earth.

Measures to minimize the likelihood of poor performance include: the careful selection of key staff members; constant training to improve individual and team or group performance; the ongoing development of reserve teams or understudies; monitoring and learning from audience or crowd response and from reviews; and frequent internal drug checks on sportspeople to detect doping fraud and disqualify violators.

Crowd-Related Crises (Stampedes)

One of the biggest challenges of organizing a mass cultural or sporting event is the management of large crowds to prevent stampedes [288].

Coronation Celebration Tragedy at Khodynka Field (Russian Empire, 1896)
Half a million people attended a celebration after the coronation of Tsar Nikolay II, running ahead of the last Russian emperor, at Khodynka Field in Moscow on May 18, 1896. The event was put on for the benefit of the general public and involved the distribution of free gifts, free food, and beer and several entertainment activities. Because they had underestimated the number of visitors and the police presence required to marshal them, the security services could not stop the crowd when they rushed forward after rumours had spread of a shortage of gifts and beer. In the end, 1389 people were caught in the stampede and more than 1300 injured. Many Russians saw this tragedy as a bad omen for the reign of Tsar Nikolay II. Konstantin Balmont, a Russian anti-monarchist poet, declaimed "*Who began to reign by Hodynka, will be finished by scaffold*" [289]. His words became prophetic: in March 1917, the Tsar was forced to renounce the crown after a liberal revolution in Russia, and in July 1918, he and the whole royal family were executed by the

---

[287]Reconfirmation of doping samples from the Games-2008 and 2012 revealed more than 100 violations, Interfax, May 10, 2017, http://www.interfax.ru/sport/561762

[288]M. Moussaïd, D. Helbing, and G. Theraulaz, How simple rules determine pedestrian behavior and crowd disasters, Proceedings of the National Academy of Sciences, 2011, 108 (17), pp. 6884–6888.

[289]Konstantin Balmont, Our Tsar, 1907

Bolsheviks who had swept away the chaotic and unsuccessful liberal administration in 1917.

Similar crowd surges for gifts, which also led to stampedes, happened in 1883 during the distribution of free toys at Victoria Hall in Sunderland (England, 183 children perished) and in 2006 during the anniversary of a popular TV show in Manila (Philippines, 74 died).

1955 Le Mans Disaster (France, 1955) 83 spectators and a driver died and more than 180 were injured at the 24 h of Le Mans car race, after two racing cars collided sending debris flying over a protective wall, right into a huge crowd of spectators. The cause of the collision was the 30-year-old racing track, which was not designed to withstand the speed of the latest racing cars.

Lima Football Disaster (Peru, 1964) During a match between Argentina and Peru, some of the Peruvian fans disagreed with the referee's decision to disallow a goal, and they tried to invade the pitch. According to standard policy, the attempt was prevented by releasing tear gas. But this caused immediate panic and thousands of spectators flooded towards the exit gates, which were closed during the match; more than 300 people were killed and 500 injured.

Disastrous stampedes also occurred at Ibrox Stadium in Glasgow (UK, 1971, 66 deaths at exit gates); at Luzhniki Stadium in Moscow (USSR, 1982, 66 deaths at exit gates); at Heysel Stadium in Brussels (Belgium, 39 causlties); at Hillsborough Stadium in Sheffield (UK, 1989, 96 deaths and more than 700 injured due to weak police control over the drunk fans); at the national stadium in Guatemala City (Guatemala, 1996, 83 deaths); at Ellis Park Stadium in Johannesburg (South Africa, 2001, 43 deaths). At Ohene Djan Stadium in Accra, Ghana in 2001, 127 people died in the crowd panic caused when police overreacted to the behavior of some fans. During the Khmer Water Festival, famous for its traditional canoe competition, in Phnom Penh, Cambodia in 2010, 347 died and more than 700 were injured in a stampede on a bridge. And during an open-air live screening of the UEFA Champions League final in a square in Turin, Italy, in 2017, there was a loud bang from an exploding firecracker. Some people in the middle of the crowd panicked and tried to run away, assuming a bomb had gone off. The panic spread and led to a massive stampede; only one person died but more than 1500 were injured. Stampedes can also occur during the riots that commonly break out between opposing fans.

Stampedes During Pilgrimages On July 2, 1990, a stampede inside the Al-Ma'aisim tunnel leading out from Mecca towards Mina in Saudi Arabia led to the deaths of 1426 Muslims taking part in the annual Hajj pilgrimage. In January 12, 2006, during the stoning of the Devil on the last day of the Hajj in Mina, a stampede killed at least 346 pilgrims. And on 24 September 2015, 2000 pilgrims suffocated or were crushed during the Hajj.

Stampedes During Music Concerts and Festivals Tragic accidents took place at a concert by British band The Who (US, 1979, 11 dead); during celebrations for the Chinese New Year in Hong Kong (Hong Kong, 1993, 21 deaths); during an open-air concert in Minsk (Belarus, 1999, 53 deaths); at Roskilde Festival (Denmark, 2000, 9 dead); after a firework display in Akashi (Japan, 2001, 11 deaths); at E2 nightclub in Chicago (USA, 2003, 21 dead); and at the Love Parade electronic music festival (Germany, 2010, 21 dead).

Avoiding stampedes requires: advance planning of customer movements and careful calculation of infrastructure loading; safety and customer care training for security and service staff; coordination on advance safety measures with suppliers and local authorities, police, firefighters and medics; and the installation of loud-speakers at key points to direct crowds if necessary. Optimal design of infrastructure, scheduling, flow monitoring and adaptive rerouting can improve the traffic of dense crowds and reduce the risk of stampedes [290].

Other Crises

Fires The most dangerous infrastructure-related accidents are fires in buildings full of people, which all too often have a massive death toll. Thus in 1903, 602 people perished from fire and smoke inhalation, or were crushed in a stampede, at the Iroquois Theatre in Chicago. The fire was started by a short circuit in a floodlight, but neither the theatre's infrastructure nor staff were ready for an emergency evacuation. In 1944, 168 people died and more than 700 were wounded in a fire and the ensuing stampede at a circus marquee in Hartford, Connecticut. In 1961, 323 people perished and 600 were injured in a fire at a circus in Niteroi, Brazil. In 1963, 74 people were crushed in the panic after a leaking propane cylinder was ignited by a popcorn machine at Indianapolis Coliseum during an ice show; more than 400 were injured. In 1973, the Summerland Holiday Center in the Isle of Man, UK—at that time one of the biggest indoor entertainment centres in the world—caught fire because young people were smoking. The limited fire resistance of some of the materials used in the building, a poorly coordinated evacuation, panic and stampede together cost the lives of more than 50 vacationers. In 1985, windy conditions at the Bradford City stadium (UK) allowed an isolated fire on part of the wooden stand to turn into a conflagration, provoking panic among spectators and resulting in more than 50 deaths. In 2018, 60 visitors including 41 children died in a fire at a shopping and entertainment center in Kemerovo (Russia), due to negligence of safety on the part of the center administration.

Measures to reduce the risk of fire include: design or refurbishment of service points using flame-retardant materials; installation of fire extinguishing systems; ongoing training of staff in how to confine fires and evacuate large crowds quickly; close coordination with emergency services on mutual drills and the communication and independent assessment of risks; a strictly enforced no-smoking policy inside and outside service points; and clear fire instructions for clients.

---

[290]Dirk Helbing, Pedestrians, Crowds, Stampedes, and Mecca, http://webarchiv.ethz.ch/soms/teach ing/ModelingSpring2014/lecture6_Pedestrians_2014.pdf

Structural Collapse of Entertainment Buildings In 1992, temporary grandstand seating at the Stade Armand Césari in Corsica (France) collapsed during a match, killing 18 spectators. The seating had been constructed in a rush, with a shortage of appropriate materials and in violation of safety rules. In 2004, as we have described in the construction subchapter above, the roof of the Transvaal Waterpark in Moscow (Russia) collapsed, probably due to design errors and structural deficiencies, resulting in the death of 28 visitors. And in 2006, the roof of an ice rink in Bad Reichenhall (Germany) collapsed under the pressure of heavy snow, claiming 15 lives.

Proper design and adequate operation of infrastructure are recommended.

Terrorist Acts Concerts, sporting events or even theatrical performances can become a target for terrorists because of the high density of people in a limited space, which ensures that a terrorist act will be broadcast nationally or even internationally because of the large number of victims.

In 1972, Palestinian terrorists took hostages from the Israeli team at the Munich Olympics (then West Germany) eventually killing 11 of them. In 1978, "Islamist" militants set fire to a cinema in Abadan (Iran) and closed the exit doors, causing the deaths of 422 people. In 1996, during the Olympic Games in Atlanta (USA), an American anti-abortion activist planted several bombs, one of which killed one person and injured more than 110. In 2002, Chechen terrorists took more than 900 spectators and actors hostage during a performance of the musical "Nord-Ost" in Moscow. Russian special forces pumped soporific gas into the ventilation system in order to neutralize the terrorists without the need for a bloody battle to seize control of the theatre. But the gas obviously put the hostages to sleep as well, and some of them never woke up, either succumbing to existing chronic conditions or suffocating because those evacuating them were not aware of this risk when moving an unconscious person. A total of 130 hostages perished in the crisis. In 2003, Chechen suicide-bombers blew themselves up during the "Wings" rock festival at Moscow's Tushino airfield, killing themselves and at least 12 others. In 2013, two Chechen brothers detonated bombs at the Boston Marathon (USA) causing the death of three people and injuring more than 200. In November 2015, there were coordinated attacks by Islamist radicals at several locations in and around Paris, including a football match at the Stade de France in the suburb of St Denis, and a rock concert at the Bataclan theatre; a total of 130 people died, 89 of them at the Bataclan [291]. In 2017, an Islamic State supporter, born in Britain to parents who had fled the Gaddafi regime, detonated a suicide bomb in Manchester (UK) at an Ariana Grande concert; he killed 22 people, mostly teenage fans or their parents. In October 2017, there was prolonged shooting from a nearby high rise block down into a crowd at a country music festival in Las Vegas (USA); at least 58 died and more than 500 were injured.

Preventive actions include: close collaboration with state security agencies to identify and assess possible risks; serious monitoring and appropriate action to

---

[291]Paris attacks death toll rises to 130, RTÉ, November 20, 2015

control high risk individuals that are known to special security agencies (for instance, 12,000 people in France who are categorized as "fichés S" [292,293]); coordinated action in the event of early warnings or changes in the terrorist threat level; rapid response from special forces to any incoming attack; and mutual drills and training for service staff with representatives of security agencies, to give staff and subsequently their guests a better chance of surviving a hostage incident.

Key Risk Mitigation Measures in Entertainment and Sports Activities

- Providing top-quality performance to satisfy audiences at every event.
- Managing large numbers of invited guests in safety and proper maintenance of infrastructure.
- Close cooperation with authorities at all levels to protect mass-audience events from the actions of malicious people, from hooligans to terrorists.

3.11 Other Services

Information and Communication (58—Publishing activities; 59—Motion picture, video and television programme production, sound recording and music publishing activities)

General Description and Key Features of Industries The creation and development of unique creative solutions: classic literary or musical works, Oscar-nominated Hollywood movies that became part of world cinema heritage, TV talk shows that make their hosts a national or international household name, and so on. Creative and highly skilled authors and specialists are crucial to the development of this content. Business within these industries is project–oriented activity. The ease with which content can be copied and transmitted all around the world in matters of seconds makes government authorities an important player in these industries as global copyright controllers.

Critical Success Factors for an Organization Within the Industries The ability to create unique content; the recruitment and retention of highly qualified and motivated staff; the ability to control the value-added distribution of content.

Stakeholders in these industries:

- Employees—30%,
- Customers—30%,

---

[292]These are persons who, because of their individual or collective activity, may be detrimental to the security of the State and to public safety, through the use or active support of violence.

[293]Michel Cabirol, Fiches S: les services de renseignement ont recensé 12.000 personnes, LaTribune, April 24, 2017, http://www.latribune.fr/entreprises-finance/industrie/aeronautique-defense/fiche-s-les-services-de-renseignement-ont-recense-12-000-personnes-693258.html

- Authorities—20%,
- Distributors—15%,
- Other—5%.

Typology of Common Risks Within the Industries The most dangerous type of crisis in these industries is the failure of content: cinema flops due to an unexciting scenario, weak acting performances or poor cinematography or special effects; new books that are a disappointment to readers; unmemorable new albums with no real hits by popular rock groups; inaccurate or fake reports from prominent journalists or media channels, and so on. The situation could be exacerbated if the authors or writers are accused of plagiarism. Some industries also have infrastructure-related risks such as fire in cinemas—where hundreds of customers could be gathered—or recording studios. Where content is being developed concerning violent political and national security questions, there are challenges ensuring the safety of journalists, responsibility for the protection of national secrets, and the protection of newsrooms and content production centers. Finally, the financial stability of an author or profitability of artistic work could be damaged by digital piracy.

Real Estate Activities (68—Real estate activities; 81—Services to buildings and landscape activities)

General Description and Key Features of the Industries The real estate business is about the administration of owned or leased property, and functions as an intermediary in transactions on a fee or contract basis between owners of buildings and customers. In general, it is a project-oriented activity. The industries are highly dependent on salespeople, engineers and other workers responsible for the maintenance and running of buildings, and financial agents and lawyers who deal with contractual issues. The activity in the industry, much as in construction, is highly cyclical and volatile because business depends on the national economic situation, on the central bank discount rate, which determines banking interest rates and thus the size of mortgage payments, and on the uncertain prospects of future economic growth. There is high competition within the industry because it is easy for new intermediary services companies to enter the market.

Critical Success Factors for an Organization Within the Industries To succeed in this subsector, a company must: find staff with advanced sales skills and a desire to provide good quality service to all kinds of people, who can maintain cordial relationships between the various participants of real estate deals and their customers; be able to find and retain customers, and convince them to commit to large investments that will have a big impact on their lives; find owners of buildings who are interested in selling or renting them at affordable prices and willing to pay a reasonable rate of commission to salespeople; and be able to maintain committed infrastructure properly and on time.

Stakeholders in these industries:

- Customers—40%;
- Partners (owners of buildings, financial initiations, lawyers, contactors, etc.)—30%;
- Employees—30%.

Typology of Common Risks Within the Industries These businesses do not generate large-scale crises, because the majority of real estate deals are unique and personal. There are small scale client-related conflicts concerning the conditions of deals or mortgage contracts, legal problems with a property, or where the disadvantages of a property have been concealed during a sale; these can be solved in negotiations or in court. Because of high competition in the market and the ease with which the reputation of a real estate company can be damaged by a negative review, effective players try to mitigate client claims and solve crises in favor of the customer. The other key issue is the safe and efficient running of infrastructure—fire and structural safety, cleaning, security and fast repair in the event of any incident.

Public Administration and Defence (84—Public administration and defence; compulsory social security)

General Description and Key Features of Industries In spite of the automation of some public administration and defense processes, the majority of governmental functions and key decision-making processes depend on the manual work of public servants.

Critical Success Factors for an Organization Within the Industries The ability to attract and retain well educated, patriotic and unselfish people willing to serve the public interest with integrity, sometimes at the expense of their lives; the ability to motivate these public servants to treat citizens fairly and in keeping with legislation, and maintain a balance between the interests of individual citizens and those of society.

Stakeholders in these industries:

- Employees—50%;
- Customers (citizens)—20%;
- Contactors—20%;
- Other—10%.

Typology of Common Risks Within the Industries Mismanagement by state representatives in performing their duties has a great influence on the scale of a potential disaster, and on the speed and quality of emergency response, because these representatives are key decision-makers in a country. The worst mistakes in world history in terms of casualties and financial losses were caused by the misjudgment of governments or rulers concerning the best strategy for social and economic development, or with military defeats that brought the losing side, and sometimes all the states involved, to the brink of collapse. The corruption of a state representative is also common, and is one of the most damaging accidents in the subsector, provoking public indignation and damaging national security.

Arts (91—Libraries, archives, museums and other cultural activities)

General Description and Key Features of Industries Libraries, archives, museums are founded to safeguard the national and universal cultural heritage. The quality of this stewardship depends on the quality of service infrastructure and on the protective action of keepers and security staff.

Critical Success Factors for an Organization Within the Industries The proper maintenance and upkeep of the structures where artistic or cultural artefacts and important documents are kept; and the recruitment and retention of honest and assiduous personnel to protect these treasuries.

Stakeholders in these industries:

- Employees—30%;
- Government and national and international law machinery—30%;
- Customers—30%;
- Other—10%

Typology of Common Risks Within the Industries The most critical risk within the industries is that of fire at a cultural heritage site, which could completely destroy archives of documents, books or pieces of art. There is also a serious threat of theft of priceless cultural objects. Another challenge for the keepers of books or cultural objects is to ensure safe handling by staff or customers, without which these treasures can gradually deteriorate or be destroyed. During riots and wars, archives and museums can become targets for different participants in the conflict, either because they want to destroy crucial documents or symbolic objects, or simply because they see such things as little more than booty.

The following services were excluded from a detailed description because they do not constitute a significant share of global economic activity [294].

---

[294]77 - Rental and leasing activities
  78 - Employment activities
  80 - Security and investigation activities
  82 - Office administrative, office support and other business support activities
  92 - Gambling and betting activities
  94 - Activities of membership organizations
  95 - Repair of computers and personal and household goods
  96 - Other personal service activities
  97 - Activities of households as employers of domestic personnel
  98 - Undifferentiated goods- and services-producing activities of private households for own use
  99 - Activities of extraterritorial organizations and bodies

Chapter 4
Conclusion

4.1 Differences in Risk Profiles and Risk-Mitigation Measures

We now summarize briefly our findings concerning the dissimilarity of the critical risk profiles and key risk mitigation measures in each of the subsectors we have considered, before coming to a final conclusion about sector differences in risk management. A special table was prepared for a better understanding of the sector differences in risk management. The table includes (i) the stakeholders, (ii) the critical success factors, (iii) the common risks and (iv) the relevant risk mitigation measures for the different sectors and subsectors. The table is available for online download as Electronic Supplementary Material.

Production (Including Agriculture)

Mining, Oil and Gas Extraction To produce resources, an extraction company needs **access to low-cost and resource-rich deposits,** for which it must maintain good relations with national authorities and local communities. Stable production depends on the **reliability and safety of production sites** without serious accidents—which could become disasters, causing massive contamination of the environment, destroying relations with regulators and locals and incurring tremendous losses to the company. In order to control the critical risks, executives in this subsector should mainly focus on good relations with governments, and on vigilant oversight of the principal safety and environment issues throughout the production

---

Electronic Supplementary Material The online version of this chapter (https://doi.org/10.1007/978-3-030-25034-8_4) contains supplementary material, which is available to authorized users.

process, from design and construction to the operation and eventual decommissioning of critical infrastructure.

Manufacturing To survive in this very competitive subsector, a manufacturer needs to **produce quality goods** at low production cost and maintain good long-term relations with customers so that they remain **loyal to the brand**. The **production processes of some goods are hazardous** and pose a threat to local communities. Executives should mainly control the R&D of innovative technical solutions, the design stage and key production processes in order to avoid mistakes and deviations. Other measures to mitigate critical risks, which should be under the direct control of executives, are in the realm of customer care, including marketing and branding, crisis communications, when and how to make urgent recalls of defective goods, customer feedback, the network for repairing goods, etc. Where there are hazardous industrial facilities, management should focuses on ensuring the reliability of production processes, maintaining high standards of occupational safety and developing strong relations with local communities.

Utilities Providing a **non-stop 24-7-365 supply of utility products for customers** at reasonable prices, and fast recovery in the event of disruption, are the main success factors in this subsector. Such continuity is possible only with **reliable infrastructure and proper investment** into both production and distribution. The construction and operation of utilities is seen by authorities as part of critical infrastructure for national security, so this subsector is strongly regulated in many aspects of its activity. And as well as the huge potential impact of interruption to the supply, the production of civil nuclear and hydroelectric power poses a threat to the environment: any serious disaster at such plants could become a national or even international catastrophe. Executives should focus mainly on ensuring the reliability of production at utility plants, lobbying for the interests of the subsector with the authorities, and being prepared for operative recovery from blackouts and for working with national emergency services to address the consequences of utility accidents.

Construction Maintaining the highest quality of work over the whole life-cycle of a construction project, and managing the available time, money, people and materials, are the critical success factors for the sustainable development of a construction company. Executives should focus on any challenge that threatens a **project's timeline, performance, budget or quality of work**.

Agriculture To succeed in this subsector, a producer must have **access to fertile land, productive sea or forest** and receive **diversified government support** to compete in an international market. **Quality and safety of agricultural production** are crucial for customers. **Fast distribution of produce to customers and proper storage of harvests** is also critical to minimize the loss of agricultural production. Executives should focus on developing strong relations with authorities, controlling key cultivation processes and efficient management of production and timely distribution to customers at all levels.

Services

Trading The ability **to sell products at the lowest possible prices, find reliable suppliers and provide excellent client service** are the main competitive advantages of a successful trading company. Executives should find a balance between (I) reducing costs by finding the lowest possible wholesale prices from suppliers, (II) paying service staff a reasonable salary and (III) allocating adequate expenditure on maintaining service infrastructure. Because this subsector depends on large numbers of modestly paid shop staff, labor turnover is one of the highest in the whole service sector. Because of this, management should establish a system to continuously recruit and retain undemanding staff who can serve clients promptly, accurately, without cheating and with heartfelt care.

Transportation **Safe and high-quality transportation services** are crucial to the success of a company in this very competitive subsector. To achieve these goals, an organization has to **recruit, train and retain skilled personnel, properly invest in the repair and sustainable use of transportation infrastructure** and provide **high-level customer care**. Because the majority of transportation accidents occur through human error, executives should focus on developing comprehensive human resources management for the selection, training and development of a large number of pilots, technicians, and attendants across a geographically highly diversified network. Senior managers should keep under their control the key processes of thorough and timely maintenance of transportation. They should also keep their attention on the challenges of excellent client service and marketing: staff training, schedule management (including minimizing and dealing with delays), service pricing, control of food quality, detailed analysis of customer feedback, crisis management in case of serious delays caused by accidents, natural disasters, staff strikes, promotion of the brand, and coordination with other transportation providers.

Hotels and Restaurants Providing **excellent client service, proper operation of service infrastructure** in all geographical areas and effective **promotion of the brand**—these are the main components of success for a HORECA (**Ho**tel/**Re**stau-rant/**Ca**fé) network. Because the quality of service (and most conflicts with customers) depend on the various levels of service personnel, executives should focus on establishing a system for non-stop selection, training and motivation of huge numbers of employees in all regions. It is important to note that this subsector has one of the highest level of staff turnover of any economic activity. To solve this challenge management must develop and promote combination of tangible and intangible incentives to attract people who take genuine pride in looking after others, and are willing to work in stressful conditions for a modest salary. Careful selection of suppliers of food, equipment, and other services is also a management priority—because like customer conflicts with staff, substandard quality in anything supplied to guests threatens the reputation of the whole brand. Management should establish a system to ensure the regular maintenance of service infrastructure in all regions, with customer safety as the top priority. Finally, the unique competitive advantages of an organization should be promoted constantly, to maintain the position of the brand and ensure continuing appeal to customers on the very competitive HORECA market.

Information Services **Proper design and adequate maintenance of software** generally depends on the **expertise of programmers** and the **competence of software project managers**. Thus, executives need to focus on recruiting people of inventive genius and retaining highly qualified and motivated staff, in order to reduce errors during programming and ensure prompt updates for any bug revealed.

Telecommunications Providing a **non-stop supply to customers 24-7-365** with a fast and innovative services determines the reputation of a telecommunication service. Executives are responsible for the design and maintenance of reliable telecommunication networks. Ensuring fast recovery after inevitable failure of some parts of the network, good customer service and swift response to client feedback should also be the remit of the managerial team. Because telecommunication infrastructure is a matter of national security for many states, management should coordinate their strategic decisions on the development of infrastructure with national authorities.

Financial Services **Trust, financial stability and good client service** are key to the reputation and therefore the survival of financial institutions. To respond to the wider context outside the institution, executives should focus on tracking and predicting economic developments worldwide, and maintaining cooperative relations with the authorities: because the activity of financial institutions influences the national and even international economy, it is generally highly regulated. Cyber risks are also critical to the stability of financial institutions, thus, protecting the institution's IT systems should be a management priority. Within the institution, the key executive concerns are to establish a system to select and retain honest and highly skilled staff, and maintain complex and detailed oversight over them, to define clear criteria for the selection of borrowers and clients, and to have some control over the financial conditions of key borrowers.

Professional Services **Unique ideas and solutions developed by highly qualified employees** are the main competitive advantage of a professional services organization. Thus, the risks in this subsector stem from staff making ill-judged recommendations, providing ineffective solutions or even deliberately misleading clients (white-collar crime). Executives should continuously assess the quality and adequacy of advice and solutions provided by their staff. They also need to recruit the best minds in the field and motivate them to develop unique and unbeatable ideas and solutions.

Scientific Research and Education Providing **high-quality services in this subsector mainly depends on brilliant researchers and teachers**. The growing influence of artificial intelligence and e-learning has not diminished the role of highly qualified staff in the performance of academic institutions. Thus, the influence of staff on the business process in this subsector is one of the highest within the service sector. Executives should continuously assess the quality of the research undertaken and the education provided by staff. They should also ongoingly recruit and retain brilliant researchers and teachers. Because both scientific research and education usually take place in specialized laboratories, maintaining scientific and educational infrastructure in a safe and clean condition is an important task for executives.

Healthcare   The **quality of medical care is highly dependent on the qualification and client service skill of medical workers**. 70–80% of healthcare incidents occur through human error, usually involving interpersonal interactions within medical teams. Therefore, executives should focus on the continuous recruitment and retention of well-qualified medical staff, and on training to prioritize vigilance in preventing medical mistakes and improving customer care, to reduce as much as possible the suffering of patients. Maintaining health care infrastructure in a safe and clean condition is essential for quality of medical services.

Entertainment and Recreation   The success of **performing artists or sportspeople mainly depends on the skills of these individuals** and of the service staff who work behind the scenes to deliver the unique quality of an artistic or sporting performance. Generally, such activity requires the gathering of a large number of people at an exact place and time. Managing executives should continuously search for individuals with exceptional talent, invest in increasing their craftsmanship and motivate them to provide top-quality performance to satisfy audiences at every event. In order to prevent fires, collapses of entertainment buildings, stampedes or criminal activity by some clients or third parties, organizers should plan every event thoroughly: maintenance of infrastructure, crowd management, cooperation with authorities at all levels to protect mass-audience events, and so on.

A comparative analysis of risk profiles in different sectors allows us to draw the following conclusions. The majority of risks in production (and agriculture) are found within product development departments and at production sites. Key risks are in the poor design or low quality of product, and in an industrial disaster that could have an impact, not only on the local community but, also on a global scale if there is contamination of the environment due to decades of production waste emissions, and so on. Once the product leaves the factory or farm and is distributed to customers, the number and severity of risks are reduced; there remains the risk of poor repair or post-sale customer care. Risks in the service sector, on the other hand, extend to every transaction within the whole service network—because most services are provided by humans, whose quality of service varies from one day to another. Therefore in the production sector, executives should focus on risk mitigation measures at the stage of product development and manufacturing, while a service sector executive should be constantly monitoring service processes at the operational level and reacting promptly to any observed deviations. In the production sector, large-scale crises with a product usually mean that the manufacturer must recall a whole batch of goods (or even remove an entire product or brand from the market) and make significant changes in production processes; in the service sector, such crises generally occur only in one chain of a service network, and the consequences can be solved promptly with limited damage if the problems revealed were not systemic.

Having analyzed the risk management picture within each of the different industry groups, we are clearly in no position to conclude that there is a set of universal risk mitigation solutions, which could be comprehensively applied across all production and service sectors: any solution would need to be adapted to the distinctive management features, critical success factors, stakeholder structure and specific types of risk characteristic of each group, subsector or even industry. In particular, it has become clear that the major incompatibility in risk management solutions is between the production sector (including agriculture) on one side and the service sector on the other.

4.2   From Differences to Similarities

Nevertheless, it is possible to find a similar set of significant challenges faced by companies from all the different subsectors within each sector. These were usually between companies or organizations that cater for the same type of customer: either retail businesses (sometimes referred to as business-to-customer or B2C) or corporate businesses (referred to as business-to-business or B2B). To clarify these similarities in risk management, we will divide production and services between B2B and B2C activities, and summarize the risks they face and how they tackle them.

Production: B2C Most industries in this group face similar critical challenges: **product-related risks that may damage customer loyalty to the brand.** For industries that manufacture products for retail customers—food, beverages, pharmaceutical products, motor vehicles, electronics, and so on—the critical factors for success are the promotion of a global or nationwide brand, non-stop innovation in product design, maintaining high product quality at a reasonable price, moving manufacturing to countries with the lowest production cost in order to offer affordable prices to customers in a competitive market, and working effectively with global logistics networks.

The most harmful crises are those that **undermine a company's relationship with customers or reflect badly on product quality**; among the notorious cases we have discussed are the Ford Pinto, the Perrier recall, the Toyota pedal crisis, VW Dieselgate and the Samsung Galaxy Note 7. Crises in this subsector are caused by errors in product design, the use of unproven or harmful components, poor product assembly or by a company ignoring or failing to respond satisfactorily to customer complaints. Any major problem with a defective product could ultimately force a recall of goods and impact the whole global or national brand.

Production: B2B Industries in the production subsectors of mining, heavy manufacturing, and utilities use simpler production processes, and generate raw products or goods for corporate customers, such as oil/gas/petrochemicals, metals and other minerals, and power. Most of the industries in this group face similar types of critical risks: **regulation-related risks, potential threat to the wider environment due to hazardous technological processes,** and **occupational safety challenges.**

These industries are generally recognized as part of the national infrastructure and make a huge contribution to the GDP and tax revenue of a government. Therefore, **their actions are strictly regulated by government and regulators**, in particular regarding national security concerns, the development of shale formations by hydraulic fracturing, open pit mining, discharge of hazardous substances during petrochemical and chemical production, construction of hydropower and nuclear plants, and approval for massive construction projects that could completely change the lives of nearby local communities.

The **production process could get out of control,** or be **poorly designed,** in ways that **damage the natural environment** through the **leakage of highly hazardous substances:** surface or marine oil spills, leakage of waste water in mining, discharge of chemicals, emission of harmful gases, metals and dust during metal smelting or thermal power plant production, fallout of radioactive materials, widespread deadly flooding after accidents at hydropower plants, and so on. The surrounding environment can be laid to waste after the construction of a major industrial facility. Serious accidents at hazardous production sites—such as those at Bhopal, Chernobyl, Deepwater Horizon, or Fukushima-Daiichi—usually become national or even international disasters because of the toxicity of the production process, the threat to the lives of thousands of people, the huge scale of the enterprise as well as the reputation repercussions. Other industries will often depend on the supply from a plant, and there may be no alternative producers, so large numbers of employees may depend on the plant's activity. Moreover, it is likely that high insurance coverage exists, and dealing with a major disaster may require government support.

Occupational safety is a challenge for every industry in this group because the production processes involved are potentially highly dangerous: drilling of oil wells, underground coal mining, smelting ofs metal, and even the loading and transport of dangerous materials or the presence of workers at hazardous production sites.

More advanced or sophisticated manufacturing for corporate customers (machinery, equipment, electronic and optical products, etc.) in many ways resembles retail manufacturing: there are high expenses on R&D, product quality is critical, there is high competition within the industry, branding is very important, distribution is through a global network, and production lines have been relocated to places with low production and labor costs. And this group of industries differs from raw product production: there is little threat to the environment from the production process because the raw materials have already been processed elsewhere, and occupational safety has less of an influence on the production process than it does in raw manufacturing, since production does not involve intensive manual labor, and the growth of robotics has drastically reduced the human workforce.

Services: B2C The majority of retail service industries face two types of extremely important risk: **poor quality of service damaging brand loyalty** and **safety problems with service infrastructure.** These industries are under constant pressure to maintain nationwide or global service networks and to ensure consistent quality of service for millions of customers in different regions. To provide retail services, thousands of people are hired as service staff.

Thus the success of a retail service business usually depends on the quality of service provided by staff, and the most dangerous crises in these industries are connected with **conflicts between customers and service staff**: specifically when employees violate working rules as in the Costa Concordia or numerous cases of pilot error, cheat customers, have a poor attitude to ensuring client satisfaction, ignore or neglect risks, are slow or careless, fail to maintain their own hygiene or that of the premises, and so on. To minimize mistakes in the manual work of staff and ensure a consistent quality of service, companies try to standardize some of the services they offer by investing in automation. This has mainly occurred through IT and communication solutions that allow customers to serve themselves without the mediation of service staff. Such trends are common in finance, telecommunication, and the remote sales of services based on information transmission (with content ranging from arts and entertainment, books and event tickets to education and public services).

Several industries have **infrastructure-related risks**, when the services they provide could threaten the lives of customers: retail trade, transportation, hotels, restaurants, night clubs, cinemas and casinos, education, repair of motor vehicles, human health care, sports events, recreation activities, museums, services to buildings, postal services, etc. Examples of this kind of risk include transportation vehicle crashes, fires in shopping malls, cinemas, theatres or schools, buildings collapsing, stampedes in poorly designed stadiums and inadequate physical protection against criminal or terrorist attacks—such as poorly protected airplane cockpits or the absence of surveillance systems and metal detector frames.

Because services usually need to develop national or global networks under one united brand, and provide a similar quality of service in any geographical location, **a crisis at a single service point can harm the whole network and brand**: there will inevitably be media coverage of the crisis, and customers—in that country and probably worldwide—may assume that the poor service demonstrated is typical of the company as a whole.

Services: B2B The industries in this group face one common critical risk, in this case the **risk of damage to their reputation through poor quality of professional services or advice.** For industries providing professional services—IT, consulting, accounting, advanced financial services, architectural and engineering activities, scientific research, and so on—the critical success factor is the training and professionalism of the staff, which enables them to solve the unique business problems of their corporate clients and thus build a reputation as a corporate service provider.

Thus the riskiest situations in these industries are those that cast **doubt on the professional competence** of a company's staff. These situations include, for example, mistakes in the solutions provided: bad advice from a consultant, blunders in an architectural design, or the critical misjudgment of liabilities that led to the demise of Lehman Brothers. Worse still is **white-collar crime**: the conscious and sometimes systematic cheating of customers through misleading and/or criminal actions, notoriously demonstrated by Enron and Arthur Andersen or in the Barings collapse.

In conclusion, we can certainly advise that risk specialists should avoid blindly and hastily taking accident response experience, and risk mitigation measures, from other sectors or subsectors to apply within their own, because of the differences we have described. Nevertheless, the similarities we have found during the research can help risk specialists to extrapolate risk mitigation experience from one industry to certain others.

Printed in the United States
By Bookmasters